Low Energy Cooling for Sustainable Buildings

Low Energy Cooling for Sustainable Buildings

Ursula Eicker

Stuttgart University of Applied Sciences, Germany

A John Wiley and Sons, Ltd, Publication

This edition first published 2009

Registered office
John Wiley & Sons Ltd, The Atrium, Southern Gate, Chichester, West Sussex, PO19 8SQ,
United Kingdom

For details of our global editorial offices, for customer services and for information about how to apply for permission to reuse the copyright material in this book please see our website at www.wiley.com.

Library of Congress Cataloguing-in-Publication Data

Eicker, Ursula.
Low energy cooling for sustainable buildings / Ursula Eicker.
p. cm.
Includes bibliographical references and index.
ISBN 978-0-470-69744-3 (cloth)
1. Sustainable buildings–Air conditioning. 2. Buildings–Energy conservation. I. Title.
TH880.E43 2009
697.9′3–dc22

2008052226

A catalogue record for this book is available from the British Library.

ISBN: 978-0-470-69744-3

Typeset in 11/13 pt Times by Thomson Digital, Noida, India

Contents

Preface ix

About the Author xi

1 Energy Demand of Buildings 1
1.1 Residential Buildings 4
1.1.1 Heating Energy 4
1.1.2 Domestic Hot Water 6
1.1.3 Electricity Consumption 7
1.2 Office Buildings 9
1.2.1 Heating Energy 9
1.2.2 Electricity Consumption 10
1.2.3 Air Conditioning 13
1.3 Conclusions 19

2 Façades and Summer Performance of Buildings 21
2.1 Review of Façade Systems and Energy Performance 23
2.1.1 Single Façades 23
2.1.2 Double Façades 23
2.1.3 Modelling of Ventilated Façades 27
2.2 Experimental Results on Total Energy Transmittance 30
2.2.1 Laboratory Experiments 30
2.2.2 Building Experiments 36
2.3 Cooling Loads through Ventilation Gains 40
2.3.1 Double Façade Experiments 40
2.3.2 Parameter Study Using Simulation 43
2.4 Energy Production from Active Façades 47
2.4.1 Thermal and Electrical Energy Balance of the Façade 53
2.5 Conclusions on Façade Performance 58

3 Passive Cooling Strategies 61
3.1 Building Description and Cooling Concepts 62
3.1.1 Lamparter Building, Weilheim 62
3.1.2 Rehabilitated Office Building in Tübingen 64
3.1.3 Low-energy Office Building in Freiburg 65

3.2 Passive Night Ventilation Results 65
3.2.1 Internal Loads and Temperature Levels 65
3.2.2 Air Changes and Thermal Building Performance 68
3.2.3 Simulation of Passive Cooling Potential 71
3.2.4 Active Night Ventilation 74
3.3 Summary of Passive Cooling 79

4 Geothermal Cooling **83**
4.1 Earth Heat Exchanger Performance 88
4.1.1 Earth to Air Heat Exchanger in a Passive Standard Office Building 88
4.1.2 Performance of Horizontal Earth Brine to Air Heat Exchanger in the ebök Building 93
4.1.3 Performance of Vertical Earth Brine to Air Heat Exchanger in the SIC Building 95
4.1.4 Modelling of Geothermal Heat Exchangers 102
4.1.5 Conclusions on Geothermal Heat Exchangers for Cooling 108

5 Active Thermal Cooling Technologies **111**
5.1 Absorption Cooling 113
5.1.1 Absorption Cycles 113
5.1.2 Solar Cooling with Absorption Chillers 117
5.2 Desiccant Cooling 125
5.2.1 Desiccant Cooling System in the Mataró Public Library 129
5.2.2 Desiccant Cooling System in the Althengstett Factory 132
5.2.3 Monitoring Results in Mataró 133
5.2.4 Monitoring Results in Althengstett 137
5.2.5 Simulation of Solar-Powered Desiccant Cooling Systems 145
5.2.6 Cost Analysis 152
5.2.7 Summary of Desiccant Cooling Plant Performance 155
5.3 New Developments in Low-Power Chillers 155
5.3.1 Development of a Diffusion–Absorption Chiller 156
5.3.2 Liquid Desiccant Systems 175

6 Sustainable Building Operation Using Simulation **197**
6.1 Simulation of Solar Cooling Systems 198
6.1.1 Component and System Models 201
6.1.2 Building Cooling Load Characteristics 207
6.1.3 System Simulation Results 211
6.1.4 Influence of Dynamic Building Cooling Loads 216
6.1.5 Economic Analysis 219
6.1.6 Summary of Solar Cooling Simulation Results 225
6.2 Online Simulation of Buildings 226
6.2.1 Functions and Innovations in Building Management Systems 227
6.2.2 Communication Infrastructure for the Implementation of Model-Based Control Systems 228
6.2.3 Building Online Simulation in the POLYCITY Project 229
6.3 Online Simulation of Renewable Energy Plants 238
6.3.1 Photovoltaic System Simulation 239
6.3.2 Communication Strategies for Simulation-Based Remote Monitoring 241

6.3.3 Online Simulation for the Commissioning and Operation of Photovoltaic Power Plants 242
6.3.4 Summary of Renewable Energy Plant Online Simulation 245

7 Conclusions **249**

References **253**

Index **263**

Preface

Investigations of building energy use have often concentrated on heating energy, which dominates total primary energy consumption in moderate or cold climatic zones. Today, there are built examples available for both residential and office building types, which demonstrate the feasibility of reducing heating energy demand almost to zero at little extra cost. Even rehabilitation of buildings to passive energy standards is possible using highly efficient glazing, excellent insulation and heat recovery for ventilation needs. Calculation methods for heating demand are available and standardized on a monthly energy balance level.

In contrast, energy consumption for electrical appliances and lighting and for summer cooling has been less analyzed and regulated, although electricity demand has risen strongly in the last few years and high thermal loads in buildings have led to increasing installation capacities of electrical cooling equipment. The loads are partly due to the continuing attraction of highly glazed buildings, where the prevention of solar irradiance transmission is difficult, but also to rising internal loads through computer equipment, electrical lighting and other appliances. Chapter 1 describes the current status of building energy performance and ongoing standardization processes. Measured consumption data for both the residential and commercial sectors show typical consumption distributions and user influence on building performance.

The analysis then proceeds in Chapter 2 to how summer cooling loads can be reduced through sustainable building design. The performance of highly glazed façades often used in modern office building projects is analyzed in detail. The total energy transmittance of single and double façades with sun shading systems determines the external cooling loads of a building. It is shown that low-energy transmittance and a good thermal separation between the outside and inside are possible, but that ventilation gains occur if the fresh air is taken directly from a double façade.

Depending on the climate and building construction, a cooling energy demand often remains, which can be covered by low-energy or active cooling systems. The possibilities of supplying cooling energy are investigated in terms of rising primary energy consumption. First the limits and potential of passive and fan-driven hybrid night ventilation strategies are analyzed. Two well-monitored case studies of office building projects in Germany are presented in Chapter 3, which show the possibilities today for new and rehabilitated buildings to reduce primary energy consumption for cooling (and heating) to a minimum. Night ventilation results demonstrate the need for very high air exchange rates for effective cooling.

In Chapter 4, near-surface geothermal heat exchangers are then described as very energy-efficient heat sinks to precool ambient air or directly cool thermal masses in buildings. Different thermal cooling technologies are compared in Chapter 5, which can be supplied by solar thermal energy or waste heat. Diffusion–absorption systems and air-based liquid sorption systems were developed at the University of Applied Sciences in Stuttgart for the low-power range below 10 kW thermal. The monitoring and simulation of two plants with desiccant rotors and evaporative cooling systems provide experience for air-based cooling systems. To achieve an energy advantage over conventional electrically powered compressor chillers, all active thermal systems have to operate with a high solar or waste heat fraction, which is due to the rather low coefficients of performance of such chillers. Simulation tools have therefore been developed to support the planning of such complex systems, and also to provide new means of online supervision of plants and building control. Chapter 6 describes the results from simulation tools and presents strategies for online simulation.

The experimental results of both demonstration projects and laboratory measurements together with the developed simulation tools provide new technology solutions for the low-power cooling range and strategies for summer comfort and sustainable climatization of energy-efficient buildings.

This work would not have been possible without the support of the Solar Energy and Building Physics Research Team in the University of Applied Sciences in Stuttgart. My thanks go especially to the simulation expert Juergen Schumacher, PhD students Dirk Pietruschka, Uli Jakob, Uwe Schuerger, Volker Fux, Eric Duminil and Antoine Dalibard, the Absorption Research Team of Dieter Schneider, Tina Paessler, Andreas Biesinger and Alexander Teusser, and the Building Research Group with Martin Huber, Aneta Strzalka, Peter Seeberger, Heiko Fischer, Lukas Weigert, Christoph Vorschulze and Uwe Bauer. Our secretary, Christa Arnold, deserves a special mention for holding the research team together during the last 10 years.

About the Author

Ursula Eicker is a physicist who carries out international research projects on solar cooling, heating, electricity production and building energy efficiency at the University of Applied Sciences in Stuttgart. She obtained her PhD in amorphous silicon thin-film solar cells from Heriot-Watt University in Edinburgh and then worked on the process development of large-scale amorphous silicon modules in France. She continued her research in photovoltaic system technology at the Centre for Solar Energy and Hydrogen Research in Stuttgart.

She set up the Solar Energy and Building Physics Research Group in Stuttgart in 1993. Her current research emphasis is on the development and implementation of active solar thermal cooling technologies, low-energy buildings and sustainable communities, control strategies and simulation technology, heat transfer in façades, etc. Since 2002 she has been the scientific director of the research centre on sustainable energy technologies (zafh.net) in Baden Württemberg. She also heads the Institute of Applied Research of the University of Applied Sciences in Stuttgart, where building physicists, geoinformation scientists, mathematicians, civil engineers and architects cooperate.

During the last 10 years Professor Eicker has coordinated numerous research projects on sustainable communities with renewable energy systems and highly efficient buildings. The largest projects include the European Integrated POLYCITY Project, a demonstration project on sustainable buildings and systems in Germany, Italy and Spain, and the European PhD school CITYNET on information system design for sustainable communities.

1

Energy Demand of Buildings

Buildings today account for 40% of the world's primary energy consumption and are responsible for about one-third of global CO_2 emissions (24% according to IEA, 2008; 33% according to Price *et al.*, 2006). The energy-saving potential is large, with 20% savings expected until 2020 in the European Union alone. The cost efficiency of building-related energy savings is high, as shown in a recent study for the Intergovernmental Panel on Climate Change (Ürge-Vorsatz and Novikova, 2008). In the industrialized countries, between 12 and 25% of building-related CO_2 emissions can be reduced at net negative costs, mainly through heat-related measures. In the developing countries, electricity savings through more efficient appliances and lighting are more important with 13 to 52% of the measures being economically feasible until 2020. As published in the Green Paper on energy efficiency by the European Commission, end energy-consumption in 2005 reached 12×10^9 MWh per year, 40% of which can be attributed to buildings (see Figure 1.1). In the USA, 36% of the total energy consumption occur in buildings. Especially in urban areas, building energy consumption is typically twice as high than transport energy, for example by a factor of 2.2 in London (Steemers, 2003).

Under the Kyoto Protocol, the European Union has committed itself to reducing the emission of greenhouse gases by 8% in 2012 compared with the 1990 level and buildings have to play a major role in achieving this goal. If building energy efficiency is improved by 22%, 45 million tonnes of CO_2 can be saved, nearly 14% of the agreed total savings of 330 million tonnes.

Low Energy Cooling for Sustainable Buildings Ursula Eicker

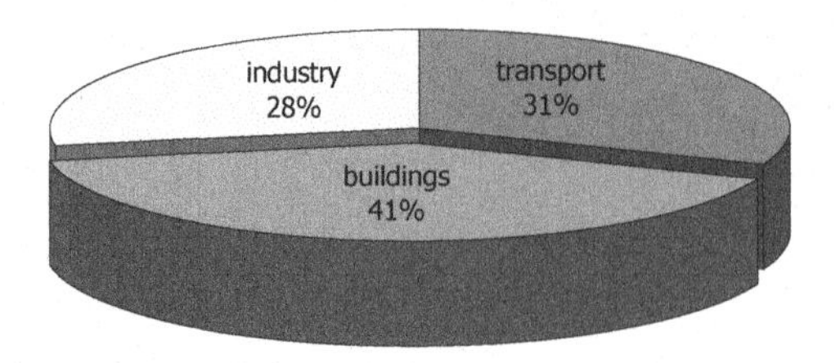

Figure 1.1 Distribution of end energy consumption within the European Union

The European Directive for Energy Performance of Buildings, signed by the European Parliament and Council in 2002, was created to unify the diverse national regulations and calculation methods, to define minimum common standards on building energy performance and to provide certification and inspection rules for a building and its heating and cooling plants. Although the performance directive only defines a common methodology for energy certification, most European countries have now increased their requirements to limit new buildings' energy demand. On average, allowed building transmission losses are now 25% lower. The heat transfer coefficient (U-value) is defined as the reciprocal sum of heat transfer resistances between room and ambient air and is today on average between 0.3 and 0.4 $\mathrm{W\,m^{-2}\,K^{-1}}$ for a building.

The reduction of energy consumption in buildings is of high socioeconomic relevance, with the construction sector as Europe's largest industrial employer representing an annual investment of 910×10^9 euros (2003), corresponding to 10% of gross domestic product. Almost 2 million companies, 97% of them small and medium enterprises, directly employ 11.8 million people.

The total primary energy consumption in Germany is about 4×10^9 MWh, corresponding to 13 878 PJ (2007 data), and is estimated to decrease by 15% until 2030 (EWI/Prognos, 2005). The main efficiency gains are expected through the reduction of transformation losses, which today are responsible for 3984 PJ and are due to decrease by 37% until 2030. In the building sector, on the contrary, the final energy consumption of 2599 PJ (2000) is only estimated to decline by 4% until 2030, which is due to the slow rate of rehabilitation.

In moderate European climates such as Germany's, about 80% of the total energy consumption is used for space heating, 12% for warm water production and the rest for electrical appliances, communication and lighting. The dominance of heat consumption, almost 80% of the primary energy consumption of households, is caused by low thermal insulation standards in existing buildings. They dominate the residential building stock with 90% of all buildings. Even in 2050, 60% of residential space will be located in existing buildings (Ministry for Transport and Buildings, Germany, 2000). Since the 1970s' oil crises the heating energy demand, particularly of new buildings, has been continuously reduced by gradually intensified energy legislation. With high heat insulation standards and the ventilation concept of passive houses, a low limit of heat consumption has meanwhile been achieved, which is around 20 times lower than today's average values. A crucial factor for the low consumption of passive buildings was the development of new glazing and window technologies, which enable windows to be passive solar elements and at the same time cause only low transmission heat losses.

In new buildings with low heating requirements, other energy consumption in the form of electricity for lighting, power and air-conditioning, as well as warm water in residential buildings, is becoming more and more dominant. Electricity consumption in the European Union is estimated to rise by 50% by 2020. Renewable sources of energy can make an important contribution to the supply of electricity and heat. Cooling and refrigeration account for about 15% of total electricity consumption worldwide, and as much as 30% in highly developed countries with a warm climate such as Hong Kong (Government Information Centre, Hong Kong, 2004). Peak electricity loads in many countries now occur in summer rather than in winter. In South Australia, for example, cooling and refrigeration were reported to account for 46% of total electricity consumption on a hot summer's day.

Urban energy management systems should include demand predictions, databases of consumption as well as strategies for operational control and optimization. Consumption data is rarely available on an urban scale, which makes projections of energy requirements difficult. Often there is no strategic energy management plan and demand and supply are not properly matched. Surveys on energy consumption patterns in communities are therefore often based on calculated demand, for example using the appliances used in residential buildings and estimated hours of operation (Zia and Devadas, 2007). A similar demand simulation approach was chosen to analyse the energy efficiency and CO_2 reduction potential in the commercial sector in Japan until 2050 (Yamaguchi *et al.*, 2007). Assuming relatively low increases in insulation thickness (from zero in the year 2000 to 60 mm in 2050), the main efficiency gains were expected through improvements in appliance electrical efficiency. This led to the surprising fact that heat demand even rises, as internal loads due to equipment were supposed to drop. A case study in the UK town of Leicester obtained energy savings of 20% by more efficient lighting in residential buildings, based on measured electricity load curves from the energy supplier (Brownsword *et al.*, 2005).

The chapter aims to contribute information on how much energy is consumed in its different forms in the building sector and which reductions are possible in best case examples. Embedded energy in the building materials and construction process is not included in the analysis, although several studies indicate a rather high importance of material and resource use during building construction and maintenance: for example, Pulselli and colleagues calculated that 49% of all energy is needed for the building manufacturing process, 35% for maintenance and only 15% for use (Pulselli *et al.*, 2007).

1.1 Residential Buildings

1.1.1 Heating Energy

Due to the wide geographical extent of the European Union covering nearly 35° of geographical latitude (from 36° in Greece to 70° in northern Scandinavia), a wide range of climatic boundary conditions are covered. In Helsinki (60.3 °N), average exterior air temperatures reach −6 °C in January, when southern cities such as Athens at 40 °N latitude still have averages of +10 °C. Consequently, building construction practice varies widely: whereas average heat transfer coefficients (U-values) for detached houses are about 1 $W\,m^{-2}\,K^{-1}$ in Italy, they are 0.4 $W\,m^{-2}\,K^{-1}$ in Finland (see Table 1.1). The heating energy demand calculated from monthly energy balances (according to European Standard EN 832) is comparable in both cases at about 50 $kWh\,m^{-2}\,a^{-1}$.

If existing building standards are improved to the so-called passive building standard, heating energy consumption can be lowered to less than 20 $kWh\,m^{-2}\,a^{-1}$. Studies in Switzerland showed that additional investment costs for passive residential buildings are about 14% (Minergie P label). For buildings with a low energy standard (reaching the Swiss Minergie label) investment costs were about 6 to 9% higher (Binz, 2006). Depending on the assumptions made for energy price increases, the additional investment costs can be compensated by lower energy costs during operation. In Germany with its high number of passive building projects, additional investment costs for the high standard are only 3 to 5%.

Since the implementation of the European Building Performance Directive in 2003, nearly all European countries have significantly increased the requirements to reduce transmission heat losses. The European Performance Directive asks for the establishment of a calculation methodology for energy demand and an energy certification process, whereas limits on energy demand are regulated by national laws.

Average U-values for new buildings are about 25% lower than in 2003. The required U-values to achieve passive building standards are listed in Table 1.1 for some cities from different European climates. For these insulation standards, heating energy consumption is between 15 and 20 $kWh\,m^{-2}\,a^{-1}$. By comparison, today's residential buildings in Germany with low energy standards have annual heating energy

Table 1.1 U-values for residential passive buildings (Truschel, 2002)

U-values $W\,m^{-2}\,K^{-1}$	Rome	Helsinki	Stockholm
Wall	0.13	0.08	0.08
Window	1.40	0.70	0.70
Roof	0.13	0.08	0.08
Ground	0.23	0.08	0.1
Mean U-value	0.33	0.16	0.17

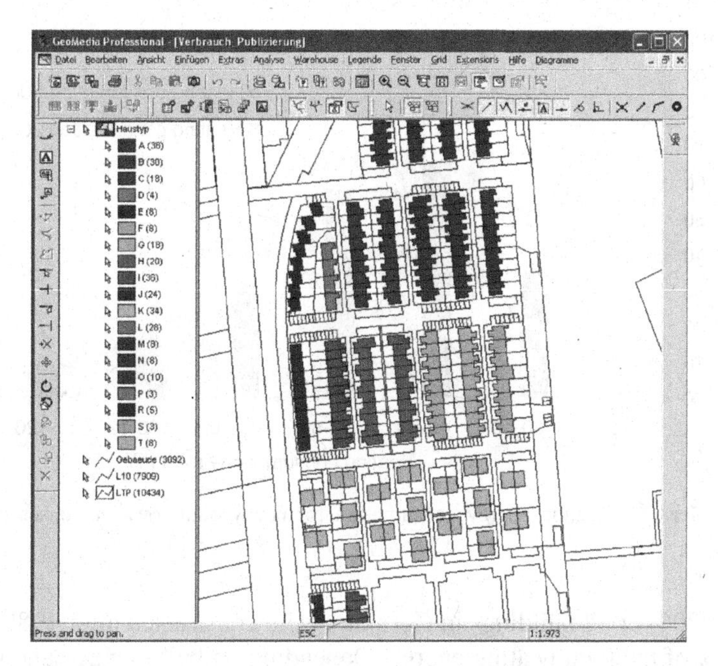

Figure 1.2 Rows of houses analysed in the POLYCITY project, visualized in a geoinformation system

consumption values of around 70 kWh m^{-2} a^{-1}. Several hundred houses in rows constructed after the year 2000 in the town of Ostfildern were measured within the European POLYCITY demonstration project (see Figure 1.2). The consumption varies strongly even for the same building type and standard deviations are about 35% of the mean value (Figure 1.3). The distributions for two years of measurement are shown in Figure 1.4.

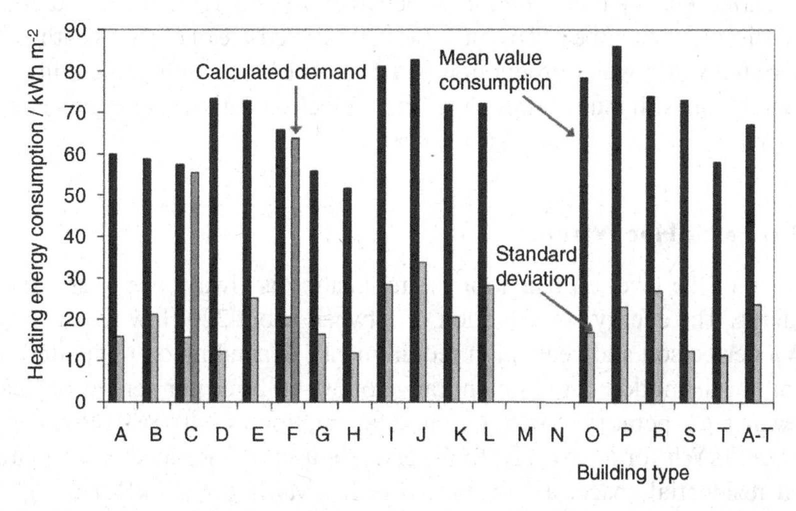

Figure 1.3 Mean measured heating energy consumption and standard deviation of different, newly built rows of houses in Ostfildern, Germany

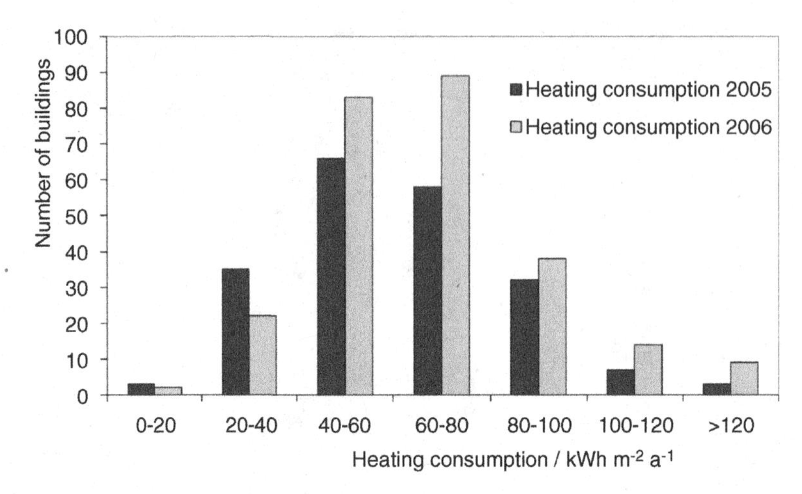

Figure 1.4 Measured heating energy consumption for newly built rows of houses in Ostfildern, Germany

Although 20% of all buildings in Germany were constructed after 1980, they only consume 5% of the total heating energy. Depending on building age and type, older buildings' heating energy consumption varies between 100 and 400 kWh m^{-2} a^{-1}. The main challenge of the next decades will therefore be the reduction of heating energy consumption for existing buildings.

Within the POLYCITY project the existing building stock of the town of Cerdanyola near Barcelona in Spain was analysed with over 6000 buildings: 44% of them are single storey buildings, 28% have two floors. More than 90% of the buildings were constructed after 1960 and 65% of the apartments are between 60 and 90 m^2. The average heating energy consumption is between 90 and 100 kWh m^{-2} a^{-1} and has increased slightly during the last decade due to increased use of central heating systems with integrated warm water production (see Figure 1.7). In comparison, in the urban area of Barcelona with multi-family apartment blocks, heating energy consumption is only 34 kWh m^{-2} a^{-1} on average (Reol, 2005).

1.1.2 Domestic Hot Water

Independent of the level of insulation, water heating is always necessary in residential buildings. The energy consumption is between about 220 (low requirement) and 1750 kWh per person and year (high requirement), depending on the pattern of consumption. For the middle requirement range of 30–60 litres per person and day, with a warm water temperature of 45 °C, the consumption is 440–880 kWh per person or 1760–3520 kWh for an average four-person household. Related to a square metre of heated residential space, a rather low average value of 12.5 kWh m^{-2} a^{-1} is for example used in German legislation. In Switzerland, a fixed value of 14 kWh m^{-2} a^{-1}

is used. To increase the share of renewable energy for water heating, some local or regional governments have introduced legislation to cover typically 60% of the warm water demand by solar thermal energy. In Catalunya in Spain, about 75% of the local communities have so-called local ordinances to oblige building constructors to implement solar thermal energy. In Mediterranean climates, the energy need for warm water heating is of the same order of magnitude as heating energy consumption, even for the given building stock. An investigation for urban housing in Barcelona in Spain showed that from a total end energy consumption of 8310 kWh a^{-1} for residential housing of 90 m^2 average size, 29% or 26 kWh m^{-2} a^{-1} was used for warm water production and 38% for heating. Cooling energy need on the other hand is less than 10 kWh m^{-2} a^{-1} (Reol, 2005).

1.1.3 Electricity Consumption

The average electricity consumption of private households is around 3600 kWh per household and year in Germany. Related to a square metre of heated residential space, an average value of 31 kWh m^{-2} a^{-1} is obtained. An electricity-saving household needs only around 2000 kWh a^{-1}. Measured electricity consumption for several hundred newly built houses in Ostfildern showed average annual consumption values between 30 and 50 kWh m^{-2} a^{-1} (see Figure 1.5). The highest number of buildings was in the class between 40 and 50 kWh m^{-2} a^{-1} (see Figure 1.6). In a passive building project in Darmstadt (Germany), consumptions of between 1400 and 2200 kWh per household per year were measured, which corresponds to an average value of 12 kWh m^{-2} a^{-1}.

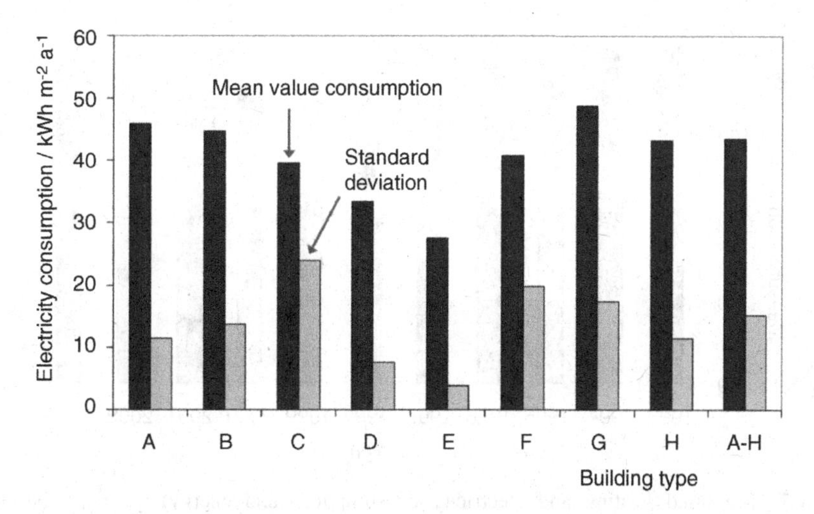

Figure 1.5 Measured electricity consumption of newly built rows of houses in Ostfildern, Germany

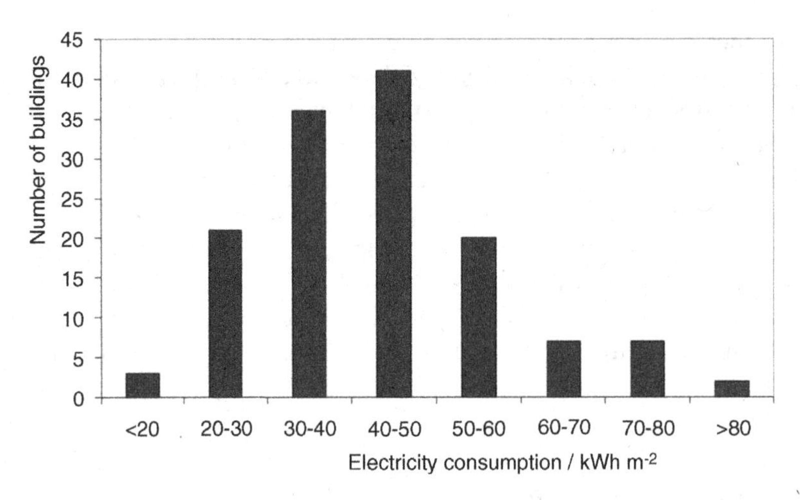

Figure 1.6 Distribution of electricity consumption in newly built rows houses in Ostfildern, Germany

Within the urban housing study in Barcelona, the average electricity consumption was 2160 kWh per household, which corresponds to 24 kWh m^{-2} a^{-1}. In the nearby town of Cerdanyola, the measured electricity consumption for mainly single or two-storey buildings was 70 kWh m^{-2} a^{-1}, with decreasing consumption during the last decade due to the replacement of electric water heaters (Figure 1.7). The high average consumption can be mainly attributed to a housing stock which is partially heated and cooled with electricity. Apartments without electrical heating systems have electricity

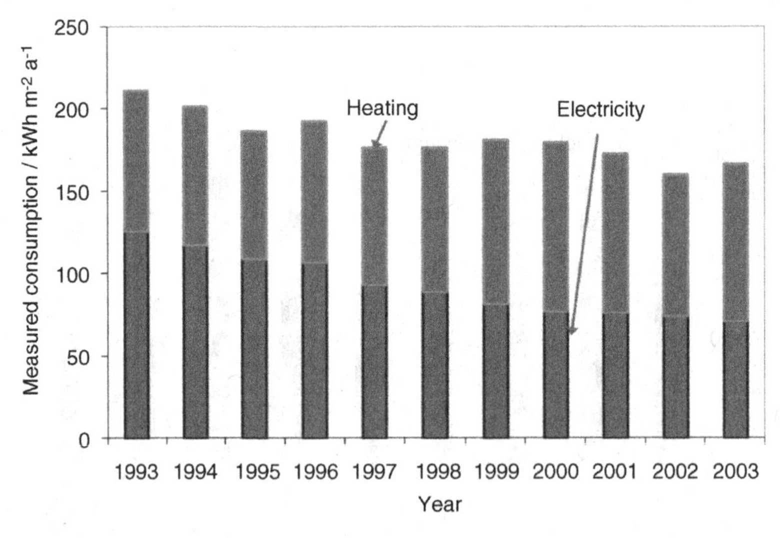

Figure 1.7 Measured heating and electricity consumption (end energy) for 6238 buildings in Cerdanyola, Spain

consumption values between 40 and 50 kWh m^{-2} a^{-1}. Measurements of electricity consumption were also taken in a social housing district with 2500 inhabitants in Turin, Italy, within the POLYCITY project: 622 apartments within 30 building blocks were analysed. Here the average electricity consumption per household is low, about 1750 kWh per year, which corresponds to a specific consumption between 14 and 20 kWh m^{-2} a^{-1}.

1.2 Office Buildings

1.2.1 Heating Energy

Existing office and administrative buildings have approximately the same consumption of heat as residential buildings and most have a higher electricity consumption. According to a survey of the energy consumption of public buildings in the state of Baden-Württemberg in Germany, the average consumption of heat is 217 kWh m^{-2} a^{-1}. The specific energy consumption of naturally ventilated office buildings in the UK is in a similar range of 200–220 kWh m^{-2} a^{-1} (Zimmermann and Andersson, 1998). From the commercial sector in Japan, values of 59 kWh m^{-2} a^{-1} have been reported (Yamaguchi *et al.*, 2007). Measured heating energy data from a variety of the author's projects (Lamparter office in Weilheim, Germany, Town Hall Ostfildern, Germany, Isbank Tower, Istanbul) and case studies literature has been gathered by the author and is shown in Figure 1.8. Heat consumption in administrative buildings can be reduced without difficulty, by improved thermal insulation, to under

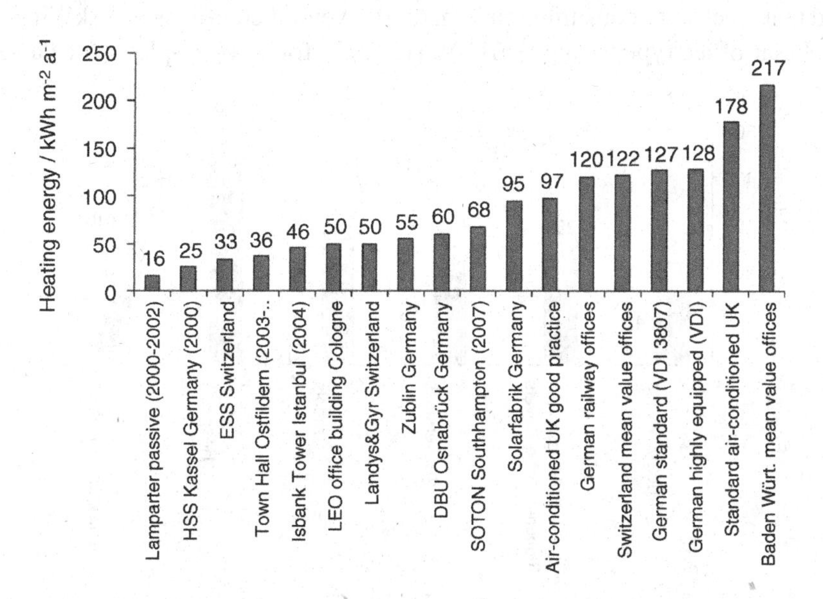

Figure 1.8 Office building projects with measured heating energy consumption in Europe

100 kWh m^{-2} a^{-1}, and even to a few kilowatt hour per square metre and year in a passive building.

1.2.2 Electricity Consumption

Total Electricity Consumption

Both heat and electricity consumption depend strongly on the building's use. In terms of the specific costs, electricity almost always dominates. A survey carried out in public buildings of the German state of Baden-Württemberg found an average electricity consumption of 54 kWh m^{-2} a^{-1}, in the UK values between 48 and 85 kWh m^{-2} a^{-1} were measured (see Figure 1.9).

When comparing the energy costs of commercial buildings with the remaining current monthly operating costs, the relevance of a cost-saving energy concept is apparent: more than half of the running costs are accounted for by energy and technical services. A large part of the energy costs is due to ventilation and air-conditioning.

Electricity consumption dominates total energy consumption where the building shell is energy optimized and can be reduced by 50% at most. Even in an optimized passive energy office building in southern Germany, electricity consumption remained at about 35 kWh m^{-2} a^{-1}, mainly due to the consumption by office equipment such as computers (see Figure 1.10).

While the measured values for heat consumption correspond well with the planned values, the measured total electricity consumption exceeds the planned value of 23.5 kWh m^{-2} a^{-1} by 42%. A survey of good practice office buildings in the UK showed that electricity consumption in naturally ventilated offices is 36 kWh m^{-2} a^{-1} for a cellular office type, rising to 61 kWh m^{-2} a^{-1} for an open-plan office and up to

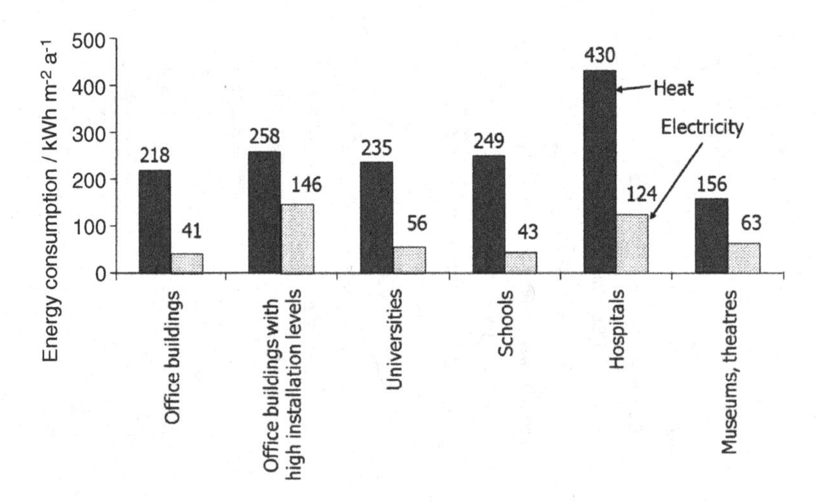

Figure 1.9 Final energy consumption by building type in Baden-Württemberg

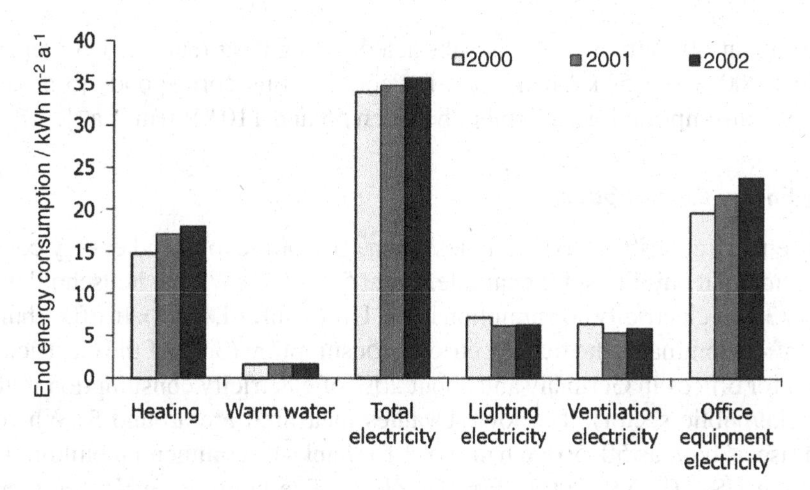

Figure 1.10 Measured consumption of electricity, heat and water heating in an office building with passive house standard in Weilheim-Teck, Germany

132 kWh m^{-2} a^{-1} for an air-conditioned office (Zimmermann and Andersson, 1998). Recent measurements in a university building in Southampton (Figure 1.11) within the European-funded Ecobuilding project SARA gave electricity consumption for equipment alone of 64 kWh m^{-2} a^{-1}. A study on commercial buildings in Osaka, Japan, indicated an average electricity consumption of 75 kWh m^{-2} a^{-1}, but was not split into different uses (Yamaguchi *et al.*, 2007). A survey on 6000 commercial buildings in the USA showed a range of total electricity consumption for offices and commercial buildings between 165 and 220 kWh m^{-2} a^{-1} (Energy Information Administration,

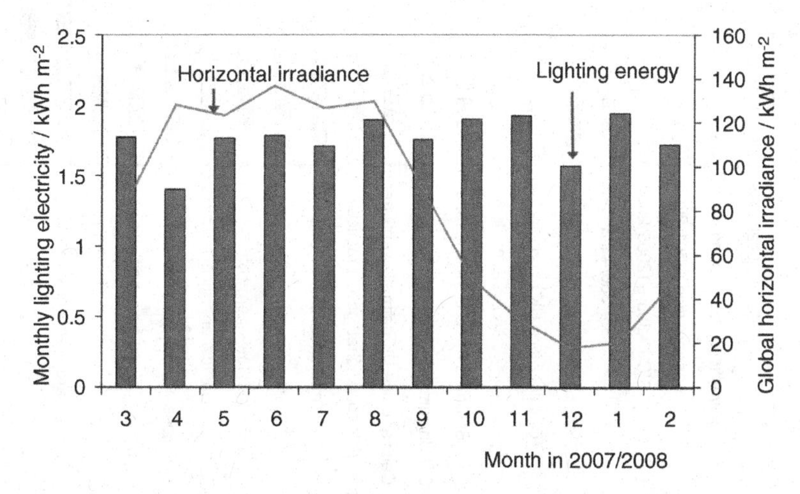

Figure 1.11 Monthly lighting energy and global horizontal irradiance for a newly constructed university office building in Southampton

2006). If about 60 kWh m^{-2} a^{-1} is subtracted for lighting (survey, US Department of Energy, 2002) and 50 kWh m^{-2} a^{-1} for cooling, this corresponds to an average electricity consumption for equipment between 55 and 110 kWh m^{-2} a^{-1}.

Lighting Energy Consumption

Lighting energy contributes typically less than 10% of the total end energy consumption in residential buildings: for example, about 500–770 kWh per household in Spain and about 3% of electricity consumption in the UK (Ashford, 1998). In office buildings lighting often dominates the total electricity consumption (36% of the electricity consumption for offices in Germany and about 20% of electricity consumption in the UK commercial/public sector). The lowest values measured are around 5 kWh m^{-2} a^{-1} and can rise as high as 50–70 kWh m^{-2} a^{-1} for banks or commercial buildings in the UK and the USA (BINE, 2000). For the ZUB office building in Kassel, Germany, with façade high windows, a low room depth of 4.6 m and daylight-dependent artificial lighting control, average lighting electricity consumption of 3.5 kWh m^{-2} a^{-1} was measured during an intensive monitoring exercise (Hauser *et al.*, 2004). A survey of lighting electricity demand or consumption has been carried out by the author and is shown in Figure 1.12. It includes measured data from office building projects like the Lamparter building, the ZUB Kassel and several case studies from the UK, the USA, Germany, etc.

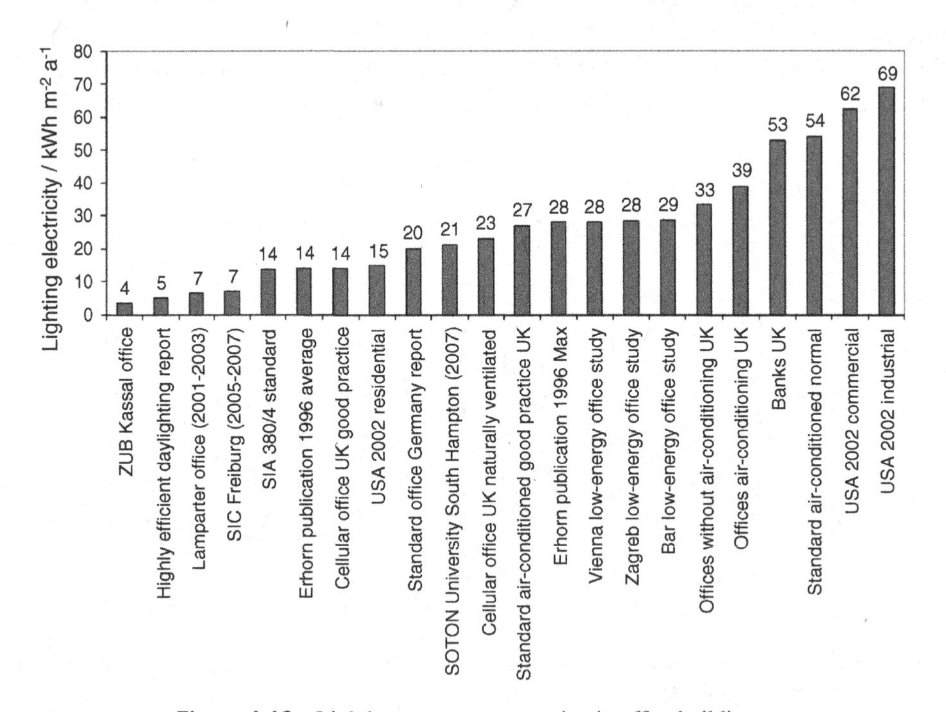

Figure 1.12 Lighting energy consumption in office buildings

Table 1.2 Approximate values for nominal flux of light, and specific connected power of energy-saving lighting concepts

Room type	Required illuminance level/ lux	Specific electric power/ $W\ m^{-2}$
Side rooms	100	3–5
Restaurants	200	5–8
Offices	300	6–8
Large offices	500	10–20

Even in newly constructed eco-buildings, electrical energy consumption for lighting can be high, especially if the users do not effectively reduce artificial lighting. A recently built office building at the University of Southampton was intensively monitored within the European SARA demonstration project (construction completed in 2005 with a $2600\,m^2$ surface area). A total annual lighting electricity value of $21\,kWh\,m^{-2}\,a^{-1}$ was measured. The daily and monthly values showed that there was no seasonal influence on lighting energy consumption (see Figure 1.11). Only on weekends was there a reduction of lighting consumption, by a factor of 10.

Worldwide, 20% of the electricity produced is used for lighting. Studies show that the introduction of market-available and highly efficient light-emitting diode technology could reduce this consumption by 30% until 2015 and by 50% until the year 2025 (European Commission, 2006).

If energy-saving lighting concepts are applied, the connection power in office rooms can be as low as $6–8\,W\,m^{-2}$, while standard values are still between 10 and $20\,W\,m^{-2}$ (see Table 1.2).

1.2.3 Air Conditioning

Annual sales of electrical room air-conditioning units are about 43 million units with the main markets in China (12 million) and the USA (11.8 million). The market penetration in Europe is smaller (about 2.8 million units in 2002), but has a high growth rate. Per thousand inhabitants, there are 0.6 installed units in Germany, 14 in Spain and 12 in Italy. The growth of climatized building area in Europe is estimated to rise from $3\,m^2$ per inhabitant (year 2000) to $6\,m^2$ per inhabitant (year 2020). The rising air-conditioning sales are due to increased internal loads through electrical office appliances, but also to increased demand for comfort in summer. Summer overheating in highly glazed buildings is often an issue in modern office buildings, even in north European climates. This unwanted and often unforeseen summer overheating leads to the curious fact that air-conditioned buildings in Northern Europe sometimes consume more cooling energy than in Southern Europe where there is a more obvious architectural emphasis on summer comfort. According to an analysis of a range of office buildings, an average of $40\,kWh\,m^{-2}\,a^{-1}$ was obtained for southern climates, whereas $65\,kWh\,m^{-2}\,a^{-1}$ was measured in north European building projects (Santamouris, 2005). Today in German office buildings the energy demand for air-conditioning

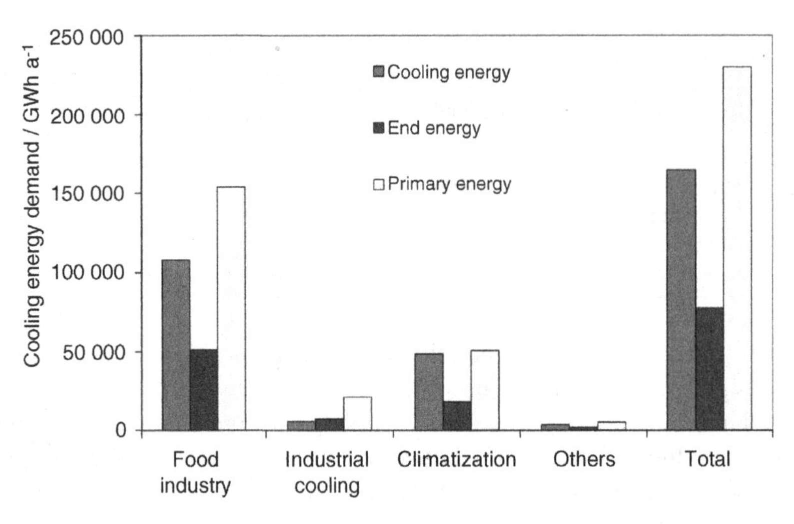

Figure 1.13 Cooling energy demand for different branches in Germany (DKV, 2002)

is estimated to be between 40 000 (Nick-Leptin, 2005) and 50 000 GWh per year (DKV, 2002). The total electricity consumption for cooling in Germany has reached 14% of the total electricity consumption, which corresponds to 5.8% of the total primary energy demand. The main cooling energy demand is in the food industry, followed by climatization applications and industrial cooling (see Figure 1.13).

The largest European air-conditioning manufacturer and consumer is Italy, accounting for nearly half of all European production (Adnot, 1999): 69% of all room air-conditioner sales are split units, with total annual sales of about 2 million units. In 1996 the total number of air-conditioning units installed in Europe was about 7 500 000. In Southern Europe, the installed cooling capacity is dominated by the residential market. Although less than 10% of homes in Spain have air-conditioning systems, 71% of the installed cooling capacity is in the residential sector (Granados, 1997).

Between 1990 and 1996, electricity consumption for air-conditioning in the European Union rose from about 24 000 GWh per year to 44000 GWh per year and further increases up to 123 000 GWh per year in 2010 and nearly 160 000 GWh in 2020 are predicted (Adnot, 1998). Spain and Italy together are responsible for about two-thirds of the total cooling energy demand (Adnot *et al.*, 2003, see Figure 1.14). About 25% of the total electricity consumption is caused by room air-conditioners, the rest by central air-conditioners. The average coefficient of performance (COP) for all cooling technologies is currently about 2.7 (cooling power to electricity input) with a target of about 3.0 for 2015. Energy labelling is now obligatory in the EU for cooling devices of less than 12 kW power. The best label (A class) is obtained for units with COP's above 3.2.

Cooling energy is often required in commercial buildings, with the highest consumption worldwide being in the USA. In Europe the cooling energy demand for such buildings varies between 3 and 30 MWh per year. Not much data is available

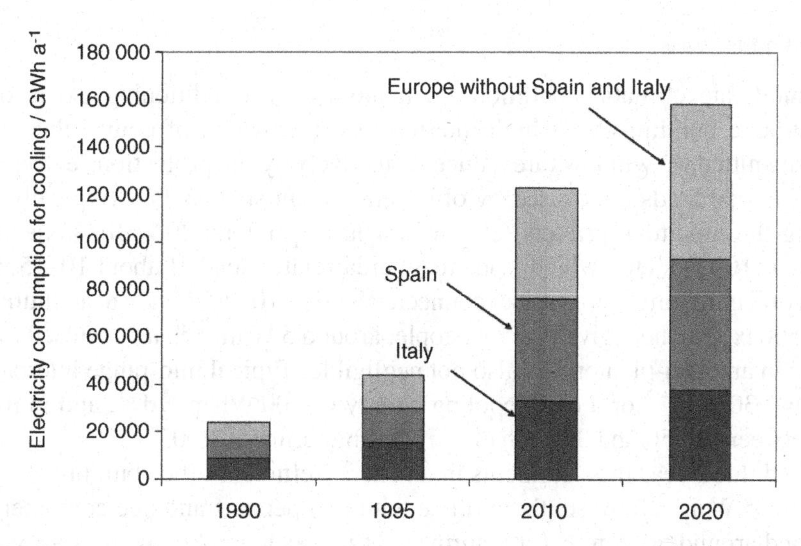

Figure 1.14 Projected electricity consumption for cooling in Europe with the two main consumer countries, Italy and Spain

for area-related cooling energy demand. Breembroek and Lazáro (1999) quote values of $20\,\mathrm{kWh\,m^{-2}\,a^{-1}}$ for Sweden, 40–$50\,\mathrm{kWh\,m^{-2}\,a^{-1}}$ for China and $61\,\mathrm{kWh\,m^{-2}\,a^{-1}}$ for Canada. The author's own survey of area-related cooling demand is shown in Figure 1.15. For residential buildings, the Catalan building research centre, Institut Cerdá estimated a typical cooling energy demand in Catalunya of $13\,\mathrm{kWh\,m^{-2}\,a^{-1}}$.

In the food trade the energy consumption for cooling is significantly higher, with a total electricity demand between 82 and 345 kWh per square metre of sales area, about 50–60% of total electricity consumption (O.Ö. Energiesparverband, 1996). The demand is dominated by food cooling equipment.

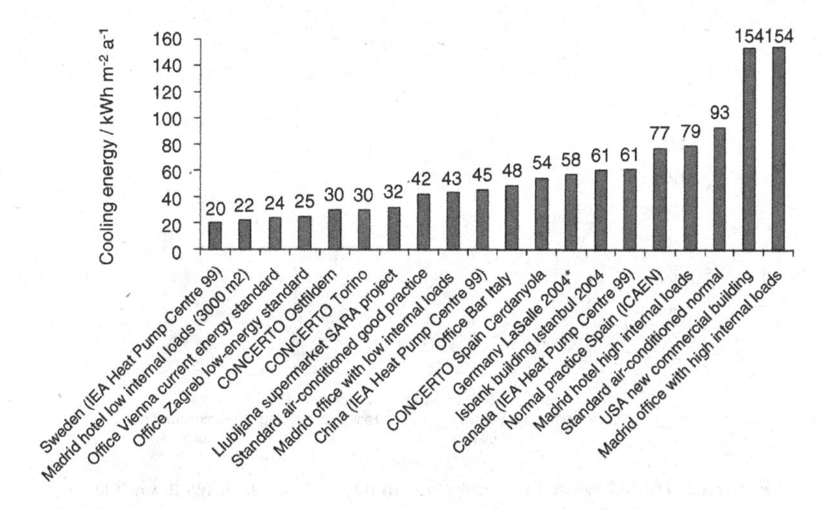

Figure 1.15 Cooling energy demand for new commercial buildings

Internal Loads

Under moderate climatic conditions, demand for air-conditioning exists only in administrative buildings with high internal loads, provided of course that external loads transmitted via windows are reduced effectively by sun protection devices. About 50% of internal loads are caused by office equipment such as PCs (typically 150 W including the monitor), printers (190 W for laser printers, 20 W for inkjets), photocopiers (1100 W), etc., which leads to an area-related load of about 10–15 W m^{-2}. Modern office lighting has a typical connected load of 10–20 W m^{-2} at an illuminance of 300–500 lx. The heat given off by people, around 5 W m^{-2} in an enclosed office or 7 W m^{-2} in an open-plan one, is also not negligible. Typical mid-range internal loads are around 30 W m^{-2} or a daily cooling energy of 200 Wh m^{-2} d^{-1}, and in the high range between 40–50 and 300 Wh m^{-2} d^{-1} (Zimmermann, 2003).

Detailed three-year measurements in an energy-efficient office building varied between 30–35 W m^{-2} for a southern office with two persons and one computer workstation and around 50 W m^{-2} for a northern office occupied by two persons with two computer workstations. This resulted in daily internal loads in the south-facing office of around 200–300 Wh m^{-2} d^{-1} and 400–500 Wh m^{-2} d^{-1} for the heavier equipped northern office. Detailed monitoring results from four other office buildings in Germany showed that appliances always dominate the internal loads, which were between 92 and 188 Wh m^{-2} d^{-1} (Voss *et al.*, 2007, see Figure 1.16).

External Loads

External loads depend greatly on the surface proportion of the glazing as well as the sun-protection concept. On a south-facing façade, a maximum irradance of 600 W m^{-2} can occur on a sunny summer's day. The best external sun-protection can reduce this

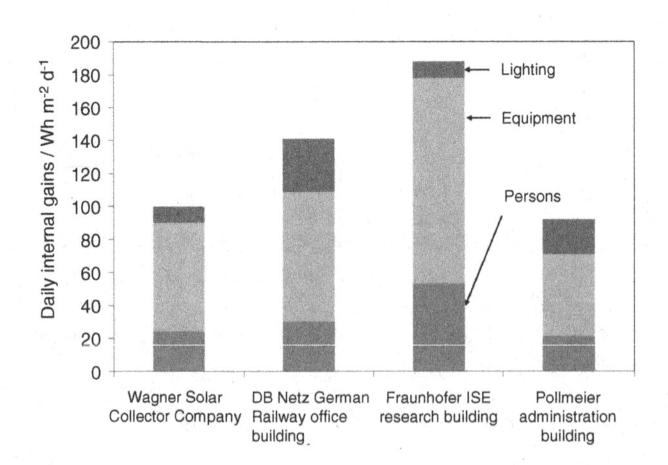

Figure 1.16 Measured internal gains in new office buildings in Germany

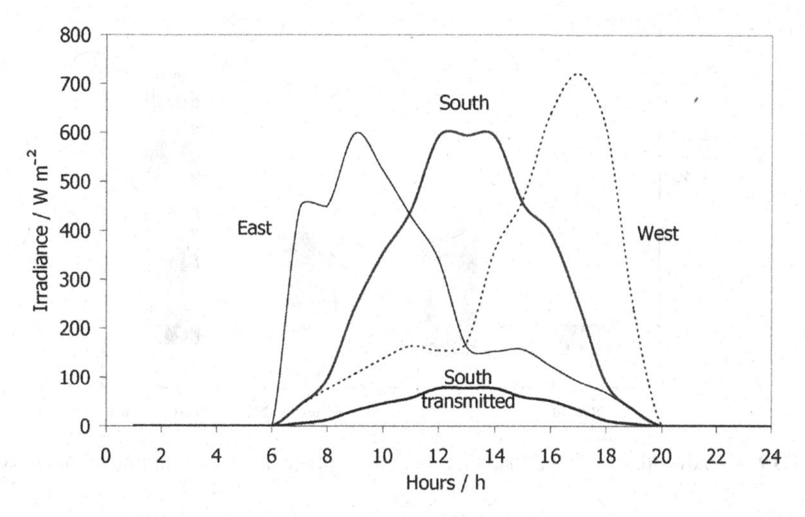

Figure 1.17 Diurnal variation of irradiance on different façade orientations and transmitted irradiance by a sun-protected south façade on a day in August (Stuttgart)

irradiation by about 80%. Together with the total energy transmission (g-value) of a coated double glazing of typically 0.65, the transmitted external loads are about 78 W per square metre of glazing surface. In the case of a 3 m^2 glazing surface of an enclosed office, the result is a load of 234 W, which creates an external load of about 20 W m^{-2} based on an average surface of 12 m^2. This situation is illustrated for south-east-and west-facing façades in the summer in Figure 1.17. The total external and internal loads lead to an average cooling load in administrative buildings of around 50 W m^{-2} (see Figure 1.18). The loads are typically dominated by office equipment and external loads (see Figure 1.19). External loads depend on the ratio

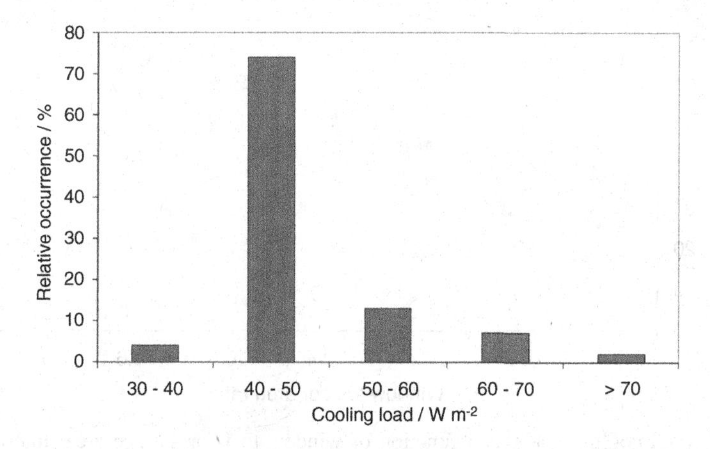

Figure 1.18 Occurrence of typical loads of office buildings in Germany

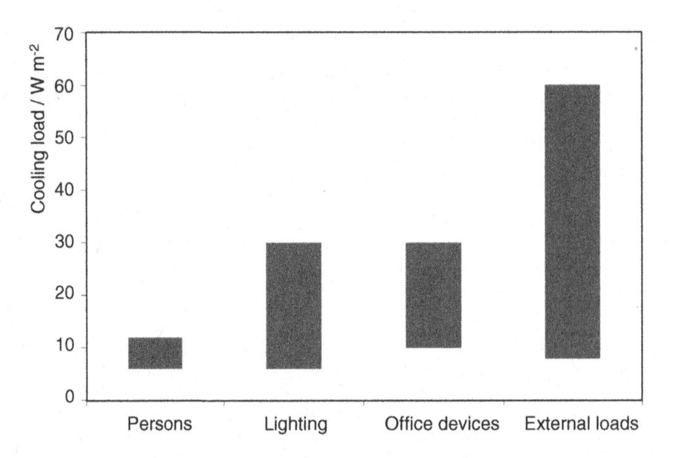

Figure 1.19 Bandwidth of cooling load distributions in office buildings (minimum and maximum)

between window and floor surface area and the shading system chosen. For surface ratios between 0.1 and 0.7, typical external loads are between 8 and 60 W m^{-2} (Arsenal Research, 2007). With internal loads varying between 17 and 29 W m^{-2} this results in typical total cooling loads between 40 and 90 W m^{-2} (see Figure 1.20). If buildings have very energy-intensive functions (such as computer centres, server rooms, etc.) the cooling load can be as high as 1000 W m^{-2} (see Figure 1.21).

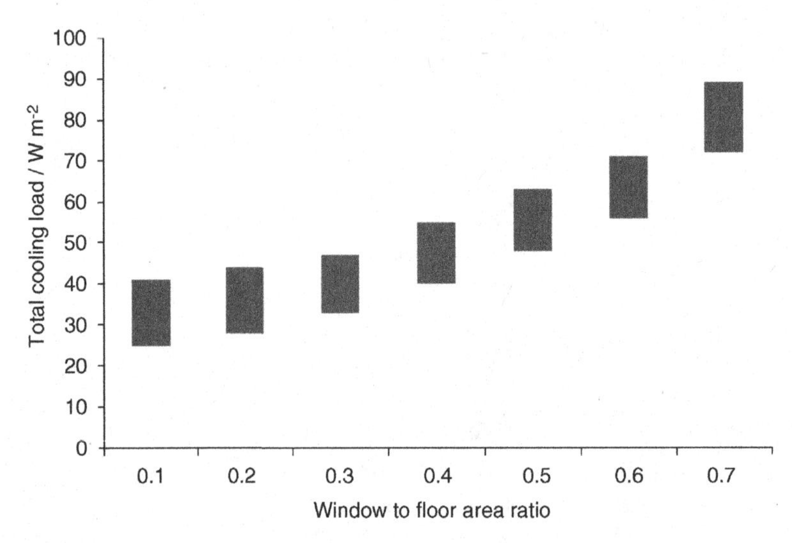

Figure 1.20 Total cooling loads as a function of window to floor surface area. Internal loads are between 17 and 29 W m^{-2}

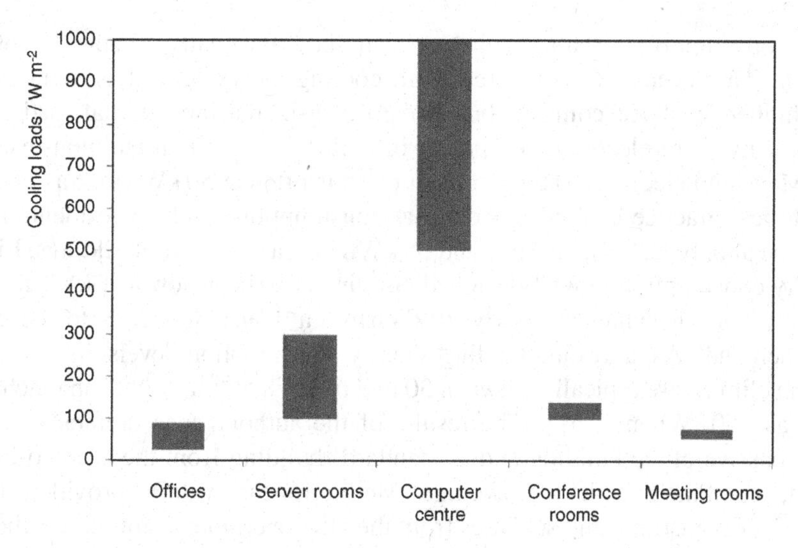

Figure 1.21 Range of cooling loads for different building and room types

1.3 Conclusions

The energy demand of existing buildings in moderate climates is dominated today by heating energy due to low levels of insulation standards. In warm climates cooling dominates the total consumption. However, as building standards continuously improve, there is a clear shift from the dominance of thermal energy to electrical

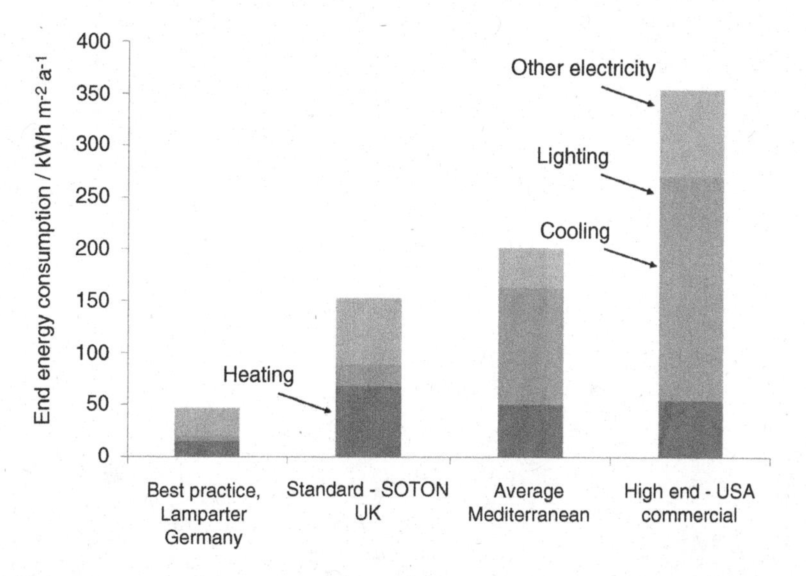

Figure 1.22 End energy consumption including thermal energy for heating and cooling and electrical energy for lighting and appliances from two case study projects and studies from Spain and the USA

energy consumption. Heating energy can be drastically reduced down to 15 kWh m^{-2} a^{-1} even in cold climates, while cooling energy demand is on the increase due to higher levels of comfort, but also to ever-increasing internal loads. These are caused by more electrical equipment in both the residential and non-residential sector. Measured electrical energy consumption is often at 50 kWh m^{-2} a^{-1} or more, although best practice examples show that consumption levels in residential buildings can be at only 12 kWh m^{-2} a^{-1} and 35 kWh m^{-2} a^{-1} in office buildings. Lighting electricity consumption in well-designed buildings can be as low as 5 kWh m^{-2} a^{-1}, but can also reach values of nearly 70 kWh m^{-2} a^{-1} and thus adds to the cooling energy demand. As a result, cooling energy consumption levels in a moderate European climate are typically between 30 and 60 kWh m^{-2} a^{-1}, but can reach values as high as 150 kWh m^{-2} a^{-1}. The results of the author's measurements in a best practice passive office building and a 'standard' building from the EU eco-building program, combined with the average Mediterranean values provided by the Catalan Energy Institute and statistics from the US government, summarize the status of building energy consumption today (see Figure 1.22).

2

Façades and Summer Performance of Buildings

Low-energy buildings in European and other moderate climates should have the following features:

- excellent thermal separation and low transmission losses between the inside and outside through a highly insulated building skin;
- high-quality glazing with U-values below at least 1.5 W m^{-2} K^{-1} and a reasonably high total energy transmittance with a g-value above 60%;
- heat recovery of ventilation air in winter for very high energy efficiency standards (passive building).

These features are all necessary to provide thermal comfort in winter at a low primary energy consumption. A good summer performance of sustainable buildings, especially offices and commercial buildings, depends on additional features:

- excellent sun protection through shading devices, if possible on the outside of the façade;
- night ventilation schemes to remove at least a part of the daily loads;
- earth heat exchangers to precool ventilation air or to directly cool part of the building mass.

Low Energy Cooling for Sustainable Buildings Ursula Eicker

Furthermore, internal loads should be as low as possible. This can be mainly achieved through optimized daylight use and energy-efficient electrical equipment.

Low glazing fractions certainly help to reduce the problem of unwanted external heat gains, but the reality is often highly glazed buildings. The role of the façade on external heat gains through transmission, and also through ventilation heat gains and secondary heat flows, is the subject of this chapter.

The current argument is that highly insulated buildings perform worse in summer, as the heat is trapped inside the building and cannot be dissipated by transmission through the building skin. For a start, this is only true if the external temperature is lower than the indoor temperature, so that there is a net heat flux from inside to outside. In hot climates, this is usually not the case. Also, under European moderate climate conditions, the mean summer temperature difference between the inside and outside is not significant, so that transmission losses are small anyway. Furthermore, highly insulated buildings are much more suited to control heat fluxes in summer: during the daytime with high external temperatures, the insulated building skin prevents unwanted transmission heat gains. During the night, heat removal can then be much more efficiently controlled through ventilation rather than transmission.

If the building is therefore highly insulated, the main external gains occur through the glazed façade part. The gains consist of two parts. The dominating part is usually related to the short-wave solar irradiance gain transmitted directly through the façade or indirectly as secondary heat flux from absorbed irradiance. This contribution can be efficiently reduced by shading devices. Experimental results on the total energy transmittance of different façades are presented in the experimental section. To analyse the main influences on the total energy transmittance, a dynamic façade model including new Nusselt correlations derived from particle imaging velocimetry and CFD (Computational Fluid Dynamics) simulations was developed and validated using temperature and heat flux measurements carried out in the laboratories of the University of Applied Sciences in Stuttgart (Fux, 2006).

The second part of external gains concerns ventilation gains, if the hygienically required fresh air is taken directly via the façade system. In naturally ventilated buildings or in buildings with only mechanical exhaust systems, this is nearly always the case. Air intake from the façade is also an issue in double façade constructions. The temperature increase through absorption on shading devices inside such double façades depends on air volume flow rates and the optical characteristics. The energy input to the building is a function of the air exchange between façade and building.

The main objective is to quantify the total summer energy input to the building through the façade system. To achieve this goal, laboratory experiments have been carried out using a large solar simulator to investigate flow and temperature conditions in different façades. The derived theoretical model was then validated on measurement results from office buildings with double façades.

2.1 Review of Façade Systems and Energy Performance

2.1.1 Single Façades

Glazed façades today are either single or double façades. In single façades, coated heat protection double glazing with low U-values between 1.0 and 1.7 W m^{-2} K^{-1} dominates the north European market, while uncoated double glazing with U-values between 3 and 3.5 W m^{-2} K^{-1} is still often used in Southern Europe. The total energy transmittance is given by the g-value, which adds the optical transmittance τ and the secondary heat flux q_i normalized by the incident irradiance G. The energy reduction coefficient F_c describes the ratio of shaded to unshaded total energy transmission and allows an easy classification of shading systems. The energy reduction coefficients of sun protection devices depend particularly on the location of the sun protection: external sun protection can reduce the energy transmission of solar radiation by 80–90%, whereas with sun protection on the inside a reduction of at most 55% is possible (see Table 2.1).

2.1.2 Double Façades

Overviews of the different double façade construction types are given by Gertis (1999), Lang (2000), Zöllner (2001) and Lee (2002), where the classification relates mainly to the type of segmentation and distinguishes between continuous air flow from bottom to top and various vertical and horizontal segmentations between the building storeys.

Gertis (1999) examined the subject of double façade systems very critically. Considering all of the physical influences (acoustic, fluid and thermal characteristics, energy, light penetration and fire protection), he concluded that double-glazed façades – except for special cases – are unsuitable for the north European climate and too expensive. Concerning summer performance, the main argument against double façades was the overheating of the façade gap and unwanted gains to the room. Gertis also considerd the simulation models very critically. In his opinion the models are not able to reproduce the thermal situation or the fluid behaviour as found in practice.

Hauser and Heibel (1996) compared measurements with simulations on different air preheating double façades with opaque and translucent outer layers and opaque

Table 2.1 Energy reduction coefficients of internal and external sun protection (Zimmermann and Anderson, 1998)

Sun shading system	Colour	Energy reduction coefficient
External sun shades	Bright	0.13–0.20
External sun shades	Dark	0.20–0.30
Internal sun shades	Bright	0.45–0.55
Reflecting glazing		0.20–0.55

inner layers. They obtained good agreement between the measurements and simulated results. The results showed that the ventilation heat losses during winter can be reduced significantly through air preheating.

Energetic comparisons of a conventional façade and a double façade system were performed by Oesterle *et al.* (1998). In their conclusions, the authors stated that the primary energy consumption of mechanical ventilation could be reduced by up to 25% by temporarily deactivating the fan and using double façade air. In these cases of no or only occasional air-conditioning, double façades may be economical.

Transient simulations of double façades were undertaken by Hausladen *et al.* (1998). Their conclusions showed that the ventilation gains during the winter period reduce the energy consumption significantly. However, especially in summertime, the problem of overheating requires an air cooling system in order to maintain thermal comfort. As a result, the marginal energetic advantage is not justified.

Similar investigations were carried out by Kornadt *et al.* (1999). The authors compared the physical characteristics of a single façade with different double façade systems. The simulations showed that buildings with double façades consume up to 40% less heating energy than those with a single façade. However, in summer the indoor temperature in rooms with a double façade system was about 8 K higher. Furthermore, the double façade system increases the total annual costs in comparison with a single façade by approximately 50%. Finally, the remaining advantage of double façades is improved sound protection.

An economic solution for a high-rise office building in a city centre was presented by Blum (1998). His indoor climate concept demonstrates that natural ventilation in combination with a ventilated cavity wall and with static heating and cooling may be an alternative to the conventional air-conditioning system. The main advantanges of a double façade system lie, in his opinion, in a controlled, natural air-conditioning especially for extreme external climatic conditions, sound protection, optimization of indoor illumination by daylight, reduction of energy consumption and reduction of yearly maintenance costs by up to 50%. The disadvantages are the slight limitation of thermal comfort in summer and the high investment costs.

Gerhardt and Rudolph (2000) suggested that window ventilation in double façades by users in high-rise buildings especially could be replaced by mechanical devices that reduce the size of the opening automatically depending on the wind speed.

Detailed investigations performed on a double-glazed façade by Kautsch *et al.* (2002) showed good agreement between the fluid flow measurements and CFD simulations. Their analysis focused on two extreme flow situations in the space between the façade layers: free, solar and wind-induced flow in the undisturbed gap between the primary and the room-side glass façade, and complex turbulent flow conditions at the double glazing surface as well as around the side elements between the façade layers. In addition to the fluid flow measurements, the thermal situation within the building was monitored and compared with TRNSYS simulations. These comparisons also show reasonable agreement.

The application of double façades in high office buildings with a high proportion of glazing in moderate climates has been investigated by Lee (2002). A detailed description of several realized façade constructions is presented. Theoretical examinations with the simulation program TRNSYS showed positive results for some of the construction types.

Jachan (2003) examined the thermal behaviour of different wall constructions with and without double-glazed façades. Experimental investigations on a double façade were compared with results obtained by the simulation program ESP-r. When comparing the investment costs as well as the costs for heating and cooling, Jachan found that the double-glazed façade is less economical than a well heat-protected conventional façade.

Different strategies for optimizing the energy efficiency of multiple-skin façades were studied and compared with the results of traditional cladding systems by Saelens *et al.* (2005). Here, three façade types (a mechanically ventilated exhaust air façade, a naturally ventilated double skin façade and a mechanically ventilated supply air façade) were analysed. Their results show that the supply air system through the double façade has the highest potential to benefit from the optimization techniques. The double skin façade is also able to control the cooling demand but is limited with respect to improvements in the heating demand. The exhaust air system is capable of significantly lowering the heating demand but is still subject to high cooling demands.

The performance of ventilated façades was numerically simulated for the Mediterranean climate of Barcelona. Most simulations were done for rather high flow velocities of $2\,m\,s^{-1}$ in an air gap of 10 cm width and without shading devices. No coupling of the façade air flow and the room was considered. Predictably, the summer heat gains were dominated by the transmission of solar irradiance (75% of total gains in July). The lowest total energy transmittance of 17% was obtained when two-thirds of the façade were closed by a shading device and an opaque zone at the bottom (Faggembau *et al.*, 2003).

A double façade thermal and air flow model was applied to a 30 m high office building in Prague with a continuously ventilated double façade (Hensen *et al.*, 2002). A maximum air temperature increase of 12 K was calculated for a shading element with 57% absorption. Compared with a single double-glazed façade with internal blinds, the cooling loads were reduced by 7–15%, depending on the floor level. The lower cooling loads are partly due to a lower optical transmission of the double façade with three glass layers.

The influence of the shading device position and colour within a double façade on the summer cooling load was evaluated using the TAS thermal simulation software (Gratia and De Herde, 2007). A surprisingly high influence of the position of the blind on cooling load was obtained. Depending on the colour of the blind, cooling energy reductions of 13.5 to 14.1% were calculated when the blind was moved from a position close to the inside glazing to the centre of the gap. Also, very high blind temperatures up to 69.8 °C were obtained for a blind with 42% absorption. For a lightly coloured blind with a 17% absorption coefficient, the temperature level at the same blind position (close to the inside) decreased to 57.8 °C. These results were calculated for a closed double façade. If the façade is open at the top and bottom, temperature levels decrease by about 18 K and the cooling energy demand by a further 6 to 9%. The main question about this work is that no experimental results are given for validation and the modelling approach for heat transfer and fluid flow is not detailed.

A special double façade solution is that of the ventilated photovoltaic façade (PV façade). A PV hybrid façade can be regarded as an architectural design concept, which might be compared with a representative granite façade and could be even more economical (Auer, 1998). By replacing the outer glazing with a semitransparent PV module, electricity and heat are generated and daylight can still be partially transmitted. Since the efficiency of PV cells drops with increasing temperature, the heat should be dissipated as much as possible and warm air can be channelled directly into the building during the wintertime. In summer natural convection should be used for heat dissipation, unless the heat can be used for active thermal cooling technologies. Due to the high absorption of the PV module, high gap temperatures in summer increase the building's cooling load.

The advantages of a hybrid PV façade can be summarized as follows:

- The transmission heat loss is reduced during the heating period.
- Ventilation losses are reduced if the preheated air is fed into the building.
- The semitransparent PV façade forms a natural shading element. The façade creates glare-free light.
- During the summer period, the building may be ventilated through the façade gap (only when there is no or low solar irradiance).
- The façade offers increased sound protection against external noise.

The disadvantages are:

- High electricity consumption for ventilation in summer, if the façade is ventilated mechanically.
- The semitransparency of the PV modules prevents maximum utilization of solar gains in winter.
- High costs.

A methodology has been developed by the author and collaborators (Mei *et al.*, 2002, 2003; Infield *et al.*, 2004; Eicker *et al.*, 1999, 2000) for calculating the thermal impact on building performance of an integrated ventilated PV façade. This is based on an extension of the familiar U-and g-values to take into account the energy transfer to the façade ventilation air. For a public library in Spain constructed with a ventilated PV façade, the energy-saving and electrical gains for different climatic conditions were compared with a conventional double glazed façade (Vollmer, 1999). A conventional double-glazed façade is more economical than a PV façade during the wintertime. However, in the special case of a library, glare-free light is an important factor and expensive shading devices would be necessary. Additional cooling loads associated with double-glazed façades during the summer also apply to the PV system, which, however, provides useful electrical gains.

Meyer (2001) analysed the influence of a hybrid double façade system on the adjacent room climate. Detailed measurements of external and internal climatic conditions were compared with numerical results from the simulation program TRNSYS. He found that winter transmission losses may be reduced by up to 50%. The required electricity for mechanical ventilation of the façade amounted to approximately 10% of the achievable gains.

The most effective application for hybrid façades may be considered in combination with solar cooling components. For example, the high façade outlet temperatures may be used to support a desiccant cooling system (Eicker *et al.*, 1998; Höfker, 2001).

Despite the controversy, in Germany approximately half of the large office buildings are presently being designed with glass double façades. At the same time, the planning and realization of double façades require a considerable effort which is often economically unjustifiable. The strong arguments between advocates and opposers of glazed double façade systems still lead to uncertainties in the planning process.

The thermal situation of glazed double façades can be summarized as follows: because of the thermal buffer present in most double façade types, transmission heat losses in winter can be reduced. Ambient air can be preheated and if intelligent control is used, running times of mechanical ventilation systems can be reduced and energy saved. However, combining air preheating from the façade and heat recovery from exhaust air is difficult, as this could only be realized decentrally in every room. It should be mentioned that in office buildings, high internal gains lead to a diminished heating demand, so that the energy-saving potential in these buildings is often not as large as might be expected. In the literature, quoted energy savings range from 'marginal' to over 25, 40 or 50%.

During the summer period, the glazed double façade increases the façade gap temperature as well as the indoor temperature, which results in an increasing demand for cooling. Shading devices within the façade gap may reduce the incoming solar radiation with the further advantage that the blinds will be protected from wind forces and other ambient influences. However, these blinds inside the double façade gap lead to an additional increase of the gap temperature. As a result, the heat transmission rate is higher than for a façade which has direct contact with the ambient air.

2.1.3 Modelling of Ventilated Façades

To model the thermal characteristics of ventilated façades, a simulation program was developed at the University of Applied Sciences in Stuttgart, which contains a new, experimentally derived empirical Nusselt correlation for the ventilated gap as well as a representation of the transient response of the whole façade system (Fux, 2006). The main motivation for studying Nusselt correlations in detail was that ventilated façades often have large gap sizes, where channel geometries for Nusselt number calculations do not give correct results and fluid flow over single plates has to be considered. Furthermore, wind-induced air flow in the façade as a forced convection term cannot be neglected, as it is often of the same order of magnitude as the buoyancy-driven flow rate. A third reason for carrying out an experimental investigation using particle imaging velocimetry and CFD simulations for parameter studies is that highly absorbing elements in the façade such as PV panels produce strongly asymmetric temperature boundary conditions, for which Nusselt correlations from the literature cannot be applied. The main results of the heat transfer analysis were:

- Nusselt correlations taken from the literature and CFD simulations can reproduce natural convection situations with small differences in surface temperatures reasonably well. For large differences in temperature, the results of the CFD simulation strongly depend on the operating density used. For plate distances of 20 cm or more, free convection over a vertical single plate fits the experimental results better than channel flow.
- For forced flow, also both approaches of flow over single vertical plates or flow between parallel plates were analysed. The best results for measurements and CFD simulations were obtained by a combination of mixed flow over a heated single plate ($Nu_{H,mix}$) and an expression for the fully developed turbulent flow within a channel ($Nu_{H,fd}$) which corresponds directly to an equation developed by Petukhov (see Equation 2.2).

$$Nu_H = \sqrt[3]{\left(Nu_{H,mix}^3 + Nu_{H,fd}^3\right)} \tag{2.1}$$

$$Nu_{H,fd} = \frac{\xi/8\, Re_d\, Pr}{1 + 12.7\sqrt{\xi/8}\left(Pr^{2/3} - 1\right)}\left[1 + \left(\frac{d}{H}\right)^{2/3}\right]\left(\frac{H}{d}\right) \tag{2.2}$$

The friction factor $\xi = 0.18/Re_d^{0.2}$ and the transforming factor H/d must be used in order to transform the Nusselt number Nu_d, which relates to the plate distance d, into the Nusselt number Nu_H, which is based on the plate height H.

The term $Nu_{H,mix}$, which includes the Nusselt correlation for a single plate, was developed for two different flow rates. For $u < 1.0\,\mathrm{m\,s^{-1}}$ (as the average velocity between both plates) $Nu_{H,mix}$ is of the form

$$Nu_{H,mix} = \left(-0.32\frac{u_{gap}}{u_0} + 0.32\right) Re_{H,res}^{c}\sqrt{Pr} \tag{2.3}$$

where $u_0 = 1.0\,\mathrm{m\,s^{-1}}$ and the exponent c is 0.63 for symmetrical plate temperatures or plate distances $d > 0.4$ m. For asymmetrical plate temperatures and plate distances $d < 0.4$ m, the exponent $c = 0.61$. For higher velocities ($u > 1.0\,\mathrm{m\,s^{-1}}$), u_{gap} is fixed to a maximum permissible value of $1.0\,\mathrm{m\ s^{-1}}$. In this case, the relation u_{gap}/u_0 is 1.0 and $Nu_{H,mix}$ becomes zero. Consequently, Equation 2.1 will be reduced to Equation 2.2.

The Reynolds number $Re_{H,res}$ is calculated from both free and forced flow conditions, where asymmetric plate temperatures for the warm and cold side of the façade gap can be used.

$$Re_{H,res} = \sqrt{Re_{H,force}^2 + Re_{H,free}^2} \tag{2.4}$$

$$Re_{H,force} = \frac{u\,H}{\upsilon};\ Re_{H,free} = \sqrt{Gr_H/2.5}$$

$$Gr_{H(WS)} = \frac{g\,\beta\,|(T_{(WS)} - T_m)|\,H^3}{\upsilon^2};\ Gr_{H(CS)} = \frac{g\,\beta\,|(T_{(CS)} - T_m)|\,H^3}{\upsilon^2}$$

$Gr_{H(WS)}$ represents the buoyancy-driven flow for the warmer plate and $Gr_{H(CS)}$ that of the colder plate.

The new correlation is able to cover a wide range of boundary conditions for mixed flow in façades: namely, plate distances between 5 and 50 cm, inlet air temperatures between −10 and +30 °C, surface temperatures between −10 and +60 °C and Reynolds numbers Re_d (for plates at distance d) between 500 and 6500 (corresponding to flow velocities analysed between 0.06 and about $2.0\,\mathrm{m\,s^{-1}}$). Compared with the simpler approaches chosen in the European standard DIN EN 13363-2 and the international standard ISO/DIS 15099, the correlation agrees excellently with experiments and CFD simulations. While the ISO/DIS 15099 method agrees quite well with the CFD results for some cases, the simplified DIN EN 13363-2 model usually leads to higher heat transfer rates.

The dynamic thermal model is based on a numerical solution of the one–dimensional heat conduction equation applied to both sides of the ventilated façade. Both sides are coupled by long-wave radiative heat exchange across the air gap, which can be up to 90% higher than the convective part. After solving the heat conduction equation on both sides of the ventilation gap in each simulation time step, the determined radiative heat flow is added as heat gain to the colder surfaces and is subtracted as heat loss from the warmer surfaces. Then the calculation of the actual time step is repeated with the new heat gains and the resulting layer temperatures are again determined. This procedure is repeated several times for each time step. The calculations have shown that three iterations usually provide a sufficient convergence in temperature. The model has been implemented both as an independent software tool (TransFact)

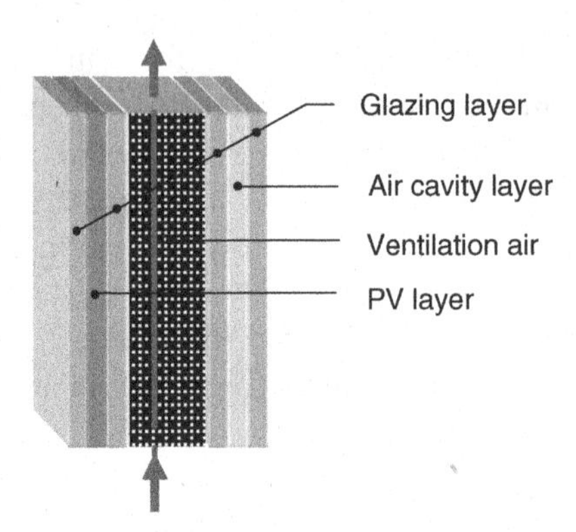

Figure 2.1 Representation of a hybrid solar façade with seven different layers for the simulation model

and as a type which can be integrated into dynamic building simulation tools (e.g. as type 111 in TRNSYS).

Generally, a ventilated façade system consists of an air space between two solid boundary walls. In the developed simulation model, the solid façade construction can be as complex as desired. An example of a possible construction for a hybrid solar façade is shown in Figure 2.1.

A PV layer embedded between two glazing layers forms the external boundary and a double glazing window the internal boundary. The air space between the outer and inner layers may be ventilated by forced convection or naturally driven flow. Apart from the construction described above, the following choice of layer types can be combined arbitrarily:

- Massive opaque layer with optional heat sources.
- Massive transparent layer (pane) with varying optical properties such as reflectance, absorptance and transmittance (the optical properties may be used optionally depending on the incident angle).
- Massive PV layer. Semitransparent layer with PV elements, considering converted electrical power depending on layer temperature.
- Ventilated air gap (massless). Consideration of long-wave radiative heat exchange between adjacent layers as well as convective heat transfer.
- Air cavity (massless). Consideration of long-wave radiative heat exchange between adjacent layers as well as convective heat transfer.

2.2 *Experimental Results on Total Energy Transmittance*

2.2.1 Laboratory Experiments

The façade causes additional summer cooling loads through direct transmission of short-wave irradiance, secondary heat fluxes from the internal surface temperature nodes and through ventilation gains, if the fresh air is supplied by the façade. The total energy transmission or g-value of a glazed façade adds the contributions of transmission and secondary heat flux and is defined as

$$g = \tau + \frac{q_i}{G} \tag{2.5}$$

where τ is the short-wave transmission coefficient of the whole façade system, G the incident external irradiance and q_i the secondary heat flux from the inner glazing surface to the room. Especially when using shading devices inside the room or in a façade gap, secondary heat fluxes are difficult to calculate. Often g-values are only estimated, which is a real problem for cooling load calculations and consequently the design of the cooling supply system. The solar simulator test rig at the University of Applied Sciences in Stuttgart provides the possibility of experimentally determining the overall energy transmission coefficient of façade systems up to 2.6 m in height.

Experimental Set-up

The simulator contains 15 Hg lamps of 1000 W electrical power each in a planar arrangement (see Figure 2.2). For special applications, the lamp field of the solar simulator can be rotated from a horizontal to a vertical position. To avoid a temperature increase in the room, each of the five lamp fields is mechanically ventilated and the heat discharged to ambient air. A ventilated air curtain between the light source and the test façade prevents long-wave heat exchange between the hot cover glass of the lamps and the outer façade surface. The temperature of the air curtain can be adapted to the surrounding temperature.

Figure 2.2 Photo of the test façade and solar simulator

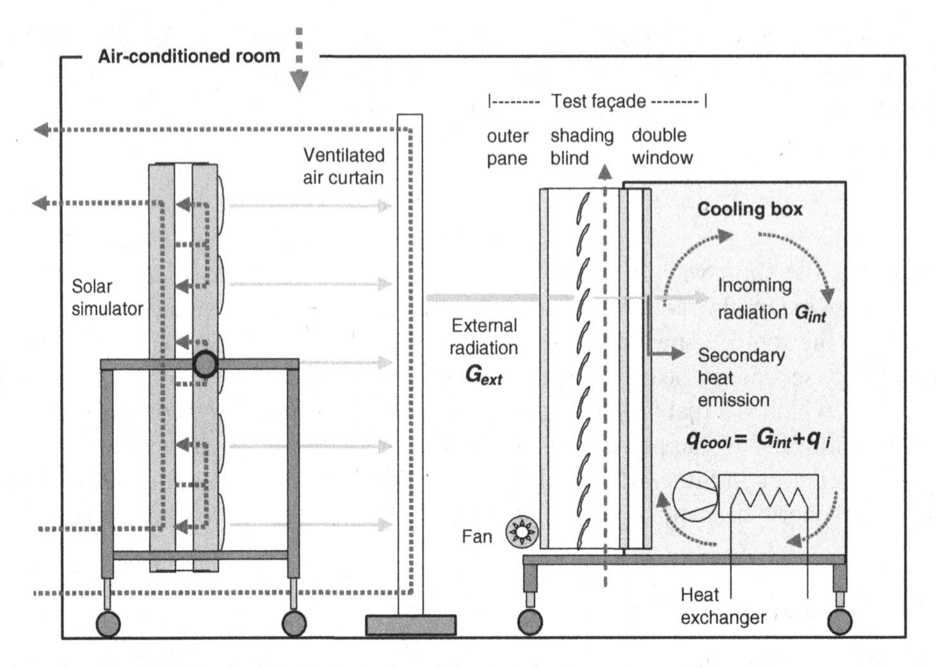

Figure 2.3 Schematic representation of experimental set-up

In the cooling box, the transmitted energy is exactly compensated by the cooling power transferred via a heat exchanger so that the box temperature stays constant (see Figure 2.3). Consequently, the required cooling energy corresponds to the transmitted radiation plus the secondary heat flux from the inner pane. To avoid transmission heat loss through the well-insulated box walls, the temperature of the ambient room is as close as possible to the temperature level within the test chamber. In order to reach convective heat transfer coefficients that correspond closely to the standard values given in the German standard DIN 4108-4 (2002) (outside 25 W m^{-2} K^{-1}, inside 7.7 W m^{-2} K^{-1}), ventilators generate a controlled flow on both sides of the façade. The set-up is based on a calorimetric reference method developed by a German research consortium (Sack *et al.*, 2001).

The dimensions of the test façade are 2.6 m × 0.7 m. In the case of a double façade system, the distance from the inner layer to the outer pane may be varied by up to 0.8 m. Surface temperatures are determined by thermocouples as well as PT100 sensors. All sensors were shielded from direct irradiance. All air temperatures are integrated over 3 m long Ni1000 sensors. The external irradiance is averaged from pyranometer measurements at 65 locations over the whole façade area. The maximum deviation from the mean value is 14% (see Figure 2.4). The system needs approximately 5 hours to reach steady-state thermal conditions.

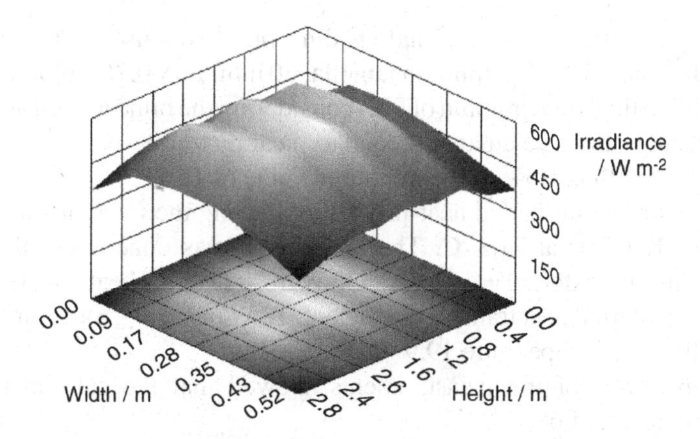

Figure 2.4 Distribution of irradiance from the solar simulator

Single Façade Energy Transmittance

The accuracy of the total energy transmission measurements was checked over a range of single façade glazing with different U and g-values.

First a low-e coated double glazing (Isolar Neutralux premium) was measured and compared with data from the manufacturer with $U = 1.4\,\mathrm{W\,m^{-2}\,K^{-1}}$, $g = 0.63$ and $\tau = 0.52$ for the optical transmission.

The experiment was carried out at boundary conditions of $475\,\mathrm{W\,m^{-2}}$ external irradiance, an external heat transfer coefficient of $25\,\mathrm{W\,m^{-2}\,K^{-1}}$, an internal heat transfer coefficient of $10\,\mathrm{W\,m^{-2}\,K^{-1}}$, a room air temperature of 21.5 °C and cooling box temperature of 22.2 °C. Measured, simulated and manufacturers g-values correspond well (g-value measured: 0.64; calculated: 0.63; manufacturers information: 0.63). The surface temperature level at the inside facing the room reached 27.6 °C (see Figure 2.5).

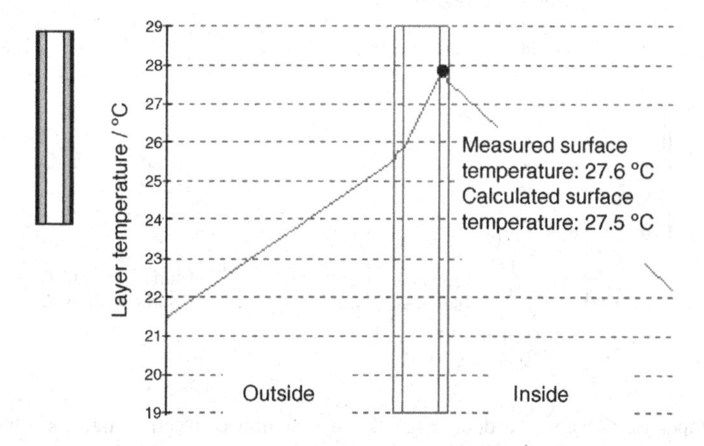

Figure 2.5 Temperature profile of double glazing without shading system

In a second experiment, an external shading blind was used which reduces the incoming radiation to 17% (aluminium lamella, 80 mm, $\rho = 0.76$, colour RAL 9010). With an overall optical transmission of 9% (considering the blind as well as the thermal protection glazing) and a calculated secondary heat flux of $30\,\mathrm{W\,m^{-2}}$, the calculated g-value is 13% and thus close to the measured value of 15%.

Compared with the unshaded heat-protecting glazing the internal surface temperature is now 6 K lower at 21.3 °C. The experiment was done under the following boundary conditions: external irradiance: $535\,\mathrm{W\,m^{-2}}$; external heat transfer coefficient $25\,\mathrm{W\,m^{-2}\,K^{-1}}$; internal heat transfer coefficient: $10\,\mathrm{W\,m^{-2}\,K^{-1}}$, room air temperature 21.9 °C; cooling box temperature 19.9 °C.

Several repetitions of the measurements showed that the calorimetric method achieves an accuracy of 6%.

Double Façade Energy Transmittance

Finally, an additional single 6 mm pane ($\tau = 0.8$, $g = 0.83$, $U = 5.8\,\mathrm{W\,m^{-2}\,K^{-1}}$) was placed in front of the shading blind corresponding to a double-glazed façade ventilated by natural convection.

The calculated overall transmission was 0.06, and the comparison between measured and calculated g-values showed similar results (measured: 0.10; calculated: 0.09). The boundary conditions were as follows: external irradiance $590\,\mathrm{W\,m^{-2}}$; external heat transfer coefficient $25\,\mathrm{W\,m^{-2}\,K^{-1}}$; internal heat transfer coefficient $10\,\mathrm{W\,m^{-2}\,K^{-1}}$; room air temperature 21.3 °C; cooling box temperature 20.5 °C. The measured temperatures agree well with the calculated results (see Figure 2.6). In this case of buoyancy-driven flow, the Nusselt correlation as well as the Reynolds number are as in Olsson's work (Olsson, 2004). Table 2.2 summarizes the measurement results

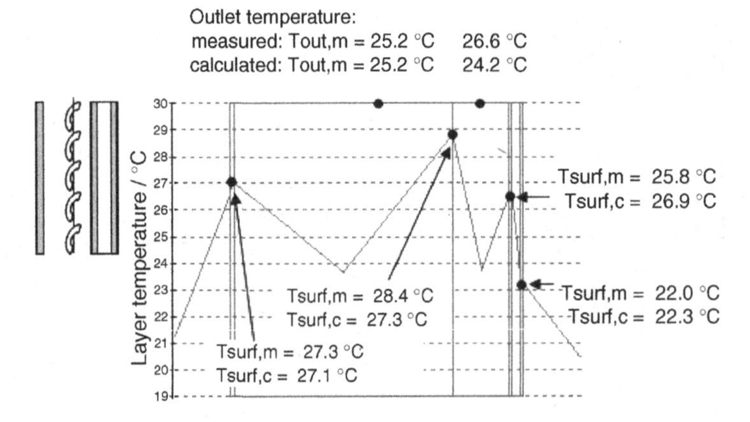

Figure 2.6 Temperature levels in a double façade with natural convection and a shading element between the double glazing

Table 2.2 Measured characteristics of single and double façades with and without blinds

	Single façade	Single façade + ext. blind	Double façade	Double façade + gap blind
g-value	0.64	0.15	0.51	0.12
Transm. int. window	0.52	0.52	0.52	0.52
Transm. ext. pane	—	—	0.80	0.80
Total transmission	0.52	0.09	0.41	0.07
q_i/G	0.12	0.06	0.10	0.05
Surface temp. ext.	—	21.3	27.4	27.7
Blind temp.	—	—	—	30.3
surface temp. gap	—	—	30.6	25.8
surface temp. room	25.3	21.3	28.9	22.6
Air inlet temp.	—	—	24.3	22.4
Air outlet temp.	—	—	28.4	26.8
Room temp.	20.6	19.9	24.0	21.0
Ambient temp.	19.6	21.9	22.9	21.4
Average irradiance	430	535	451	464

from the single façade with low-e coated glazing and a double façade with and without blinds. The free cross-section for air entry in the double façade was always 14% of the total façade surface area.

If the free cross-section of air entry is modified in the double façade, temperature levels increase with decreasing opening section and flow velocity. Whereas the air temperature increase was only 2 K for a 44% cross-section, it increased to 6 K for a smaller opening of 10% (see Figure 2.7). To calculate the free cross-section, the air entry area is related to the total surface area of the façade element. Although

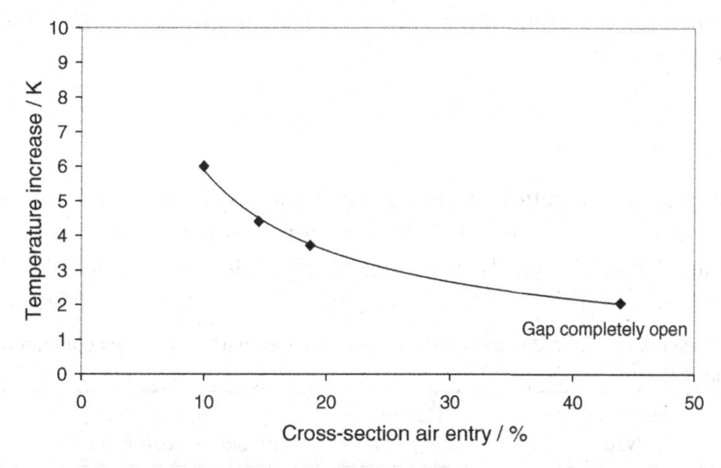

Figure 2.7 Air temperature increase as a function of the free cross-section of the air entry in relation to the total façade surface area

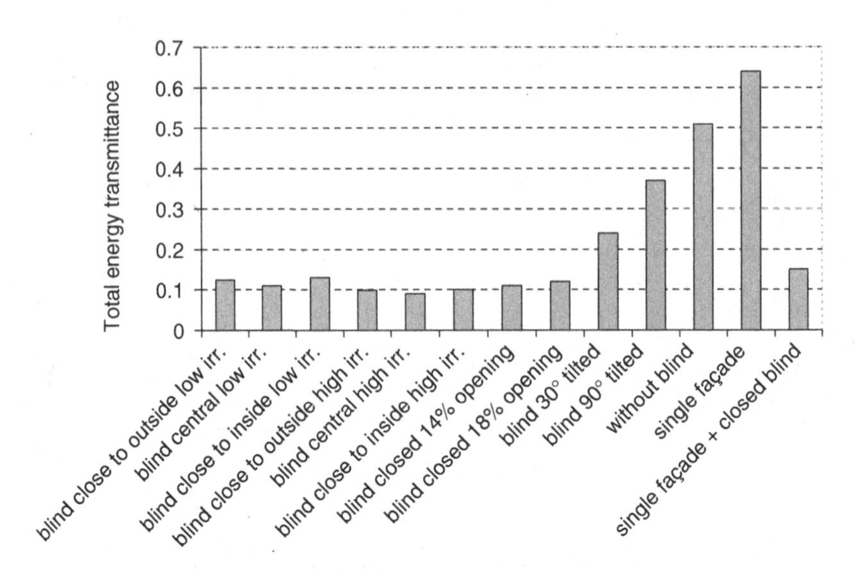

Figure 2.8 Summary of laboratory *g*–value measurements for double façades with varying blind position, changing air gap free cross-section and blind tilt angle. The last two bars show a single façade without and with a blind

temperature levels increase with decreasing air flow rates, the measured influence on the *g*-value is negligible.

Measurements were also taken on the double façade system where the sun shading position was varied within the 0.5 m air gap. The influence on *g*-value and temperature levels was negligible. Within the measurement error of the calorimetric equipment, neither blind position nor free cross-section of air entry have a significant influence on the total energy transmittance (see Figure 2.8).

The measured energy reduction coefficients F_c can now be compared with values from the literature (Table 2.3).

2.2.2 Building Experiments

An experimental investigation of a double façade system similar to the laboratory set-up was carried out at the Zeppelin Carré business centre in Stuttgart (see Figures 2.9 and 2.10). This partly historical building block was restored in 1996 and a

Table 2.3 Measured energy reduction coefficients of façades with sun shades compared with values from the literature

Façade	Sun shading system	Colour	Energy reduction coefficient	Literature values
Single	External	Grey	0.23	0.13–0.30
Double	Gap integrated	Grey	0.23	

Figure 2.9 Detailed view of double façade with monitoring equipment

double-glazed façade system was part of an integrated energy concept (AIT, 1998). The façade system consists of a 2.95 m high and 1.55 m wide outer single glazing and an inner low-e coated double glazing with a U-value of 1.4 W m^{-2} K^{-1} and a g-value of 0.62. The air space between both panes is 0.5 m, and the openings at the top and bottom are 0.15 m high, which corresponds to 10% free opening section. The external single glass pane in the façade is thicker than the laboratory glass (16 mm with $\tau =$ 0.69 instead of 6 mm with $\tau = 0.84$), resulting in 13% less optical transmission; the other optical characteristics are the same.

The sun shading system installed consists of lamellae of 0.05 m width, which are mounted at a distance of 0.175 m from the outer single glazing. Three cases with different opening positions of the shading devices were compared. Case A: without sun blind; Case B: sun blind with 45° lamella position; Case C: sun blind totally closed (see Figure 2.11).

Figure 2.10 Office building with double façade system in Stuttgart, Germany

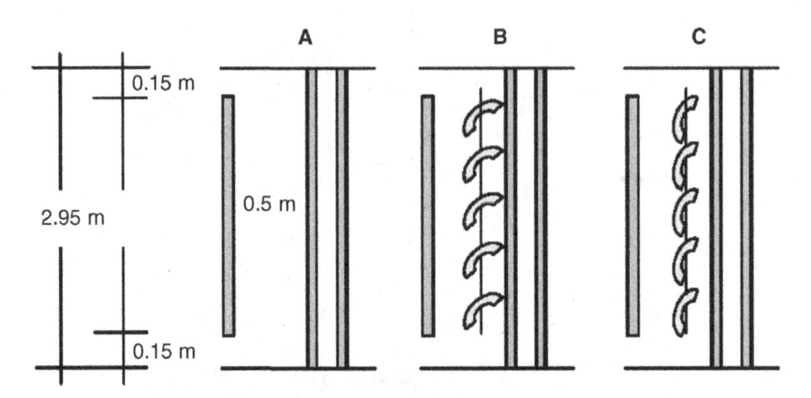

Figure 2.11 Schematic view of the façade system at the Zeppelin Carré building in Stuttgart, Germany

It is known that due to solar irradiance absorption the ambient air temperature in front of a façade is higher than a shaded air temperature sensor. Here the inlet air temperature of the double façade was already 1–2 K higher than ambient air temperature. The asymmetric placing of the sun shades in a façade gap of 0.5 m leads to different flow velocities in the narrower external gap and the wider gap towards the double glazing. Detailed measurements were taken in 2004 on temperatures at different heights of the south-west-facing double façade and on flow conditions. For a completely closed blind position (Case C), maximum blind temperatures of 44 °C were measured at an irradiance of about 600 W m^{-2}. The blind temperature difference between the top of the façade and the bottom was a maximum of 3 K. For high irradiances between 500 and 600 W m^{-2}, the blind temperature was about 10–12 K higher than the inlet air temperature (see Figure 2.12).

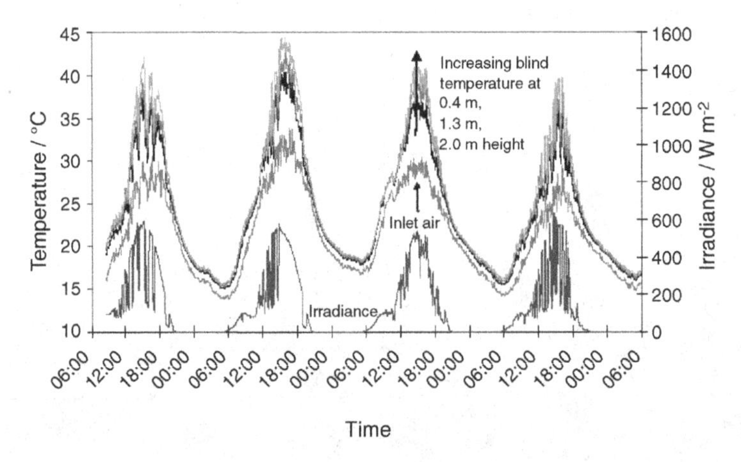

Figure 2.12 Inlet air and blind temperatures at different heights in the façade gap together with irradiance conditions over 4 days

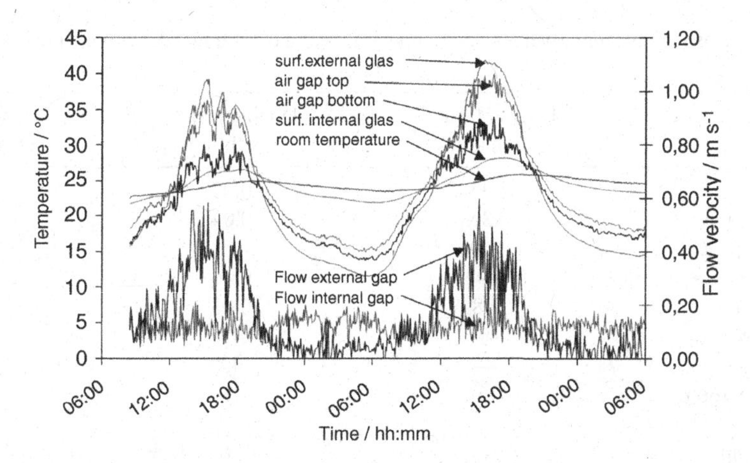

Figure 2.13 Surface and air temperatures and flow velocities in a double façade

In the external gap with a 0.175 m distance between the outer glass pane and the blind, the flow velocities varied between 0.05 and 0.6 m s^{-1}; in the inner, wider gap of 0.325 m distance the flow velocities were between 0.1 and 0.3 m s^{-1}. The maximum temperature increase of the air gap was 8 K. Due to the high temperatures in the double façade, the inner surface of the heat-protecting glazing increases from night temperatures of 22 °C to 28 °C during the day, causing additional cooling loads (see Figure 2.13). The boundary conditions correspond to the first 2 days shown in Figure 2.12.

Further measurements on the south-eastern façade gave similar results. The mean temperature increase in the air gap was 6.5 K at an average irradiance of 560 W m^{-2}, while the gap entry temperature was 2.8 K higher than ambient air. The mean flow velocity in the narrow gap was 0.41 m s^{-1} and 0.14 m s^{-1} in the wider gap. The blind temperature reached 45 °C maximum. In the laboratory with air inlet through the complete façade gap (44% cross-section), the blind temperature was only 5–6 K above the inlet air temperature. A reduction of the inlet air free cross-section down to 14%, which is close to the 10% opening in the Zeppelin Carré building, reduced flow velocities and increased temperatures. Typically, blind temperatures were then about 8 K above inlet air temperatures and maximum temperature increases of the façade air were 5 K. These slightly lower temperature levels indicate that flow conditions on site in real buildings are more turbulent than laboratory measurements with nearly undisturbed air flow entry into the façade. Finally, measurements were taken to determine the g-value of the building façade. The short-wave irradiance transmission was measured using two pyranometers inside and outside of the façade. The secondary heat flux can only be directly measured if no irradiance strikes the sensor; that is, for the completely closed blinds. From those heat flux measurements the internal heat transfer coefficient was determined at 10 W m^{-2} K^{-1} and used to calculate the other secondary heat fluxes. The averaged results are listed in Table 2.4.

Table 2.4 Measured average g-values and temperatures of a double façade with different sun shading positions

	Double façade no shading	Double façade 40° blind	Double façade closed blinds
g-value	0.43	0.07	0.03
Tot. transmission τ	0.35	0.04	0.01
Transm. ext. pane	0.69		
Energy red. F_c	—	0.16	0.07
q_i/G	0.08	0.03	0.02
Surface temp. ext.	40	41	42
Blind temp.	—	43	38
Surface temp. room	36	28	29
Air inlet temp.			
Air outlet temp.	36	36	37
Room temp.	30	25	27.5
Ambient temp.	30	27	29
Average irradiance	700	700	470

When comparing measured and calculated g-values, the results are practically identical. Whereas the simulated surface temperature levels deviate by a maximum of 1 K, the simulated outlet air temperatures are 2–4 K lower than the measured values. This again is due to the heat transfer correlations derived from laboratory experiments with lower turbulence. The total g-value of the façade with closed blinds is only 3% compared with 12% in the laboratory experiment. The main reasons are:

- a high average incidence angle of 80° in the real building with 70% less transmission;
- 13% lower transmission of the thicker external pane;
- about 10% less transmission due to dirtying of the real façade.

In conclusion it can be said that double façades with integrated sun shading elements effectively reduce external gains through solar irradiance with measured g-values as low as 3% for completely closed blinds and 7% for partly closed blinds. If low-e coated glazing is used to separate the air gap thermally from the room air, the secondary heat fluxes are also low. However, the ambient air is heated significantly through absorption in the double façade. If this air is used to ventilate the adjacent rooms, ventilation heat gains occur, which reduce energy consumption in winter, but add to the cooling load in summer.

2.3 Cooling Loads through Ventilation Gains

2.3.1 Double Façade Experiments

Ventilation heat gains through a double façade depend on the air exchange rate between the façade and the adjacent room and on the absorption characteristics of the glazing

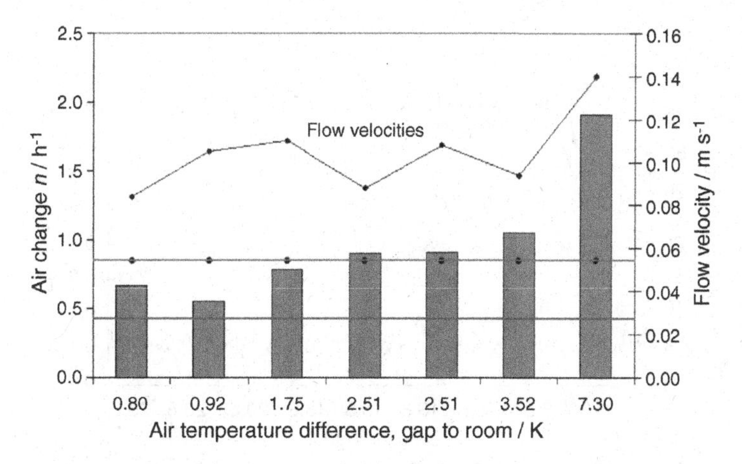

Figure 2.14 Measured air change n as a function of temperature difference between façade air and room air. The two horizontal lines indicate the minimum hygienic air exchange rate for one and two persons

and integrated blinds. The air exchange rates were measured in summer 2004 in the Zeppelin Carré building using a tracer gas system from the Dantec Company. Concentrations were measured in three locations of the room simultaneously. Both the shading blind position and the window opening size were varied. The position of the blinds, which can be tilted to six different angles between closed and fully open, did not have any influence on the air exchange rate. There is no clear dependency of air change rate on temperature difference between the façade and room air. Measurements on one tilted window showed an air change increase with temperature difference (see Figure 2.14), whereas with two tilted windows there even a decrease was measured (see Figure 2.15). The main influence is the opening cross-section of the window: if only one of the two windows is tilted, the mean air exchange is $0.97\,h^{-1}$; if both

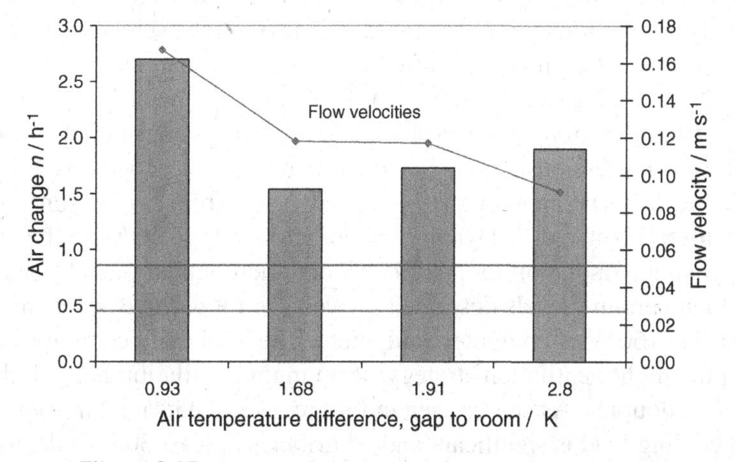

Figure 2.15 Measured air change n for two tilted windows

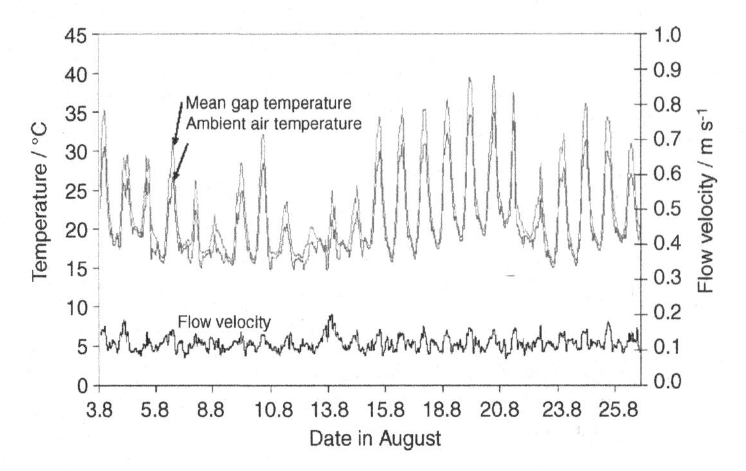

Figure 2.16 Flow velocities, ambient air and mean gap temperature

windows are tilted, the air exchange increases to 1.96 h^{-1}. Both sets of measurements were taken with the internal doors closed. If the internal doors are open, the cross-ventilation increases the air exchange rate up to 50 h^{-1}.

The measured flow velocities during the month were quite constant and reached values between 0.1 and 0.2 m s^{-1} (see Figure 2.16). This corresponds to air exchange rates with the room between 0.5 and 1.7 h^{-1}. Average daily heat gains through ventilation for a fixed air exchange rate of 0.97 h^{-1} are 36 Wh m^{-2} d^{-1}. The gain was calculated from the volume flow rate and the temperature difference between the façade outlet and room air. As the tilted window starts drawing in air from the centre of the gap onwards, it might be more realistic to use the mean gap temperature as the room inlet temperature. The ventilation gain then reduces to 25 Wh m^{-2} d^{-1}. If ambient air were used for ventilation the heat gains for the month considered would be only 11 Wh m^{-2} d^{-1}. If both windows are tilted and a fixed air exchange rate of 1.96 h^{-1} is used, the daily heat gain from ventilation using the mean gap air temperature is 54 Wh m^{-2} d^{-1}. The highest value of 86 Wh m^{-2} d^{-1} is obtained if the temperature difference between the top of the façade and room air is considered and if the empirically derived regression is used to calculate the air exchange rate for one tilted window. These values compare with typical daily gains from two persons in the office considered of 66 Wh m^{-2} d^{-1}. Dynamic building cooling loads were then calculated using the dynamic simulation tool TRNSYS and the boundary conditions of air exchange and temperature levels described above. The room loads were calculated for two persons and 160 W of computer equipment. The total daily cooling load strongly depends on the night ventilation strategy: if no night ventilation is used, the cooling load more than doubles. As can be seen in Figure 2.17, the heat gain from ventilation to the total cooling load is significant and contributes at least 30% of the total load.

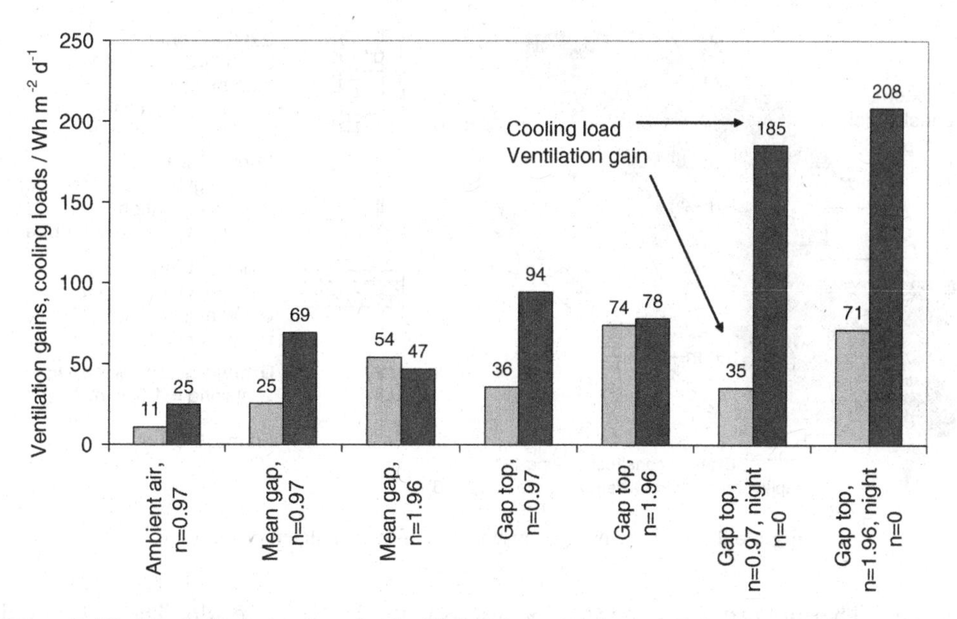

Figure 2.17 Ventilation gains from the double façade for varying ventilation strategies together with the total daily cooling load of the office

When extrapolated to the complete month of August, the total ventilation heat gain with varying air exchange rates and two tilted windows is about 2 kWh m^{-2} per month. This value is not that high in absolute terms, but three times higher than ambient air ventilation.

2.3.2 Parameter Study Using Simulation

Further studies on the total heat gain of double and active façades were done with TRNSYS, into which the newly developed façade model was integrated (type 111). A schematic description of the room in the office building described above, as well as of the wall constructions used for the building model, is given in Figure 2.18.

At first the influence of window sizing for a single façade system was investigated. The air change rate was set to 0.7 h^{-1}, and all internal walls as well as floor and ceiling areas are boundary walls where the adjacent room temperature was assumed to be identical to the zone temperature. The external façade faces south and the ambient climatic conditions used (temperature and solar radiation) correspond to the standard reference climate for Germany as defined in DIN 4108-6 (2003). The internal loads were set to zero.

In Figure 2.19, the yearly heating and cooling loads of the office are shown for different window areas (given as a percentage of wall area). The monthly heating energy balance method from the European standard EN 832 implemented in the software

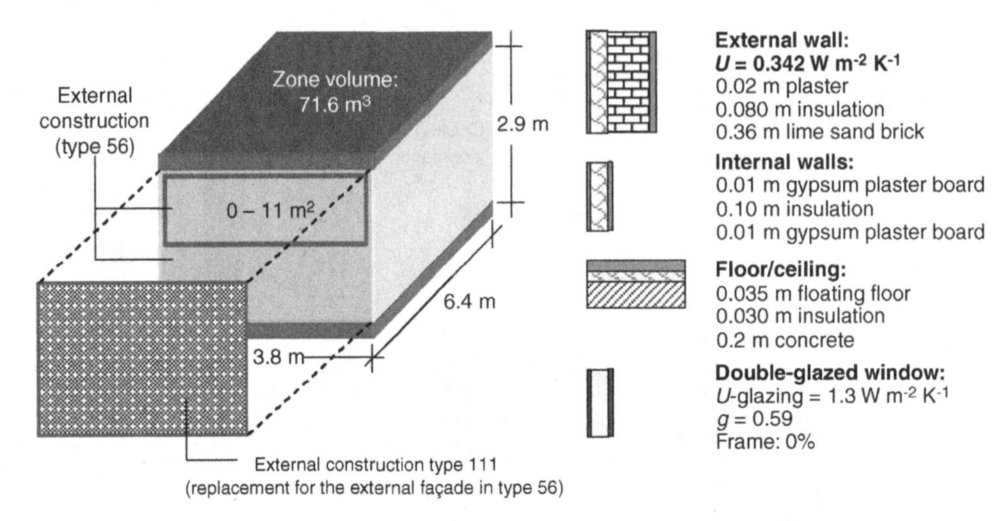

Figure 2.18 Schematic representation of the simulated office room

package Thermplan corresponds nearly exactly to the TRNSYS results. The new model (type 111) gives a higher cooling energy demand than the standard façade model (in type 56), which is mainly due to neglecting the angle-dependent reflection losses. As expected, the energy use for heating is reduced by increasing the window area due to the solar gains. The energy use for heating and cooling reaches an optimum for this office situation for a window area percentage of about 25%. This confirms the importance of low window fractions to maintain low cooling loads in office buildings. The single façade can now be compared with a double façade with and without sun shading elements. In all of the cases, a constant infiltration rate of $0.7\,h^{-1}$ was assumed. The external heat transfer coefficient was set to a more realistic value of $10\,W\,m^{-2}\,K^{-1}$,

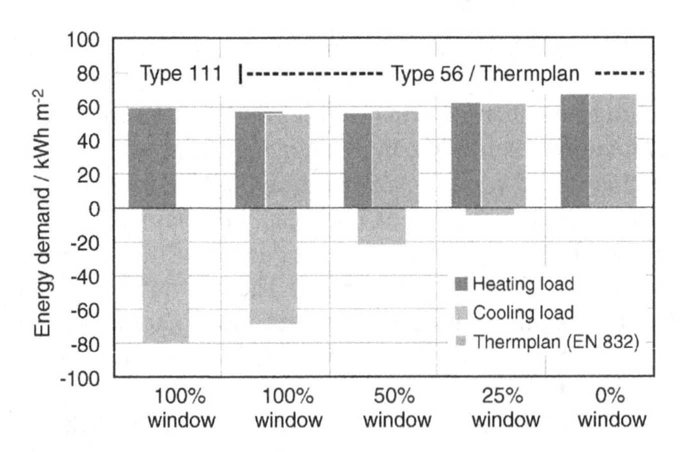

Figure 2.19 Heating and cooling energy demand as a function of window fraction calculated with dynamic and monthly energy balance simulation tools

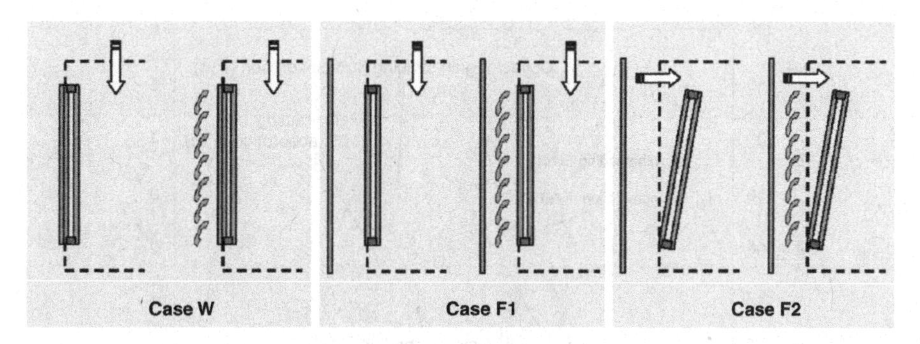

Figure 2.20 The different façade types examined

which corresponds to a low external wind speed. The distance from the inner window to the outer pane of the double façade is 0.5 m. The following cases were compared (see Figure 2.20):

- Cases W(a) and W(b): Single window façade, ambient air ventilation. Case W(a) represents a single façade consisting of a double-glazed heat protection window ($U = 1.5\,\mathrm{W\,m^{-2}\,K^{-1}}$, $g = 0.43$). In Case W(b), an additional external shading blind was considered with a shading factor of 90% for the whole year. Ventilation is done by the ambient air temperature.
- Cases F1(a–c): Double façade, ambient air ventilation. Case F1(a) represents a double façade system without shading devices. In F1(b), an additional shading blind was placed between the panes. The shading factor was also set to 90% for the whole year. The solar absorption coefficient for the blind was assumed to be 10% for Case F1(b) and 30% for Case F1(c). Ventilation is also done by the ambient air temperature.
- Cases F2(a–c): Double façade, double façade ventilation. Cases F2(a) to F2(c) have ventilation air coming from the façade air gap temperature at the outlet opening. This is symbolized by a tilted internal window.

The calculated maximum temperature increases in the double façade range from 3 to 5 K on a hot summer's day with a maximum ambient air temperature of 32 °C and $600\,\mathrm{W\,m^{-2}}$ maximum irradiance (here 2nd September is taken for detailed analysis). The corresponding volume flow rates are calculated for free convection conditions for each time step. From the temperature rise and the façade volume flow, first the total ventilation gains are calculated, which are produced by the façade acting like a solar air collector;

$$q_{vent} = \dot{V}_{total}\,\rho\,c\,(T_{outlet} - T_{inlet})\,/\,A_{façade} \tag{2.6}$$

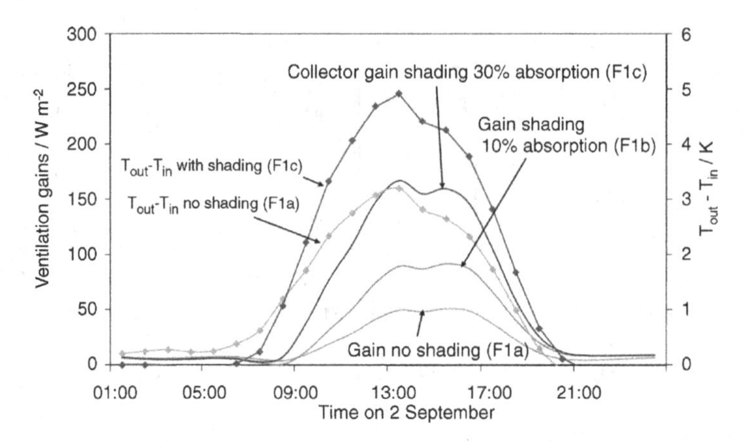

Figure 2.21 Total ventilation gains for a double façade without shading (F1(a)), with a blind of 10% absorption (F1(b)) and 30% absorption (F1(c)). The temperature increase of the façade air is also shown

Without a sun shading element inside the double façade, the low absorption leads to a maximum area related heat gain of $50\,\mathrm{W\,m^{-2}}$. With a 10% absorbing blind this heat gain increases to $90\,\mathrm{W\,m^{-2}}$ and to over $150\,\mathrm{W\,m^{-2}}$ for a system with 30% blind absorption (see Figure 2.21).

From this total ventilation heat gain, only a fraction enters the room. At an air exchange rate with the room of $0.7\,\mathrm{h^{-1}}$ only $50\,\mathrm{m^3\,h^{-1}}$ of the façade air enters the room and adds to the cooling load. The heat gain per square metre of façade is 10% of the total heat produced in the façade: that is, between, 5 and $15\,\mathrm{W\,m^{-2}}$. The secondary heat flux from the room surface temperature to room air is highest if no shading element is present:

$$q_{conv} = h_c\left(T_{surface} - T_{room}\right) \tag{2.7}$$

For the unshaded double-glazed single façade, a maximum secondary heat flux of $90\,\mathrm{W\,m^{-2}}$ is obtained. Only slightly lower values are calculated for the double façade without blinds. If sun shading blinds are present, the surface temperatures are reduced and the secondary heat flux is between 25 and $40\,\mathrm{W\,m^{-2}}$ (see Figure 2.22).

The total ventilation heat gain over a summer period varies between 50 and $100\,\mathrm{kWh\,m^{-2}_{\textit{façade}}\,a^{-1}}$ depending on the absorption of the façade. If all that air were to enter the room, a very significant additional energy input to the $24\,\mathrm{m^2}$ room of $23\text{–}46\,\mathrm{kWh\,m^{-2}\,a^{-1}}$ would occur. However, as the air exchanges with the façade are usually low, only a small fraction of about 10% of the heated air really enters the room (unless strong cross-ventilation takes place!). Furthermore, the heat gain is not necessarily equal to the required cooling energy, as often room temperatures stay below the setpoint of 24 °C despite the heat input. Using the standard meteorological data from Stuttgart, Germany, in the simulation with low average ambient temperatures,

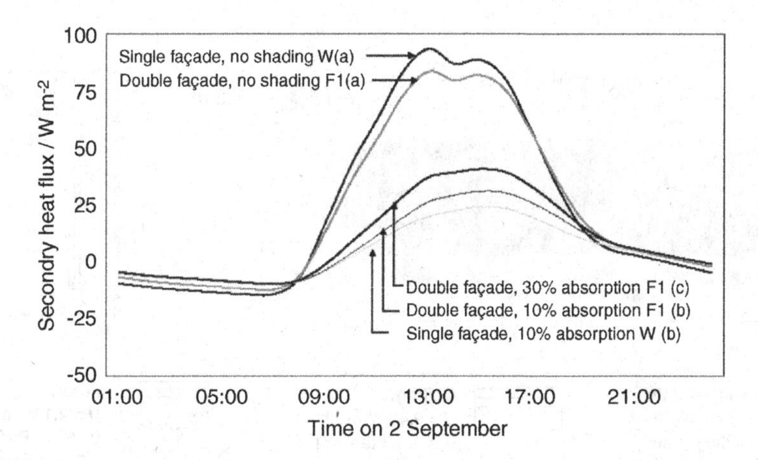

Figure 2.22 Secondary heat flux for unshaded and shaded single and double façades

ventilation with ambient air at a fixed air exchange rate of $0.7\,h^{-1}$ on average leads to heat removal during summer of $13\,kWh\,m_{room}^{-2}\,a^{-1}$. Unwanted summer gains from secondary heat fluxes are between 17 and $20\,kWh\,m_{façade}^{-2}$ for the unshaded single and double façade, respectively. When related to the room surface area, this corresponds to $8–9\,kWh\,m_{room}^{-2}$ cooling loads. It should be noted that only for the unshaded façades is net energy transferred from the surface to the room air; in all other cases the façade remains cooler than room air on average and there are even heat losses between 10 and $17\,kWh\,m_{façade}^{-2}$ during the summer period. It is also clear that effective summer heat losses from both ventilation and secondary heat fluxes are only possible in a very moderate summer climate with average ambient temperatures below the room setpoint temperature. If weather data from a warmer summer (such as in Stuttgart in 2003) is taken, both ventilation heat and secondary heat fluxes are positive and add to the room's cooling load.

2.4 Energy Production from Active Façades

Finally, the influence of an active PV ventilated façade is studied for a library building with a south-facing PV system (see Figure 2.24). Figure 2.23 shows a schematic view of the building as well as a cross-section through the façade and the construction of the external enclosures.

Measurements were done on a 6.5 m high façade collector element in the building in Spain (see Figures 2.24 and 2.26) as well as on a similar 2.2 m high element located at the test site of the University of Applied Sciences in Stuttgart (see Figure 2.25).

As the total absorption coefficient of the outer PV module is about 80% (PV module and glass spacing) and thus significantly higher than a standard sun shading element,

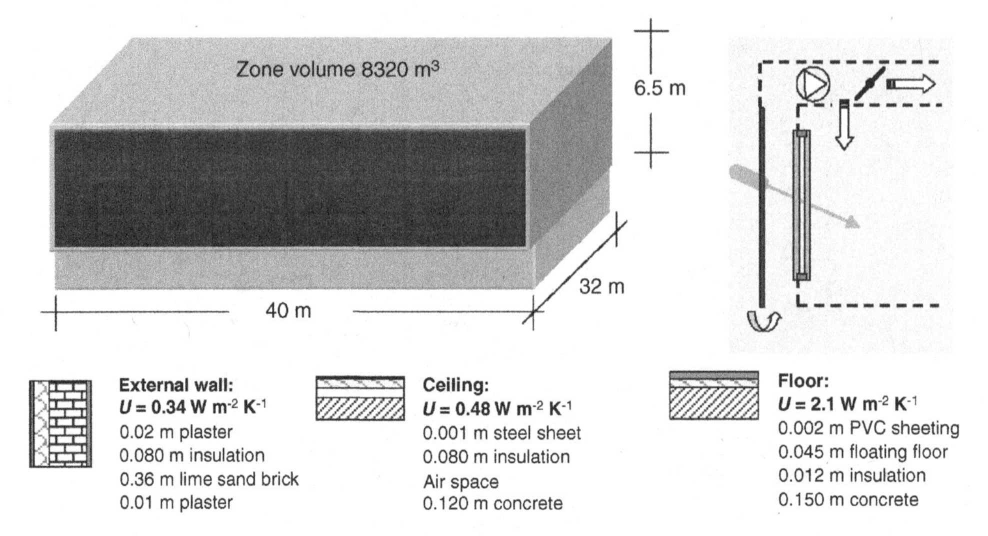

Figure 2.23 Building description of the Mataró library building with an integrated PV double façade

temperature levels are in general well above a standard façade. Temperature levels were measured on all surfaces as well at the air entry and exit (see Figure 2.27).

On a day with a high façade irradiance up to nearly 700 W m^{-2}, all temperature levels were analysed in detail and the façade simulation model was tested. The entry air temperatures in the façade were significantly higher (up to 6 K) than the air temperature measured on a shaded side of the building (see Figure 2.28).

On the outer glass pane of the PV module, temperature levels of nearly 60 °C were measured (see Figure 2.29). To model this type of façade with highly absorbing elements and asymmetric heat flow conditions, the new Nusselt correlation from Equation 2.1 was used. As can be seen in Figure 2.29, the calculated surface

Figure 2.24 Public library building in Mataró, Spain, with ventilated PV façade

Figure 2.25 PV façade element with series-connected air collector at the Stuttgart test site

temperatures on the double-glazed window (T_s-Window) correspond nearly exactly to the measured data. This is a good indication that the heat transfer rate within the gap must have been calculated correctly.

The air temperature increase reaches 15 K at average flow velocities of 0.3 m s^{-1} (see Figure 2.30). These high temperature levels in the air gap and the rather poor thermal separation from the room by uncoated double glazing then lead to inside surface temperatures up to 37 °C and consequently to high secondary heat flows to the adjacent room (see Figure 2.29).When comparing the façade outlet temperatures in Figure 2.30, different approaches for the calculated Nusselt numbers were used. The newly developed Nusselt function as well as the correlation for mixed convection over a single vertical plate from Churchill (1977) provide nearly the same temperature profile. Because of the very slow air flow rate of a maximum of 0.3 m s^{-1}, the correlation of Petukhov (1970) for fully developed forced convection is not useful for reproducing results obtained from the Mataró façade. The Petukhov correlation for fully developed flow in an air channel is proportional to the measured flow velocity and gives very low heat transfer coefficients below 2 W m^{-2} K^{-1}. The highest heat transfer coefficients over 4 W m^{-2} K^{-1} are obtained from single plate flow (see Figure 2.31).

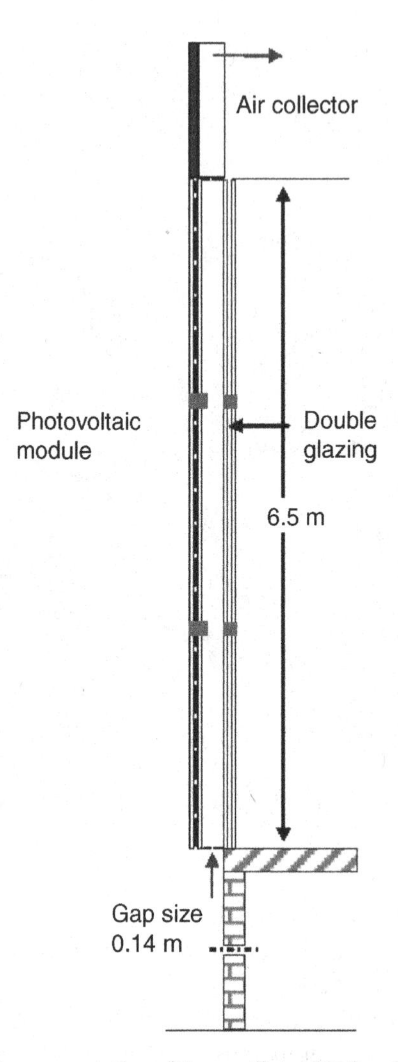

Figure 2.26 Schematic representation of the ventilated PV façade in the Mataró building

Maximum convective heat fluxes of 50 W m^{-2} were obtained from the warm PV side and 10 W m^{-2} from the colder glazing side. This corresponds to maximum convective heat transfer coefficients of 3.2 W m^{-2} K^{-1} on the warm side and 2.5 W m^{-2} K^{-1} on the cold side.

Using the new Nusselt correlation for the determination of convective heat transfer coefficients, all temperature levels can be calculated to a good precision (Figure 2.30) and annual energy balance simulations can be done for different operating strategies. Several cases were distinguished to show the influence of different gap sizes and of the chosen control strategy for façade ventilation.

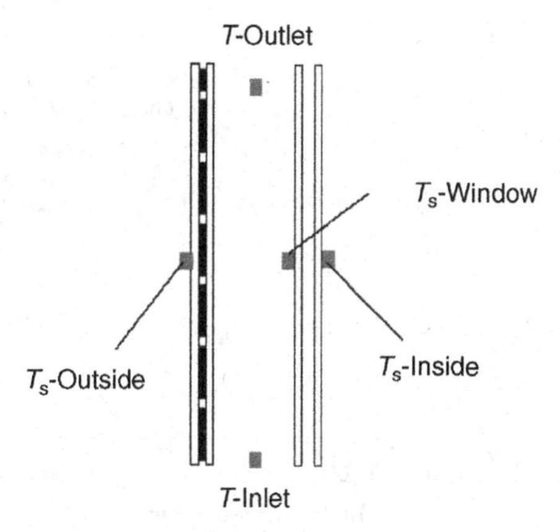

Figure 2.27 Position of sensors for temperature measurements and for calculation points

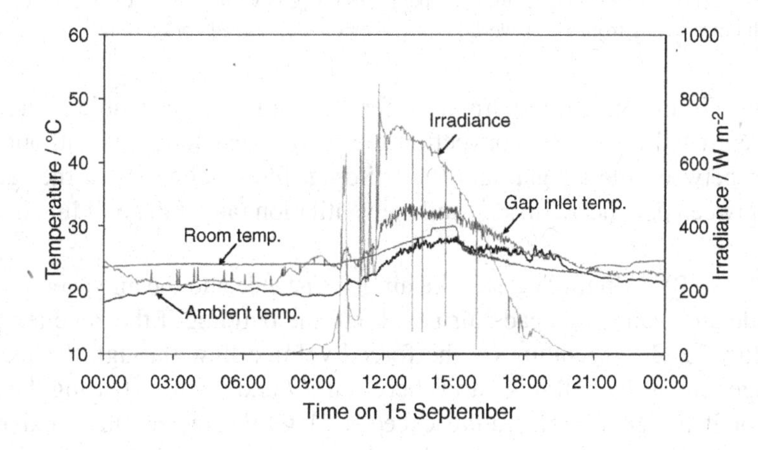

Figure 2.28 Ambient air and façade entry air temperature together with room air temperature and irradiance on a vertical south-facing surface

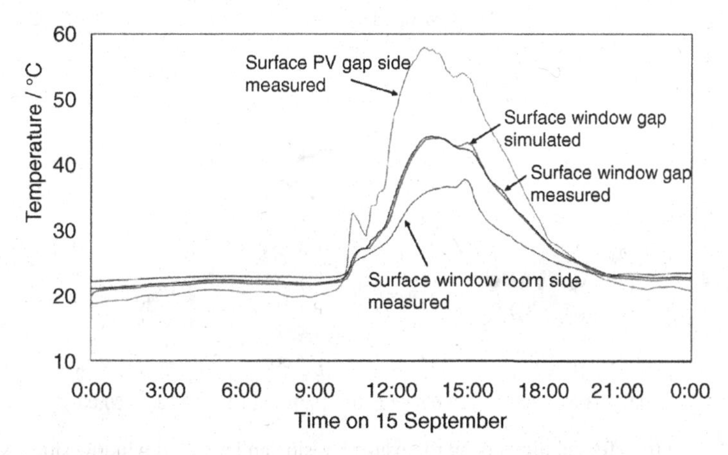

Figure 2.29 Surface temperatures on the absorbing photovoltaic element, inside the air gap and on the surface facing the room

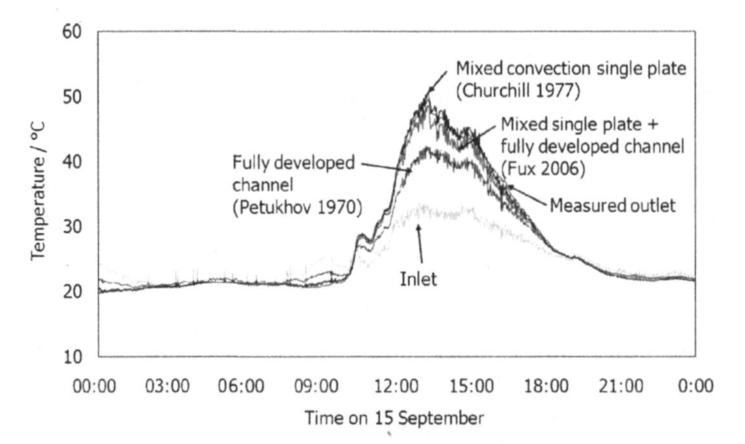

Figure 2.30 Measured air inlet and outlet temperatures together with calculated outlet air temperatures using Nusselt correlations from Fux's work and from mixed and forced convection

- Case PV0: To estimate the thermal buffer effect of the PV double façade the air gap is ventilated by forced convection the whole year long but without thermal coupling between the air gap and the building. Flow velocities in the façade are varied between 0.4 and 1.5 m s^{-1}, and the ventilation rate of the building is constant at 0.7 h^{-1}.
- Cases PV1 to PV3: In these cases, the air gap distance and the air speed rate within the façade are changed. Facade air is led into the building if the zone temperature falls below 21°C. Depending on the forced volume flow through the façade, the air change rate of the building varies between 0.7 and 2.0 h^{-1}. During the summer period, or if the zone temperature exceeds 21°C, the façade air is exhausted to

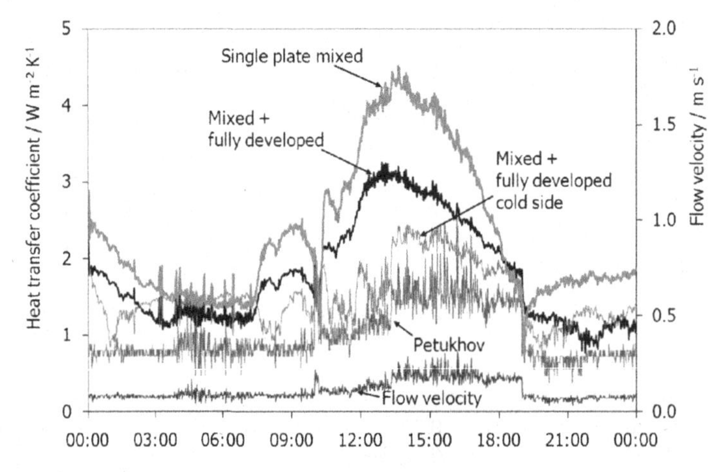

Figure 2.31 Heat transfer coefficients on the warm PV side and the cold window side using different Nusselt number correlations

the surroundings or may be used for solar thermal cooling systems. The façade is ventilated without interruption the whole year long (8760 h).

- Case PV4: This case represents a more realistic control strategy. The preheated façade air is led into the building only during winter days between 15 September and 15 May. During the night (from 18:00 to 07:00) the fan is switched off and the flow velocity within the façade gap is reduced from 0.4 to 0.1 m s^{-1}, which simulates free convection. The volume flow through the façade causes an air exchange of 0.7 h^{-1}. The building supply air temperature corresponds either to the gap outlet temperature or to the ambient temperature depending on the operation of the fan. In the summer period, the fan works only during the daytime but the hot façade air is not connected with the building and is usable for additional solar cooling systems.

In the following section the complete thermal and electrical energy balance of the façade is evaluated, including pressure drops from forced ventilation systems. The parameters for the study are summarized in Table 2.5.

2.4.1 Thermal and Electrical Energy Balance of the Façade

The thermal energy balance for heating and cooling, the electrical energy produced by the PV system as well as the necessary power of the fan to ventilate the double façade are considered in this section. The average electrical efficiency and the solar irradiance transmission through the glazed part of the PV modules are 10%. To calculate the electrical power P_{el} of the fan for ventilating only the double façade, the pressure drop Δp within the façade must be estimated over the sum of the single flow

Table 2.5 Gap size and flow velocities of the ventilated PV façades considered

Case	Gap width / m	Flow velocity / m s^{-1}	Air change / h^{-1}
PV0a	0.10	0.40	0.7
PV0b		1.00	0.7
PV0c		1.50	0.7
PV1a	0.05	0.81	0.7
PV1b		1.16	1.0
PV1c		2.31	2.0
PV2a	0.10	0.40	0.7
PV2b		0.58	1.0
PV2c		1.16	2.0
PV3a	0.20	0.20	0.7
PV3b		0.29	1.0
PV3c		0.58	2.0
PV4	0.10	0.40	0.7

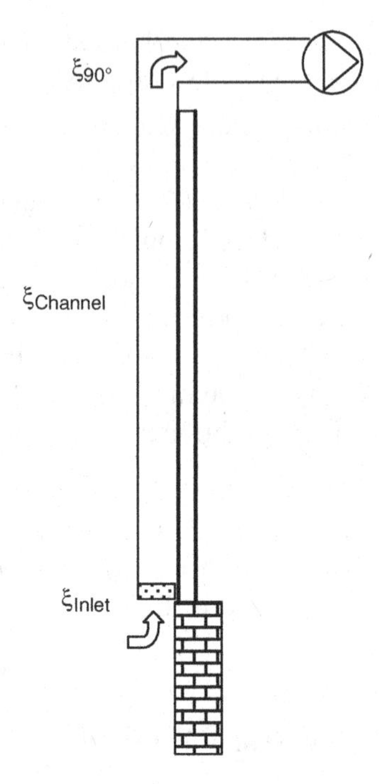

Figure 2.32 Flow resistances within the double PV facade with a 14 cm air gap

resistances (see Figure 2.32)

$$P_{el} = \frac{\dot{V}\,\Delta p}{\eta_{Fan}} \tag{2.8}$$

The efficiency of the fan η_{Fan} was assumed to be 80% and the pressure drop is calculated using

$$\Delta p = \xi\,\frac{\rho}{2}\,\frac{H}{d_h}u^2 = \left(\xi_{Channel}\,\frac{H}{d_h} + \xi_{inlet} + 1 + \xi_{90}\right)\frac{\rho}{2}u^2 \tag{2.9}$$

The resistance of a channel for turbulent flow is

$$\xi_{Channel} = \frac{0.3164}{\sqrt[4]{Re_d}}$$

where Re_d is calculated using a hydraulic diameter which is approximately twice the width of the air gap: $d_h \approx 2\,d$.

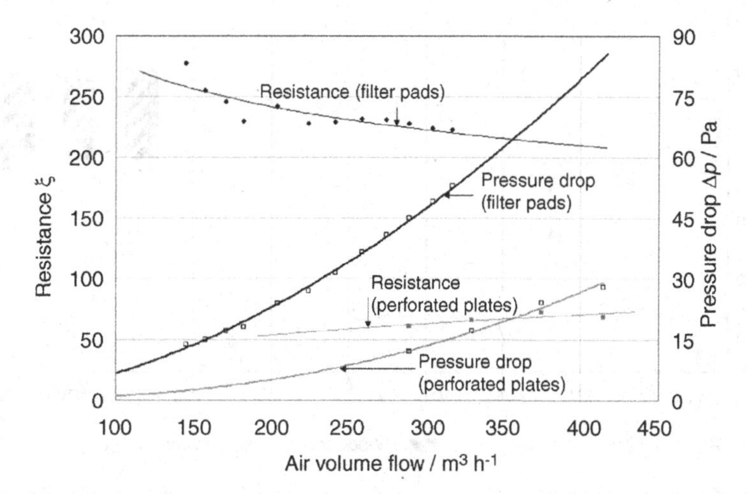

Figure 2.33 Flow resistance and pressure drop in a ventilated double façade

For the rectangular bow, a resistance of $\xi_{90^\circ} = 1.3$ was assumed. The resistance of the openings depends on the covers used: the inlet resistance may vary between 1.5 and 300, depending on the opening section of perforated plates or on the density of the filter pads used. The author's experimental investigations of a single PV façade element gave a total resistance of about $\xi = 60$ if a perforated plate was used only at the openings. A total resistance of about $\xi = 230$ was measured if the original dirty filter pads from the library in Mataró were used additionally. A flow resistance of $\xi = 200$ for the façade entrance region was assumed when using clean filter pads. The measured resistances and the pressure drops as a function of air volume flow are shown in Figure 2.33.

The yearly energy consumption of the fan is obtained from the operating time, which varies from 8760 hours for Cases 0 to 3 down to 5352 hours in Case 4. When considering the electrical energy production of the PV modules, a total value of about 15 000 kWh a^{-1} (or 58 kWh m^{-2} a^{-1}) was obtained for all cases. The calculated maximum difference between Cases PV1c and PV3a, where the air gap velocities amount to 2.3 and 0.2 m s^{-1}, respectively, is about 400 kWh a^{-1} and therefore not visible in Figure 2.34. Because of the quadratic increment of the flow resistance with the air velocity, the energy used for ventilation exceeds the generated energy when there are high-volume flows (PV0c and PV1c). For Case PV4 where the geometry as well as the flow rate corresponds to Case PV2a, the influence of the lower operating time of the fan is cancelled out by the lower PV efficiency when the fan is switched off.

The thermal energy gains from façade ventilation and secondary heat gains from the internal façade surface to the room air are higher than the electrical energy produced. The energy for the façade ventilation is deducted from this produced electrical energy, and is important for higher flow velocities. The values which refer to the façade area are differentiated into winter and summer situations. As can be seen in Figure 2.35, the

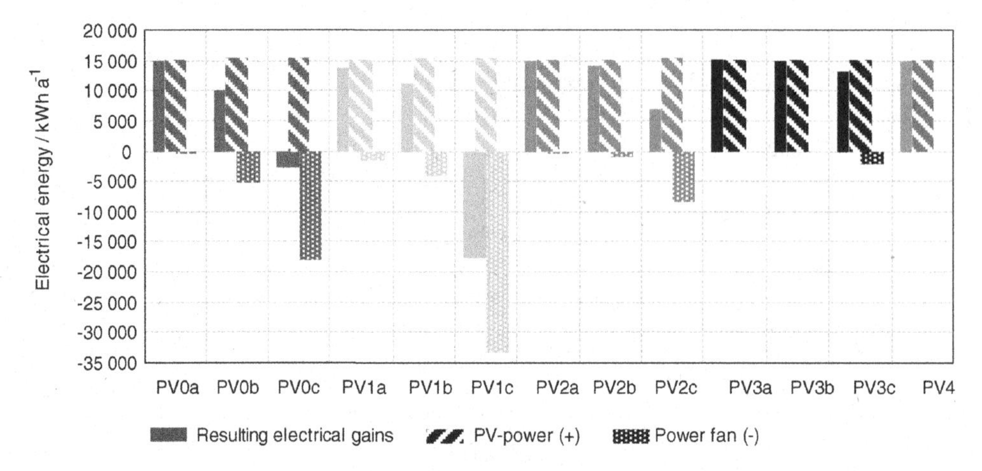

Figure 2.34 Comparison of the electrical energy production by the PV system and the energy used for forced ventilation

thermal collector gains and the secondary heat emission increase with higher volume flows through the façade. Because of the nearly constant energy production of the PV elements, the necessary energy for the fan increases with higher air speed rates and the remaining energy therefore decreases.

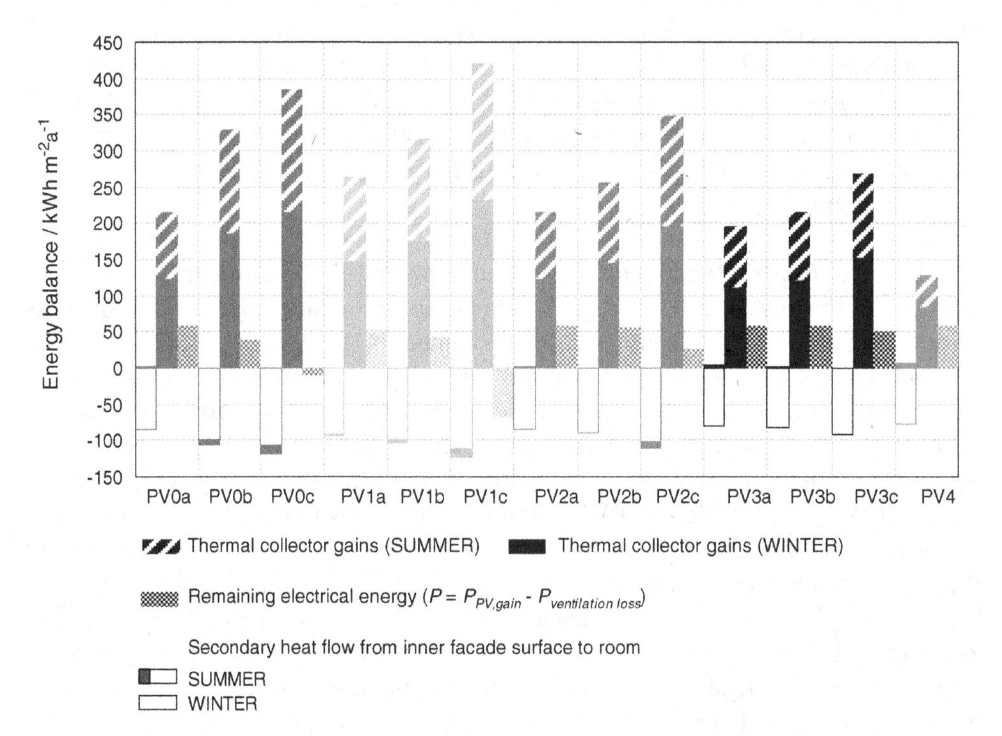

Figure 2.35 Relations of secondary heat flow, façade collector thermal gains and electrical power supplied by the PV modules differentiated into winter and summer situations

The thermal gains in summer and winter are approximately within the same range. However, the secondary heat transfer from the inner façade surface to the room air is very low in summer because of the compensation between day and night where the heat flow directions are different. Case PV4 directly shows the influence of the operating time of the fan in comparison with Case PV2a. Because of the reduced operating hours in Case PV4, the thermal gains are proportionately lower.

Higher flow velocities for lower gap sizes lead to higher thermal yield at the expense of higher pressure drops. The coefficient of performance (COP) describes the relation between the thermal energy gain of the façade collector $Q_{Façade}$ and the energy consumption of the fan E_{vent}:

$$\mathrm{COP} = \frac{Q_{Façade}}{E_{vent}} = \frac{\sum \dot{V} \rho c \left(T_{gap,out} - T_{gap,in}\right) \Delta t}{E_{vent}} \tag{2.10}$$

For the widest gap of 0.2 m with the lowest flow velocity within the air gap, and thus the lowest resistance (Case PV3a), the COP reaches a maximum value of 50. The dominating influence on the COP is clearly the low flow velocity, which reduces electrical energy consumption without greatly reducing the thermal efficiency. At more realistic operation modes (such as in Case PV4 with a 10 cm air gap) the COP reaches values of 12 (see Figure 2.36).

The thermal efficiency η of the façade air collector can be calculated directly from the relation of thermal gains $\dot{Q}_{Façade}$ to the incident irradiance G:

$$\eta = \frac{\dot{Q}_{Façade}}{G\, A_{Façade}} = \frac{\dot{V} \rho c_{air} \left(T_{outlet} - T_{inlet}\right)}{G\, A_{Façade}} \tag{2.11}$$

A façade collector, however, also interacts thermally with the building, as the transmission losses through the façade increase the air gap temperature. To compensate for

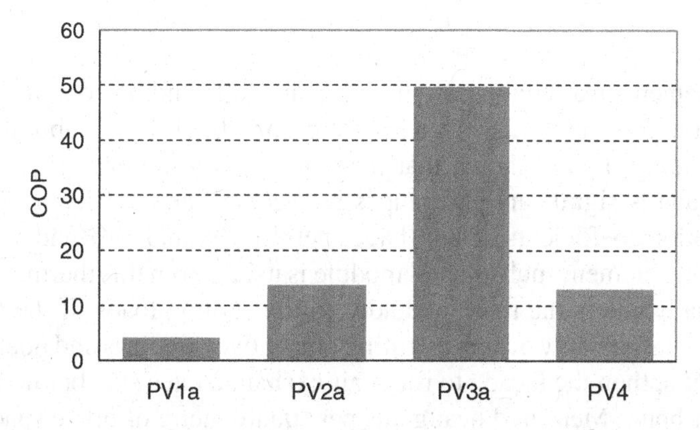

Figure 2.36 Ratio of produced thermal energy to electrical fan energy

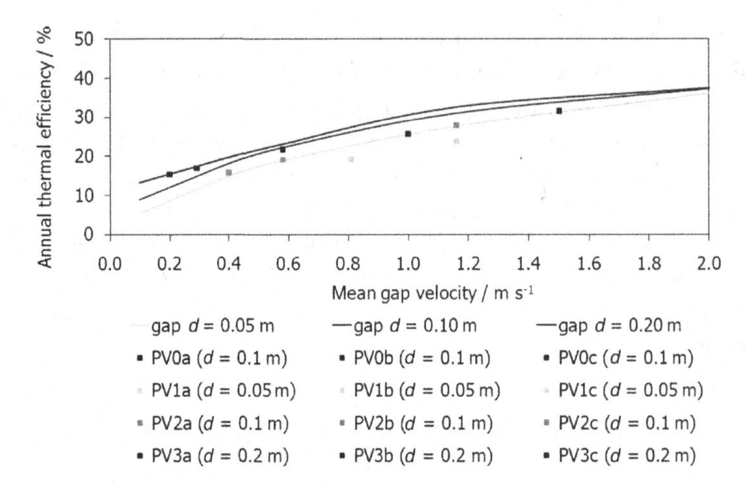

Figure 2.37 Mean annual thermal efficiencies from building simulation (points) compared with results from the façade thermal model

this effect, the thermal gains were calculated for day and night separately. With the assumption that the thermal influence of the building is identical for day and night, the remaining collector energy may be estimated from the difference between the two: $Q_{col} = Q_{day} - Q_{night}$. In Figure 2.37 the thermal efficiencies calculated by the new simulation tool TransFact (solid lines) were compared with the averaged results from TRNSYS which were calculated from the day and night differences. These calculated efficiencies correspond to a stationary thermal model of a ventilated double façade (Eicker, 2003). At the same flow velocity the thermal efficiency increases with gap size, as the total volume flow increases, the average temperature level reduces and the heat losses to ambient also decrease.

2.5 Conclusions on Façade Performance

The summer performance of single and double façades with sun shading systems or active solar elements was analysed both experimentally and by computer simulation. Experiments on one-storey-high façades were done both in the laboratory and in a real office building. It was shown that the summer thermal energy production in a ventilated façade is significant and ranges between 50 and 100 $\mathrm{kWh\,m_{façade}^{-2}\,a^{-1}}$ for typical absorption coefficients of sun shades between 10 and 30% and can double if a highly absorbing element such as a PV module is used. From this thermal energy, only a fraction usually enters the room and adds to the cooling loads. If the air exchange with the double façade only occurs by tilting one or two windows and not allowing any cross-ventilation, then the façade to room air exchange rates stay below 1–2 room air exchanges per hour. Measured heat gains per square metre of office space in August are between 36 and 86 Wh m^{-2} d^{-1}, which corresponds to about 2–5 kWh m^{-2}_{room} a^{-1} of additional cooling loads. Unwanted summer gains from secondary heat fluxes only occur if no shading system is used and the internal glass pane heats up significantly. In all other cases, even with an integrated PV module, the inner façade surface remains cooler than room air on average and there are even heat losses between 10 and 17 kWh $m^{-2}_{façade}$ during the summer period. Secondary heat gains are between 17 and 20 kWh $m^{-2}_{façade}$ for an unshaded single and double façade, respectively. Related to the room surface area, this corresponds to 8–9 kWh m^{-2}_{room} cooling loads.

Measured air temperature increases in a one-storey-high double façade were between 3 and 5 K, with peak values of 8 K. The blind temperature measured in the façade was 10–12 K above the air inlet temperature, slightly higher than for the laboratory measurements. The temperature levels increase by about 4 K when the cross-sections of the air inlet and outlet are reduced from 44% to 10%.

The total energy transmittance or *g*-value was determined by a calorimetric method for single and double façades in the laboratory. Energy reduction coefficients of 23% were measured for both a single façade with external shading and a double façade with sun shades in the air gap. The *g*-values were thus reduced from 64% to 15% for the single façade and from 51% to 12% for the double façade. On the real building, less precise measurements with pyranometers for optical transmission and heat flux measurements were taken, which resulted in *g*-values of 43% for the unshaded double façade and a very low value of 3% for the completely shaded façade.

The position of the sun shades within the double façade relative to the outer pane had no measurable influence on the *g*-value. Also the size of the façade opening cross-section does not influence the *g*-value (within measurement uncertainty), although air temperature levels increase by a few Kelvins. Apart from the energy transmittance, the ventilation rate determines the additional cooling load and has to be analysed when evaluating the summer performance of a façade.

In conclusion, it can be stated that externally shaded single and internal gap shaded double façades can effectively reduce the total energy transmittance to a building in summer. The highest energetic priority is always the reduction of short-wave solar irradiance, which is the dominant energy flow. With measured *g*-values as low as 7% with even slightly open blinds, this condition can be fulfilled by both single and double façades. Secondary heat flows only play a role if no shading system is used and are otherwise negligible. If the façade is used for providing fresh air to the room, additional cooling loads of the order of 10–30% of typical office cooling room loads occur. All results were obtained for moderate German summer climatic conditions and the increase in cooling load from façade ventilation and secondary heat flux will be even stronger in warmer climates.

For the special case of ventilated PV façades, thermal energy can be produced and delivered to the building by forced ventilation at high COPs. Temperature increases are usually between 10 and 15 K and thus mainly useful for preheating the air. If a building's fresh air supply is delivered through the PV façade, thermal gains are typically higher than the delivered electrical energy from the façade. For the Spanish case study building, a ventilation rate of 0.7 h^{-1} gave winter thermal gains between 80 and 150 kWh m^{-2} a^{-1}, increasing with flow velocity. The summer gains, which can be used to drive a thermal cooling machine (see section 5.2), are of the same order of magnitude. In total, the thermal gains are 2.5 to 5 times higher than the electrical energy produced. If higher ventilation rates can be used in the building, increasing volume flows improve the heat transfer and thus the collector efficiency, which can reach 15–35 % as an annual mean value. However, as the pressure drop increases with the square of flow velocity, care has to be taken to minimize ventilation power. The best COPs were obtained for larger gap sizes, where flow velocities were below 0.5 m s^{-1} and COPs over 10 can be obtained.

3

Passive Cooling Strategies

Cooling of buildings can be achieved at very different energy consumption, ranging from zero energy for purely passive over low-energy consumption for earth heat exchange up to high electrical energy requirements for active compressor chillers. The application of different systems depends strongly on the cooling load, which has to be removed.

Ambient air can be directly used to cool office buildings through night ventilation or daytime ventilation using earth heat exchangers. Night ventilation can rely solely on buoyancy or wind-driven natural forces, whereas fans are required to drive the volume flow through the earth heat exchanger or to support the night ventilation volume flow control. The energy which can be removed at night depends on the air exchange rates, the convective heat transfer from all surface areas and the temperature swing of ambient air.

Zimmermann states that daily cooling loads should not exceed 150 Wh m^{-2} for night ventilation to be efficient, and that night-time ambient temperature should be at least 5 K below room temperature for more than 6 h at air exchange rates of 5 h^{-1} (Zimmermann 2003). If summer nights are very cool with ambient temperatures below 16 °C, loads up to 250 Wh m^{-2} d^{-1} can be removed. Shaviv and colleagues claim that achieving 20 air changes per hour is important for locations in Israel and recommend forced night ventilation strategies, if natural ventilation does not reach this air exchange rate (Shaviv *et al.*, 2001). Measured night air changes in low-energy office buildings such as the Fraunhofer Institute of Solar Energy in Freiburg, Germany, and the railway building DB Netz AG Hamm on the other hand are only 2–5 h^{-1}(Pfafferott, 2003

Low Energy Cooling for Sustainable Buildings Ursula Eicker

Pfafferott *et al.*, 2004). This led to only a 1.2 K room temperature decrease during working hours, if passive cooling alone was used.

If fans support the air exchange, care has to be taken to limit the additional electrical energy consumption: in the DB Netz AG building the electrical consumption was $0.46\,W\,m^{-3}\,h$, so that quite low coefficients of performance (COPs) of 4.5 were obtained. Despite an efficient control strategy with night ventilation restricted to early morning hours, up to 400 hours were above 25 °C even in a moderate German summer.

One of the best performing office buildings today with a passive energy building standard was constructed in 1999 in Weilheim, Germany, and has been monitored by the University of Applied Sciences in Stuttgart over three years to evaluate experimentally its energy performance. The building has a user-driven passive night ventilation strategy for summer cooling. Extensive data sets were recorded to evaluate the summer performance of the building, including 170 hours of tracer gas measurements for the nightly air exchange rates in the unusually hot German summer of 2003.

As a low-energy compromise, fan-driven night ventilation strategies were experimentally analysed in two office building projects in Freiburg and Tübingen, Germany. The Solar Info Centre building in Freiburg has an exhaust air night ventilation system. In addition, monitoring results are presented from the office building of the engineering company ebök in Tübingen. This is one of the first office buildings to have been rehabilitated to the passive energy standard, and using mechanical night ventilation effectively to discharge ceilings with phase change material. Here the coefficient of performance for the night cooling energy related to the required ventilation power is the main factor for evaluating the system.

3.1 Building Description and Cooling Concepts

3.1.1 Lamparter Building, Weilheim

The Lamparter office building in Weilheim is a compact building with a net surface area of $1488\,m^2$ and a gross room volume of $5540\,m^3$, of which $1000\,m^2$ is heated and mechanically ventilated (see Figure 3.1). The average U-value of the building is $0.3\,W\,m^{-2}\,K^{-1}$. Triple-glazed windows with wooden frames have a U-value of $1.1\,W\,m^{-2}\,K^{-1}$ and the highly insulated roof, wall and floor constructions have U-values between 0.1 and $0.16\,W\,m^{-2}\,K^{-1}$. On the south-west-orientated side of the building, 46% of the façade surface is glazed, and on the northern side 32% (see Figure 3.1).

The building's heating energy is distributed via the mechanical ventilation system with heat recovery, covering an extremely low measured heating energy consumption between 15 and $19\,k\,Wh\,m^{-2}\,a^{-1}$.

Summer cooling is done with a passive night ventilation concept, whereby the user has to open manually the upper section of the windows ($4000\,cm^2$ open cross-section

Figure 3.1 Office building with passive energy standard in Weilheim, Germany

for two windows) and air flow takes places via opened corridor doors to the roof, where flaps automatically open if the internal air temperature is 2 K above ambient temperature (see Figure 3.2). This concept works because the building is occupied by one single engineering company with no internal security measures. Otherwise, openings above the internal doors would be necessary.

The room air temperature setpoint is 22°C. The control of the earth to air heat exchanger differs between the heating (ambient air below 15°C) and cooling period (ambient air above 15°C). During the heating period it is bypassed if the soil temperature, measured at 2.35 m depth, is colder than the ambient air and the room temperature is below 22°C. During the cooling period the earth to air heat exchanger is bypassed

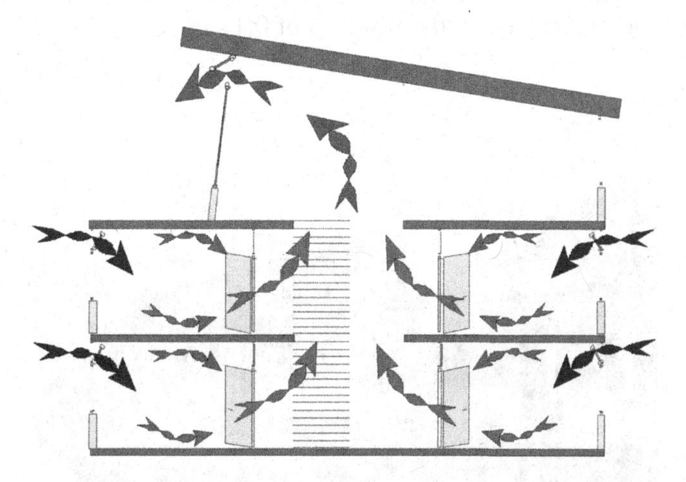

Figure 3.2 Night ventilation concept for the ground and first floor of offices with opening flap at the roof

if the room temperature is below 22°C, that is, too cool. Otherwise the earth heat exchanger is always used during summer operation of the building, regardless of soil temperature. This can have the effect that unwanted gains can occur using the heat exchanger, if outside temperatures are lower than the soil temperature. A better control strategy would be to compare inlet air and soil temperature in summer to prevent heat input.

3.1.2 Rehabilitated Office Building in Tübingen

The passive energy office building in Tübingen has been rehabilitated and is now occupied by the engineering company ebök and is part of a former military barse called the Thiepval Barracks (see Figure 3.3).

Roof and wall insulation (30 cm and 24 cm thick) and triple-glazed windows with a U-value of 0.8 W m^{-2} K^{-1} correspond to passive house standards. The floor insulation in the existing building is only 7.5 cm due to low ceiling heights so that a low cost perimeter insulation had to be chosen to prevent excessive ground losses. The building has a useful heated floor area of 833 m^2 and a gross room volume of 3724 m^3. It is mechanically ventilated with winter heat recovery and summer precooling of the ambient air through a brine-based, ground-coupled heat exchanger installed around the building perimeter. A natural ventilation concept was difficult to implement due to the low height of the building, resulting security problems with window openings and restrictions due to the protection of listed buildings. Therefore the mechanical ventilation system was sized to provide up to 4000 m^3 h^{-1} (about two air changes) for night ventilation at a low pressure drop. The specific electricity consumption for night ventilation is a total of 0.48 W m^{-3} h for both the supply and exhaust fan. This compares with the daytime ventilation with a reduced volume flow of 2000 m^3 h^{-1} and a total electricity demand for the two fans of 0.17 W m^{-3} h.

Figure 3.3 Rehabilitated passive standard office building in Tübingen, Germany

Figure 3.4 Low-energy office building in Freiburg, Germany

3.1.3 Low-energy Office Building in Freiburg

The Solar Info Centre (SIC) in Freiburg is a new office building for renewable energy companies (see Figure 3.4). Therefore great care was taken to obtain an energy-efficient building.

The net floor area is about 14 000 m^2, distributed over six floors in the northern and western wings and three floors in the eastern wing. The outline is U-shaped. The gross room volume amounts to about 58 000 m^3; 45% of the façade is glazed. The building is connected to a district heating system from the neighbouring university hospital power plant, assisted by solar collectors. To obtain a neutral CO_2 balance, a CO_2 trading investment was made in a highly efficient heat recovery system of the hospital's power plant which saves more energy than the SIC building needs. Mechanical night ventilation supplies the required cooling energy, drawing cold ambient air through open flaps in the façade and rejecting the warm air over the roof. The night air exchange rate is 2 h^{-1} from 22:00 until 06:00 in the morning. The fans are only operated if the ambient air is 3 K below room temperature and a minimum cooling power of 116 kW is achieved.

Located on the first floor is a seminar room with a floor area of 170 m^2. Air-conditioning of this room is achieved by geothermal energy. Five vertical borehole heat exchangers of 80 m each supply cooling and heating energy. This system is designed for a cooling load of 16 kW. The fan is capable of generating a volume flow of about 5100 $m^3\,h^{-1}$.

3.2 *Passive Night Ventilation Results*

3.2.1 Internal Loads and Temperature Levels

Two office rooms in the Lamparter building were analysed in detail: on the south-west side of the building a 20 m^2 office is occupied by two persons with one computer

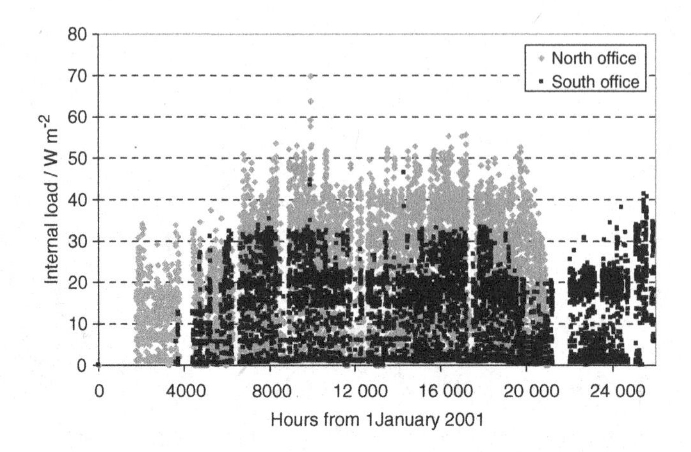

Figure 3.5 Hourly internal loads of the northern and southern offices

workstation and on the north-east side a similar office has two persons with two CAD workstations. The hourly internal loads were monitored for three years and are about 30 – 35 W m^{-2} for the southern office and around 50 W m^{-2} for the northern office (Figure 3.5).

The main part of the load is due to the computer equipment. Measurements have been taken to separate lighting energy consumption, the plug electricity consumption for all other equipment and the heat contribution of the persons using presence sensors. The lighting energy consumption in winter is about 100 Wh m^{-2} d^{-1}, slightly higher in the northern office than in the south-facing office. In summer, lighting energy consumption is mostly zero, but about once a week, 50 Wh m^{-2} d^{-1} is used for lighting. The people contribute between 50 and 100 Wh m^{-2} d^{-1}, depending on presence, and the main consumption is due to other electrical equipment, mainly computers. If both persons are present and use their electrical equipment, the daily load from electricity consumption alone is about 300 Wh m^{-2} d^{-1} (Figure 3.6).

During summer the daily internal loads in the south-facing office are around 200–300 Wh m^{-2} d^{-1} and 400–500 Wh m^{-2} d^{-1} for the heavier equipped north office. The external loads from solar irradiance and wall transmission cannot be easily measured and have been evaluated using building simulation as described below. Typical external loads per square metre of floor space on a sunny day with partly shaded southern windows are 30 Wh m^{-2} d^{-1} from short-wave transmission and 15 Wh m^{-2} d^{-1} from secondary heat flux. Transmission through the glazed façade is usually negative, as internal temperatures are higher than external temperatures during the night and morning hours and amount to about −40 Wh m^{-2} d^{-1}. In total the external loads are therefore not very significant.

Under typical German climatic conditions such as the summers of 2001 and 2002, the night ventilation concept is highly efficient with only 1.9 to 2.4% of all office hour room temperatures above 26 °C. This corresponds to only 50–60 hours above 26 °C

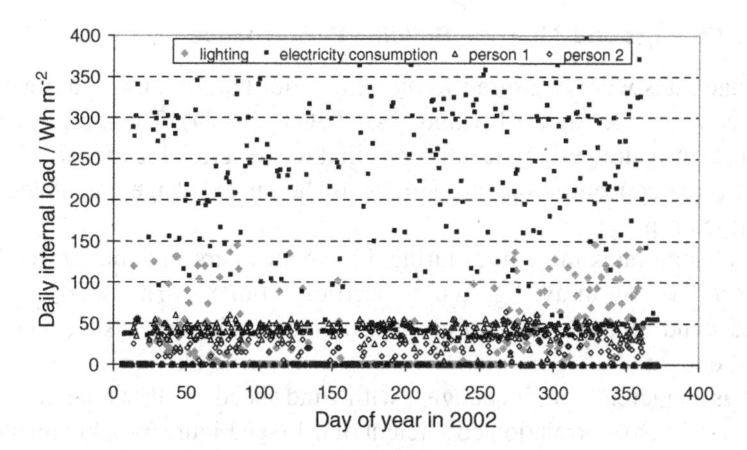

Figure 3.6 Daily internal loads from lighting, electricity consumption and the presence of two persons in the northern office

or 10–30 hours above 27 °C. In 2003, with a mean summer temperature 3.2 K higher than usual, however, 9.4% of the office hours had room temperatures above 26 °C, which is more than 230 h or about 5 weeks. This value is just below the maximum allowed value of 10% according to German standard DIN 4108.

The cumulated hours above a given temperature level show the direct correlation between ambient temperatures and high internal temperatures: only in 2003 with about 100 hours above 30°C were there a few hours above 30°C within the offices, otherwise this did not occur at all (Figure 3.7).

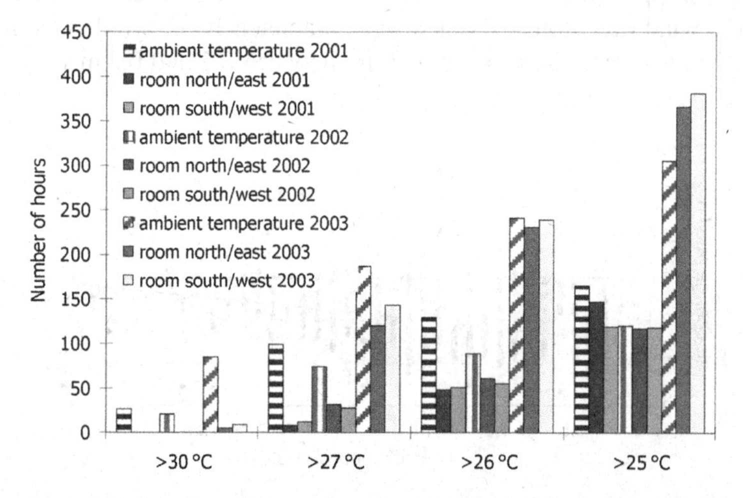

Figure 3.7 Number of hours where given temperature levels are exceeded

3.2.2 Air Changes and Thermal Building Performance

Air exchange rates were measured in the Lamparter building using a tracer gas unit from Dantec (Vivo Dosing 20H31 and Vivo TriGas 20T34). The three-channel measurement unit measures SF_6 concentrations in three different locations within a room. Only if all three concentrations are similar is the air exchange calculated from the concentration drop.

The air change rates measured during 170 night hours in summer 2003 gave an average of 9.3 h^{-1} at an average wind speed of 1.1 m s^{-1}. The wind direction was between east and south for 90% of all measurements with decreasing speed over the course of the night.

There was an increase of air change n with wind speed v following a linear correlation with a very weak correlation coefficient of 0.1 (see Figure 3.8), but no measurable increase of air change with temperature difference between the inside and outside air:

$$n = 1.8173v + 7.2544$$

The analysis of air flow within the building using artificial fog showed that the air exchange was largely wind induced and that the thermal buoyancy effect was especially small in the first-floor offices. The neutral zone within the two-storey building is at the top of the first floor and the driving pressure for buoyancy is therefore small.

During daytime with the mechanical ventilation system in operation, the flow velocity measured below the north-east office ceiling was quite constant at 0.15 ± 0.02 m s^{-1}. Higher wind speeds or wind blowing directly on the north-eastern façade only marginally increased the air velocities. At night with only passive ventilation the flow velocity dropped to 0.01 m s^{-1} if the windows were closed. When the windows are open, the wind direction is the decisive parameter. If the wind is from northern directions, the air velocities at the ceiling are between 0.2 and 0.3 m s^{-1} even at low

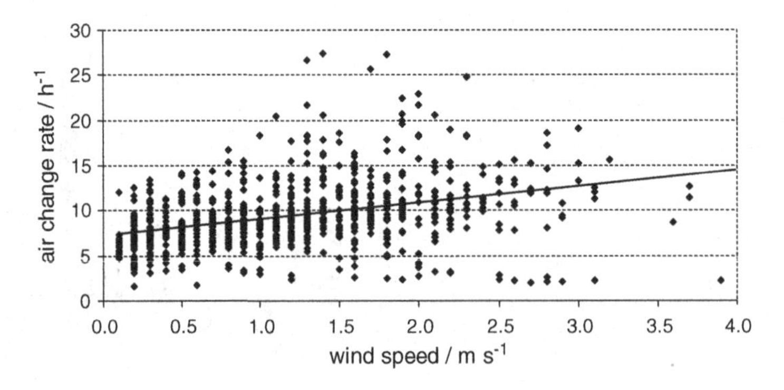

Figure 3.8 Air changes as a function of wind speed

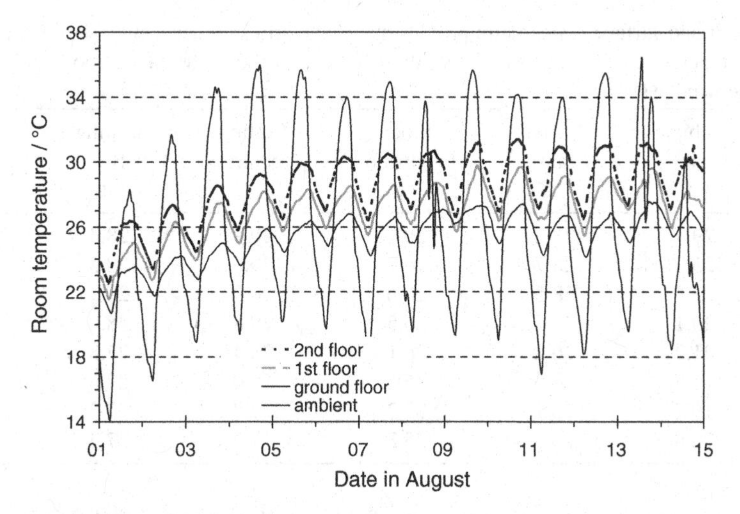

Figure 3.9 Measured temperature stratification within the building in summer 2003

external wind speeds of 1 m s^{-1}. If the wind is not from the north, the air velocity at the ceiling remains low at about 0.12 m s^{-1} even at higher external wind speeds.

There is a temperature stratification in the building, which is nearly 2 K per floor (Figure 3.9). Whereas the top floor is highly glazed and not regularly used, the first floor has the same types of office rooms as the first floor and stratification is a significant problem. All air exchange measurements were carried out in the first-floor offices.

Considering the usually suggested literature value of 5 h^{-1} for efficient night ventilation, the measured air exchange with an average of 9.3 h^{-1} seems sufficiently high during summer. However, the room air temperature levels during the hot month of August only drop by about 3 K from the daily peak temperature and remain 5 to 6 K above ambient air temperatures.

A detailed analysis of the room and surface temperature profiles showed that there is a clear temperature stratification within the room despite high air change rates. The cooler night air enters the room via the top of the window and then drops to the floor. The floor temperature therefore cools down below the average room air temperature. The ceilings on the other hand are not effectively discharged, as air movement is very small at the top of the room and the temperature difference between the ceiling surface and the temperature measured at 30 cm below the ceiling is very small (less than 1 K). Only close to the window opening does the ceiling cool down by about 1.5 K during a warm summer night, and at the room centre by only 1 K (see Table 3.1).

The air changes measured during the night varied between 6 and 14 h^{-1} and follow the external wind velocity (see Figure 3.10). Wind speeds in general decreased during the course of the night, which is very unfavourable for manually opened windows for passive cooling: in the early evening hours ambient air temperatures are mostly well above room temperature and the high air exchange rates even increase the room load.

Table 3.1 Air and surface temperature profiles during a warm summer's night (8 to 9 August 2003). The ceiling temperatures were measured near the windows, at the centre of the room and between the centre and the windows

Time	Ambient air /°C	Room air /°C	Floor to corridor /°C	Ceiling centre /°C	Ceiling between /°C	Ceiling at windows /°C
20:00	29.3	28.8	26.7	29.1	28.4	28.0
22:00	25.3	28.7	26.7	29.0	28.3	27.8
00:10	23.5	28.5	26.6	28.9	28.1	27.2
00:50	23.1	27.5	26.5	28.8	28.0	26.7
03:00	21.2	26.2	26.1	28.4	27.8	26.2
05:00	19.8	26.3	25.6	28.0	27.5	25.8
07:10	19.7	26.1	25.2	27.7	27.2	25.4
09:30	22.6	26.7	25.3	27.9	27.3	26.5

In the early morning hours with a more useful 5 to 6 K temperature difference between room and ambient air, the air changes were down to 6 to 8 h^{-1}.

Simply multiplying the measured air volume flow by the measured temperature difference between the inside and outside from 21:30 to 08:00 gives a cooling energy of 9.3 kWh. Equally distributed over the room surface areas of 94 m^2 this gives an average nightly removed heat flux of 100 Wh per m^2 and night. Heat flux measurements taken at several points of the ceiling, however, gave only low integrated values of 15–20 Wh per m^2 and night. During the day, about 45–50 Wh m^{-2} d^{-1} is taken up by the ceiling. Rather than just the ceiling, the floor and the internal walls also contribute to the night discharging of the room. There is also the effect that air flow is not unidirectional from the outside to inside; that is, it is not only ambient air entering the room, but also partly air flowing from the warm corridors. The simulation studies showed that for air change rates of 8 to 10 h^{-1}, there should be a more significant drop in room

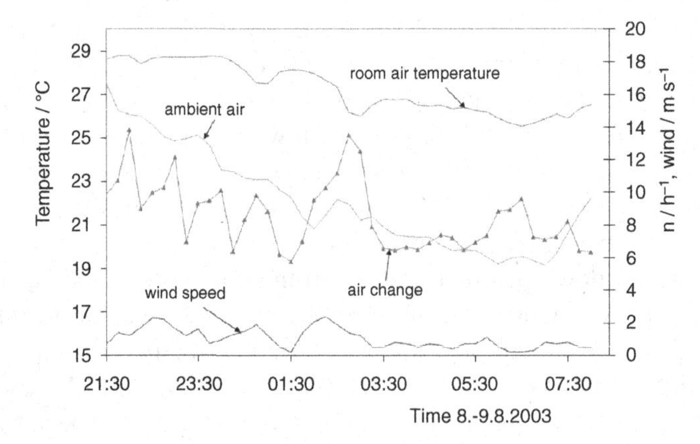

Figure 3.10 Development of room and ambient air temperature as a function of air change and wind speed

Table 3.2 Temperature profile during a cool summer's night (22 to 23 August 2003). The ceiling temperatures were measured near the windows, at the centre of the room and between the centre and the windows

Time	Ambient air /°C	Room air /°C	Floor to corridor /°C	Ceiling centre /°C	Ceiling between /°C	Ceiling at windows /°C
20:00	26.0	25.5	24.2	25.0	25.0	25.0
23:30	18.0	25.1	24.1	25.1	24.2	23.1
00:10	16.8	25.2	24.1	25.1	24.2	23.2
00:50	16.2	23.7	24.0	25.0	23.5	22.9
03:00	15.6	22.2	22.5	24.3	22.7	20.8
05:00	14.2	20.5	22.0	24.1	22.1	20.8
07:10	13.2	22.3	21.6	24.0	22.6	21.6
08:50	17.6	22.7	21.8	23.8	22.4	21.4

temperature. Due to the always decreasing temperature difference between ambient and inside air during the course of the night, about 6 kWh should be removed. These results indicate that effectively the air change between ambient and inside is less than 8 h^{-1}. Further research is necessary to quantify the amount of two-directional flow, which up to now could only be visualized qualitatively by fog experiments.

When the night air temperatures were significantly lower, ceiling temperature decreases of 3–4 K were measured for the surfaces (Table 3.2). Again the heat transfer is significantly better on the ceiling near the windows. Also during that night the air exchange rates were much more favorable for passive cooling: early in the evening with high ambient temperature levels, the air exchange rates were low, but increased to high values in the middle of the night.

3.2.3 Simulation of Passive Cooling Potential

Building simulation studies using TRNSYS were then carried out to investigate the effect of parameter changes, which could improve the passive night cooling strategies. As air change rates were not measured during the whole operation time, the simulation used constant air exchange rates whenever the windows were open. The measured room and surface temperature data during August was very well reproduced for air change rates between 4 and 8 h^{-1}. External sun shading was set constant at 50%, as the external shutters consist of two parts, of which the upper part is mostly open for daylight use. All other boundary conditions were taken from measurement data.

First the influence of air change on the nightly room temperature drop was evaluated, while keeping all convective heat transfer coefficients constant. The simulated room air temperature only approaches ambient temperature during the night if air exchange rates are very high. For a constant air change of 8 h^{-1} the room temperature after night ventilation stays at 4 K above ambient air temperature. Even at 15 to 20 air changes per hour, the room temperature stayed at 2 K above the minimum ambient air temperature. At 50 air changes per hour, the difference between room air and ambient air decreases to 1 K.

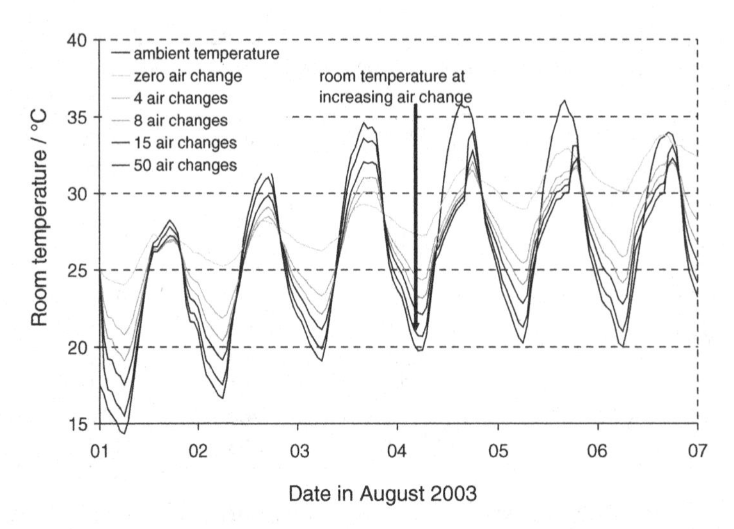

Figure 3.11 Simulated room temperatures as a function of air exchange at constant heat transfer coefficients

Using the monitored window opening times from the window contact switch information, the effect of early evening room temperature increase due to the window openings can be clearly seen in Figure 3.11. A maximum nightly cooling energy of about 9 kWh can be theoretically achieved if the air exchange started at 21:00 and a very high air exchange rate of 50 h^{-1} were available (see data on 5 or 6 August). However, as the windows are manually opened by the user at 18:00, the room first heats up through the high ambient temperature and 2.7 kWh infiltration gains occur! So, effectively, only 6.3 kWh of effective cooling energy can be provided by the night ventilation as a maximum.

At a more realistic lower air exchange rate of 8 h^{-1}, a maximum of 5.5 kWh can be removed if night ventilation starts at 21:00 or 4.7 kWh effectively if the windows are opened at 18:00. For similar measured ambient air temperature profiles (see Table 3.2), and similar measured average air exchange rates (8.6 h^{-1}), the room temperature decreased less than in the building simulation and the cooling energy of 9.3 kWh per night was about double the simulated value. This shows that the air exchange contained not just ambient air, but was partly composed of air from the hallway.

If the user does not close the windows during the daytime, which was the case for the first ten days in August, there is a very strong increase in daytime temperature. This additional heat load of more than 5–9 kWh during the day must first be removed by night cooling.

The heat fluxes from all internal surfaces were then calculated to evaluate the thermal discharging during the night: 40% of the total energy delivered by the room surfaces at night came from 39 m^2 of internal walls, although these walls are light constructions

Table 3.3 Simulated heat fluxes and cooling energy removed for given boundary conditions

Air change /h^{-1}	T_{amb} Max /°C	T_{amb} Min /°C	T_{room} Min /°C	Ceiling flux /Wh m^{-2}	Floor flux /Wh m^{-2}	Walls flux /Wh m^{-2}	Q_{cool} Max /kWh	Q_{cool} eff. /kWh
8	36	20	24.2	55	68	50	5.5	4.7
50	36	20	21.0	84	91	68	8.9	6.3

consisting of two 1.3 cm gypsum boards on both sides filled with mineral wool. About equal shares of nearly 30% were provided by the 20 m^2 of heavy 22 cm concrete ceiling and the 4 cm of mastic asphalt floor covering and a nearly negligible quantity from the lightweight external wall (Table 3.3).

Concerning the effective cooling energy for the hot days of 36°C maximum and 20°C minimum temperature to the floor surface area of 20 m^2, this results in removable loads of 235 Wh m^{-2} d^{-1} for an air change of 8 h^{-1}. The simulated ceiling surface temperature decrease of 1.5 K was obtained using convective heat transfer coefficients of 3 W m^{-2} K^{-1}.

The decrease of room air temperature through higher air exchange rates does not solve the problem of high surface temperatures, if heat transfer is not significantly improved at the same time. Doubling the convective heat transfer coefficient from 3 to 6 W m^{-2} K^{-1} resulted in an additional ceiling temperature decrease of 0.5 K. Increasing convective heat transfer to 18 W m^{-2} K^{-1} decreased the minimum night-time surface temperature of the ceiling by another degree (Figure 3.12).

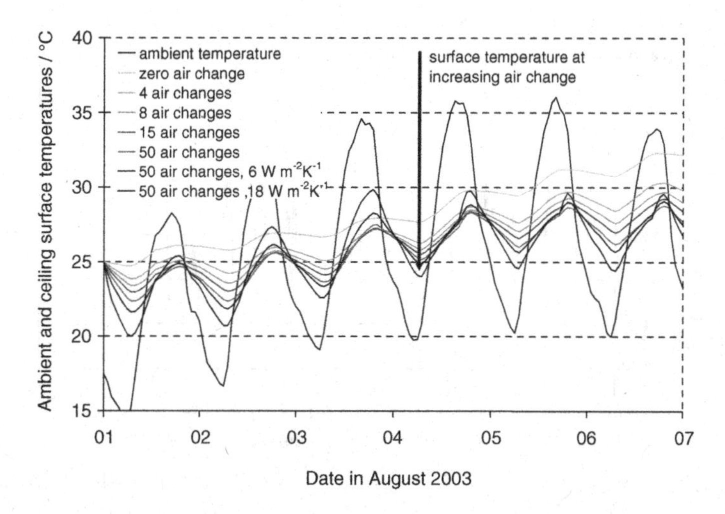

Figure 3.12 Ceiling surface temperatures as a function of air exchange rate and heat transfer coefficient

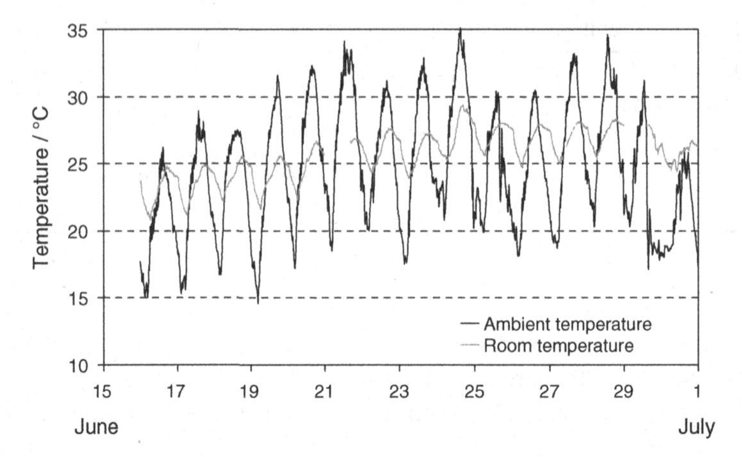

Figure 3.13 Ambient air and room temperatures in the SIC building during June 2005

3.2.4 Active Night Ventilation

Mechanical Exhaust Air Ventilation in the SIC Building

The active night ventilation system of the SIC building with its two air exchanges per hour was capable of reducing room temperature by about 3 K during the hot weeks of June 2005 (see Figure 3.13). During such extended warm periods, this leads to ever-increasing room air temperatures which reached up to 26°C in the morning and 29°C in the afternoon.

The measured average nightly cooling power for the 14-day hot period was 215 kW, which corresponds to 15 W m^{-2} or 120 Wh m^{-2} per night removed cooling load (Figure 3.14).

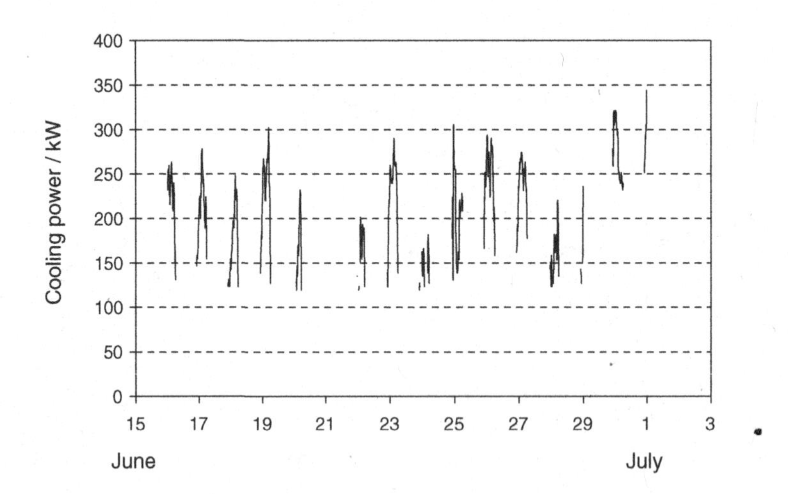

Figure 3.14 Cooling power of mechanical night ventilation in the SIC building in summer 2005

The results indicate that two air exchanges per hour are simply not sufficient to remove the daily internal loads, despite reasonably cool night temperatures of 20°C or lower. Detailed measurements of one section of the building (one of eight exhaust air fans with a volume flow of 5250 $m^3 h^{-1}$), carried out by the University of Applied Sciences in Offenburg, showed that average nightly COPs varied between 5 and 10.

Mechanical Ventilation in the ebök Building

The ebök building in Tübingen uses active ventilation for supply and exhaust air. During the night a volume flow of up to 4000 $m^3 h^{-1}$ of ambient air is injected into the building for cooling purposes, which corresponds to two air exchanges per hour as in the SIC building. As can be seen in Figure 3.15, the cooling loads of the day cannot be fully removed during night-times, resulting in room temperatures above 26°C in the morning hours of the second week. This means a drop of only 2–3 K compared with the preceding evening.

A maximum of 147 Wh m^{-2} per night of internal loads is removed during the evaluated time (last two weeks of June 2005) while the average is about 85 Wh m^{-2} per night. The peak cooling power is nearly 14 kW with an average of 7.4 kW. Mechanical night ventilation obviously requires electrical power for the fans, which is significant even if highly efficient fans are used. In the ebök building, the mean COP for night ventilation during the two-week measurement campaign in 2005 was 4.0, with maximum values of 6.0 (see Figure 3.16). The design power of the fans for summer ventilation was 300 W for a total daily volume flow of 2000 $m^3 h^{-1}$ and 1100 W for night ventilation with 4000 $m^3 h^{-1}$ total flow rate. Measurements of the installed system showed that at 2066 $m^3 h^{-1}$ supply air flow and pressure drops of 123 Pa in the supply and 84 Pa in the exhaust channel, the power consumption was 200 W for the supply fan

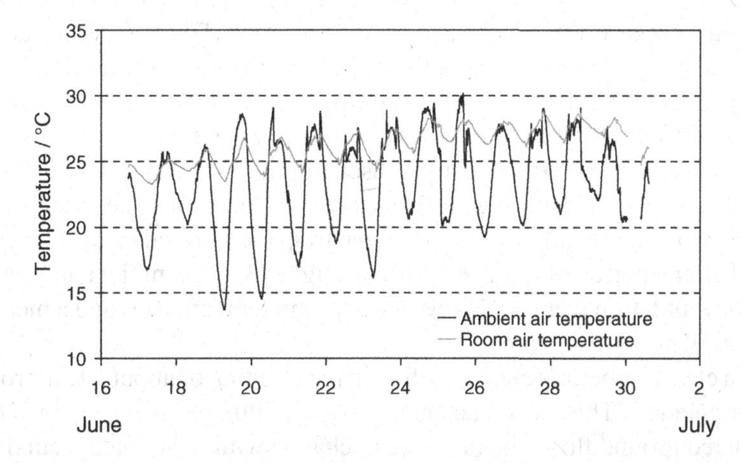

Figure 3.15 Ambient air and room temperatures in the ebök building during summer 2005

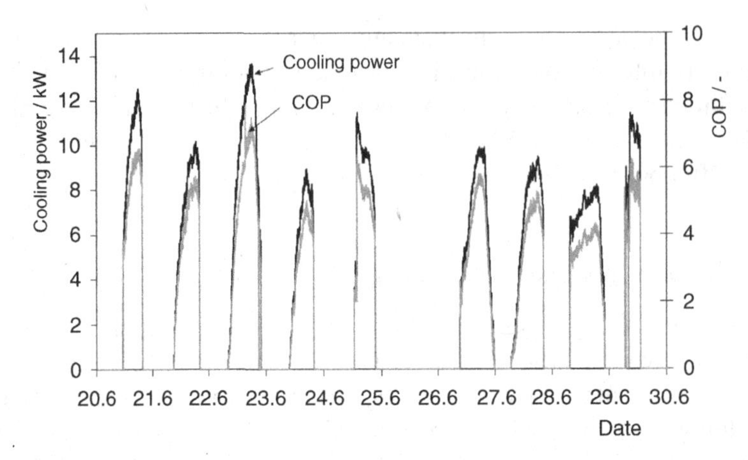

Figure 3.16 Cooling power and coefficient of performance of mechanical night ventilation system

and 155 W for the exhaust fan, closely matching the design value. The specific total consumption of 0.17 W m^{-3} h is excellent. During night ventilation with a measured supply air volume flow of 4021 m^3 h^{-1} at a pressure drop of 395 Pa, the supply air fan consumed 1072 W, while the exhaust air fan at a slightly lower flow and pressure drop (3600 m^3 h^{-1} and 356 Pa) required 850 W. The specific total consumption related to the supply air flow is 0.48 W m^{-3} h, still a good value for ventilation systems, but 75% higher than the design value.

Further measurements in several rooms were taken in 2006 during a very hot period in July. Internal loads averaged over 24 hours were only 4 W m^{-2} in the top floor and 5 W m^{-2} in the bottom floor offices. External loads including sun shading systems were 7.5 W m^{-2} on average on the ground floor and 4.6 W m^{-2} on the top floor. The ground floor has a significant heat transfer to the ground due to the rather low insulation thickness of the floor (U-value of 0.35 W m^{-2} K^{-1}). The measured heat flux via the floor to the earth was between 2 and 3 W m^{-2} on average. This is the same amount that the measured heat flux removed from the ceiling during night ventilation (2.7 W m^{-2} on average). The measured effective discharging time of the ceiling during mechanical night ventilation was about 13 hours with a ceiling temperature decrease of 1.5 K. The main limitation on a more effective night discharging of storage masses is the total air exchange rate, which is limited here to 2 air exchanges per hour. The air outlet injection to the ceiling worked effectively, as can be seen from the infrared measurements during the day and after a period of night ventilation (Figure 3.17). Ambient air temperatures during night ventilation were 4.2 K lower than room temperatures and a mean cooling power of 7.6 W m^{-2} was achieved.

There is a clear temperature stratification in the building of about 1 K for rooms with the same orientation. This can be attributed to a heat flux of 2–3 W m^{-2} via the rather badly insulated ground floor, the radiative exchange with the cooler ground floor and higher storage masses of the ground floor ceiling. Strong drops in room temperature

Figure 3.17 Infrared thermography of ground floor office after night ventilation (28 June 2005, 06:00) and during the daytime (28 June 2005, 14:00)

were observed when the user fully opened the windows in the morning (e.g. on 19 July in the ground floor west office in Figure 3.18). However, the rapid rise in temperature after closing the windows showed that only the air was cooled, and the storage masses could not be discharged during this time.

Generally the daytime temperature level at the ceiling near the window is about 1 K higher than close to the inner door, but also nearly 1 K cooler after night ventilation, as the mechanical ventilation directs the cool air towards the window at the ceiling (see Figure 3.17). The ceiling temperature difference is 3 K between day and night. The temperature at the ceiling near the window has a daily maximum at 11:00, after

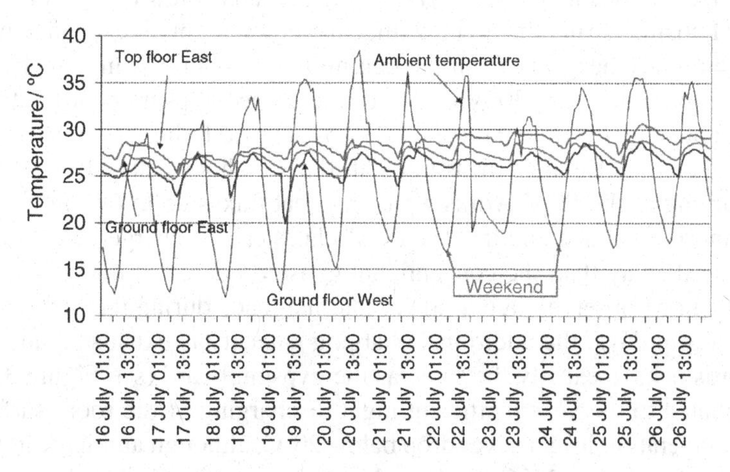

Figure 3.18 Ambient and room temperatures in the ebök office building during a hot period in July 2006

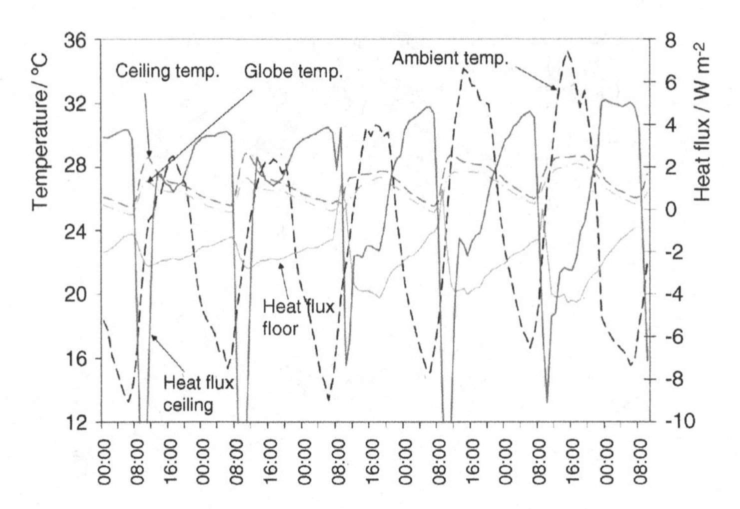

Figure 3.19 Temperature levels of room surfaces and globe thermometer and heat fluxes to and from the surfaces (positive = heat flux from surface) from 1 to 6 July 2006 in a ground floor office

which external irradiance on the east façade decreases (see Figure 3.19). The operative temperature has a maximum of 27°C for external temperatures up to 35°C. During the heating-up phase of the room the average heat flux of the ceiling is $-2\,\mathrm{W\,m^{-2}}$; that is, there is a net heat flux into the ceiling, which is not fully discharged during the night. During a cooling-down period of the room (e.g. 15 to 17 July 2005) the average heat flux is positive at $4\,\mathrm{W\,m^{-2}}$; that is, there is a net discharge of the ceiling at night.

The use of phase change materials (PCMs) in the gypsum boards of the top floor ceiling and wall (16.2 and 19 $\mathrm{m^2}$ surface area) did not significantly improve the situation. The melting point is rather high (26–28°C) and the latent heat capacity limited to about 80 $\mathrm{Wh\,m^{-2}}$. The main problem, however, is again the low heat flux for discharging the PCM boards during the night, with a maximum night air exchange of 2.6 $\mathrm{h^{-1}}$. The average nightly heat flux from the ceiling is less than $2\,\mathrm{W\,m^{-2}}$, corresponding to a total energy removed of 30 $\mathrm{Wh\,m^{-2}}$. The reference gypsum board without PCM showed a slightly higher temperature swing (0.5 K) and a slightly lower heat removal of 25 $\mathrm{Wh\,m^{-2}}$ per night. Detailed measurements during a hot week in July 2006 showed that the charging of the PCM works well with heat fluxes up to $6\,\mathrm{W\,m^{-2}}$, especially during the morning hours, and that the PCM plates can be charged for a longer time period during the day than the conventional gypsum board. At mid-day a reduction in heat flux can be observed due to lower internal loads during the lunchtime break. After three warm days, the additional charging capacity is exhausted and the PCM ceiling boards behave exactly the same as the gypsum board (see Figure 3.20). This can be attributed to the very inefficient night discharging of the room surfaces. The supply air temperature levels never drop below 20°C, although ambient air night temperatures are often below 16°C. From ambient air up to the air distribution box after the air handling unit, there is already a temperature increase of 2 K. A further 2 K

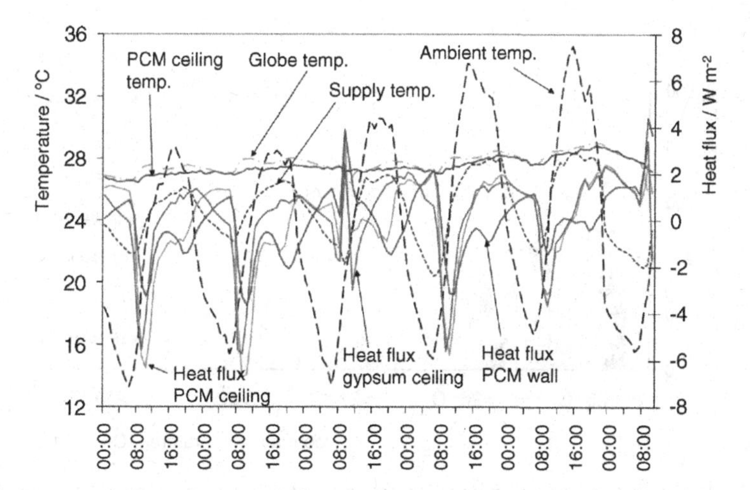

Figure 3.20 Temperatures and heat fluxes on the top floor of the ebök building with a comparison of PCM boards in the ceiling and wall and a conventional gypsum board. The measurements were taken from 1 to 6 July 2006

temperature increase then occurs until the air outlets are reached. This high night supply air temperature level combined with rather low air exchange rates lead to night heat fluxes of only 1–2 W m^{-2}. The effective measured heat storage capacity of the PCM boards was therefore only 24 Wh m^{-2} compared with 17 Wh m^{-2} for the gypsum board.

Despite the limitations in heat removal through mechanical night ventilation, summer comfort in the ebök building is generally acceptable. In summer 2005 with 220 hours of ambient temperatures above 25°C, all rooms had very satisfactory room air conditions, with 4.2% of all office hours above 26°C with a maximum of 7.7% (see Figure 3.21). The total number of working hours is 2871 h.

In the warmer summer of 2006 (e.g. with a mean ambient temperature in July of 4 K higher than average) 10.6% of all office hours had more than 26°C room air temperatures with maximum values of 16.6%. With monthly average temperatures of 23.5°C in July, a room air temperature level of 27°C is considered acceptable, but here only 6% of all office hours are above 27°C (see Figure 3.22). Although the night ventilation concept could be improved with higher night air flows and lower supply air temperatures, the building functioned well in both summer and winter and had an extremely low total primary energy consumption of 50 kWh m^{-2} a^{-1} for heating, lighting, ventilation and auxiliary electricity consumption.

3.3 Summary of Passive Cooling

New experimental results for night ventilation are presented for some of the energetically best European office buildings today.

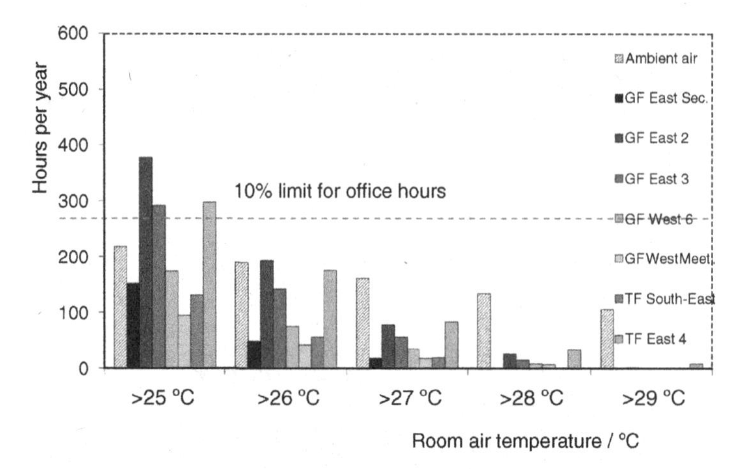

Figure 3.21 Hours in the year 2005 when the room air temperature exceeded a given value (only working days from 08:00 to 18:00 were considered). The key orders the rooms on the ground floor (GF) and top floor (TF) from left to right

Night ventilation measurements and simulations showed that complete discharging of internal surfaces during longer hot periods in summer can only be achieved with very high air change rates and high heat transfer coefficients. In real building situations, especially close to the neutral zone, the flow direction is not necessarily from the inside to outside so that the effective air changes for cooling are often lower than measured. Correct design of the neutral zone placement (as high as possible) is important for effective natural night ventilation. In the building project investigated, wind-induced

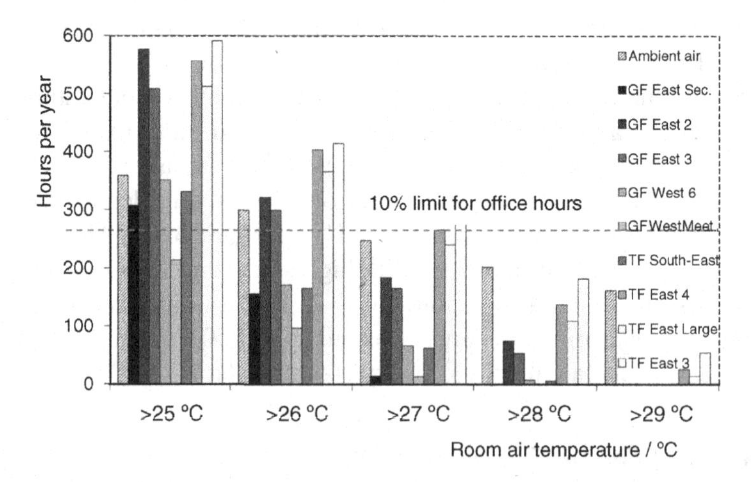

Figure 3.22 Hours in the year 2006 when the room air temperature exceeded a given value (only working days from 08:00 to 18:00 were considered). The key describes the rooms from left to right (ambient air, ground floor (GF), top floor (TF))

cross-ventilation dominated the air exchange. A significant disadvantage of manually controlled passive cooling systems is the early evening heat gain through air changes, which reduces the night cooling potential by 20–30%.

However, user-driven night ventilation concepts work very satisfactorily during moderate Central European summers with no more than 150–200 h above 25°C. Internal loads as high as 400 Wh m^{-2} d^{-1} in a north-facing office can be removed, so that less than 50 h are above 26°C. The same applies to a south-facing office with lower internal loads of about 200 Wh m^{-2} d^{-1}. During hotter summers with 300 h above 25°C, nearly 10% of all office hours are above 26°C in both south- and north-facing offices, which shows the limits of passive cooling concepts in warmer climatic regions. An energy-efficient improvement is obtained if night ventilation is restricted to the later hours of the night through the control of external or internal openings and if a fan can increase the total air volume flow.

Mechanically driven night ventilation results were obtained from two further office buildings in Southern Germany. The ratio of cooling power to electrical power required for the fans (i.e. the COP), was between 5 and 10, which is still a better performance than conventional chillers provide. The night temperature drops were only modest, about 3 K on average, which is due to the low air exchange rate of about 2 h^{-1}. The removed heat load was between 85 and 120 Wh m^{-2} per night, which is significantly less than the usually generated internal heat loads. The limit to higher air exchange rates is clearly the necessary electrical power, so that very low pressure drops at the air inlet and within the tubing system are essential. Hybrid systems combining natural and forced ventilation might also be an interesting alternative.

4

Geothermal Cooling

Research into ground-coupled cooling systems, which are more energy efficient than conventional compression cooling systems has been carried out for many years. The initial interest was in direct coupling of buildings with the earth, in order to take advantage of the fact that the sub surface soil temperature in many regions falls below or is within comfort conditions (Labs, 1989; Dahlem, 2000). Direct contact with the ground allows heat to dissipate into the earth, and stabilizes building temperature at or near the temperature of the adjacent soil. In order to be effective, most of the envelope of the building must be in contact with soil at a sufficient depth to eliminate the effect of variations in the surface temperature. A principal disadvantage of direct coupling is that the missing thermal separation leads to additional heating demand in winter.

Where direct contact with the soil is not possible (as is usually the case, due to a variety of constraints), the ground may still be used as a heat sink by means of either earth to air or earth to water heat exchangers. Examples are given by Fink *et al.* (2002). If air is used as the cooling fluid, the heat exchangers are typically pipes buried beneath the surface, through which air is drawn into the building by electric fans (Kumar *et al.*, 2003). As air moves through the pipe, energy is transferred to the adjacent soil. Since air has a much smaller heat capacity than any type of soil (by several orders of magnitude), its temperature will become equal to that of the soil after a certain distance with only a small disturbance to the earth's temperature field. The air drawn into the building may be either ambient air, in which case the system is referred to as 'open-loop', or interior air, circulating in a 'closed-loop' or multi-pass system. Argiriou (1996) presents several models for calculating the time-dependent heat transfer between the air and the ground, and the resulting temperature of the air supplied to the building.

Direct cooling of the building is only possible if the temperature of the soil is lower than the desired temperature of the indoor air. With alternated signs, this is equally true for direct heating (Rafferty, 2004). This restriction cannot be met in practice in many locations, especially in the warmer regions most in need of cooling, such as North Africa, the Middle East, parts of China, India and most tropical countries. As a result, although commercial installations of ground source cooling systems are gradually becoming more common, the application of this technology is still restricted mainly to temperate climates. Indicative of this situation is the fact that while the Association of German Engineers (VDI) has already published a series of engineering guidelines on the planning and installation of ground-coupled heat pumps (VDI, 2000), other countries do not have a single system installed to date.

Water- or brine-filled ground heat exchangers are either designed in a horizontal configuration with a shallow depth of about 2 m or installed as vertical loops. Vertical ground heat exchangers are constructed by inserting one or two high-density polyethylene U-tubes into vertical boreholes of 75 to 150 mm diameter. Vertical loops are usually connected in parallel to reduce the pressure drop. In Europe, double U-tubes are common, whereas in the USA single tubes prevail. Zeng *et al.* (2003) showed that double U-tubes reduce the borehole resistance by 30–90% and thus increase heat transfer.

In Germany, vertical pipes up to a depth of 100 m are increasingly used for closed-loop direct cooling of buildings, with water as the heat transfer fluid. The higher heat capacity of water is advantageous, as the electrical energy needed for circulating the fluid through the earth heat exchanger is considerably lower than if air is used. The water cooled through contact with the earth is then distributed in the building using either activated concrete slabs with buried pipes or an air-based ventilation system, in which the air is cooled by the water (in an additional heat exchanger). Several large office buildings in Germany now use this low-energy cooling system (the passive office building Energon in Ulm, Solar Info Centre in Freiburg, Post Tower in Bonn, with heat exchangers to air-based distribution systems; Technikum Biberach, the office building DS Plan in Stuttgart with direct water circulation in activated concrete ceilings). The main restriction of the system is the available ground temperature level, which at 15 m depth corresponds approximately to the annual average air temperature above ground and then increases again by about 3 K per 100 m (see Figure 4.1).

If this temperature level is too close to the desired room temperature, the cooling power of such a system is too low to be cost effective. This problem may be overcome by one of two methods:

- By employing indirect cooling systems, such as reversible heat pumps (chillers), which use the ground as a heat sink and are therefore more efficient compared

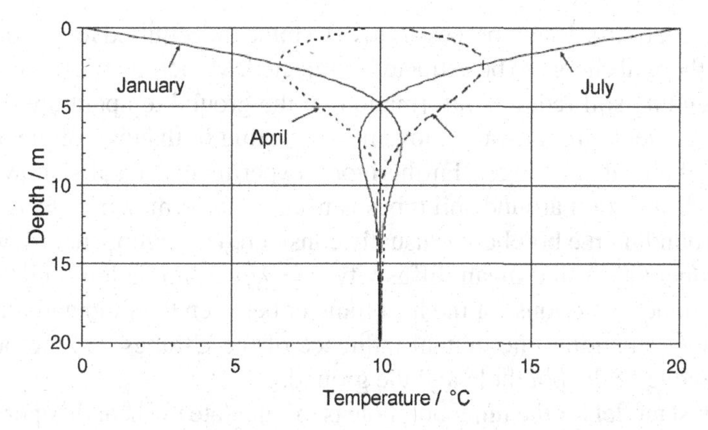

Figure 4.1 Variation of ground temperature with depth for German climatic conditions

with similar systems exposed to cooling tower temperatures or hot ambient air (Karagiorgas *et al.*, 2004).

- By cooling the soil below its undisturbed temperature prior to circulating fluid through the heat exchanger. When horizontal heat exchangers at shallow depths are used, soil temperature may be influenced by surface treatment. At greater depths, the ground may be cooled in the winter season as heat pumps in the heating mode extract energy from the soil. Li *et al.* (2006) showed that unbalanced heat extraction from the ground reduces temperature levels by 6°C within 5 years for heat pump operation only. If only heat is rejected to the ground from a cooling machine, after 13 years the soil temperature will be over 35°C and no longer suitable for air-conditioning. Only balanced heat fluxes kept ground temperatures constant over a 30-year period. Eklöf and Gehlin (1996) carried out thermal response test measurements at a borehole cooling installation for telephone switch stations in Sweden located at Drevikstrand and Bromma. In Drevikstrand, four boreholes were placed in a line with about 5 m spacing on average. The systems were designed to cool the circulating fluid from 22 to 16°C. With a measured high thermal conductivity of about 4 W m^{-1} K^{-1} and a rather low borehole resistance of 0.09 K m W^{-1} the mean capacity was determined to be 30 W m^{-1}. In Bromma, 13 boreholes at 130–160 m depth were placed in an irregular pattern. Similar soil conductivities and even lower borehole resistances were measured. The required temperature levels were lower at 20°C return and 14°C supply. The thermal capacity is significantly lower at 17 W m^{-1}.

Environmental and geothermal energy obtained via horizontal or vertical ground heat exchangers is usually used for heating (mostly via ground-coupled heat pumps), but such heat pump systems can also be used for space cooling. However, although there are over 500 000 ground source heat pumps installed worldwide (IEA, 2002), there is very little experience with such systems coupled to cooling machines.

The main task of the design process is to determine the required length of tubing for the required thermal energy to be extracted or rejected. Modelling the thermal response of the surrounding soil requires information on the ground temperature distribution, the moisture content, groundwater movement, freezing or thawing of the soil and the geometry of the heat exchanger. Furthermore, experimental results showed that the temperature distribution around soil tubes is usually not symmetric (Bi *et al.*, 2001). The soil surrounding the borehole is usually considered as homogeneous with a mean thermal conductivity λ and mean diffusivity $\alpha = \lambda/\rho c$. Simulation models are often separated in 'inner' solutions for the heat transfer between the fluid and the perimeter of the borehole, including the mutual influence of the U-tubes and the outer region between the edge of the borehole and the ground.

The simplest model for the inner borehole is to calculate the heat flux per unit length of the borehole q_b from the resistance R_b, the mean fluid temperature $T_f(t)$ and a mean borehole temperature $T_b(t)$. The thermal resistance contains the convective resistance between the fluid and pipe wall, the conductive resistance of the wall and borehole filling material:

$$q_b = \frac{T_f(t) - T_b(t)}{R_b} \tag{4.1}$$

As the steepest temperature gradients occur at the pipe–soil or backfill material interface, several authors developed detailed models for the near temperature field. Zeng *et al.* (2003) developed analytical models for a range of tube configurations within the borehole, Piechowski (1999) solved heat and moisture transport equations for horizontal pipes embedded in the soil. Piechowski states that the soil temperature around a pipe drops by 30–40% within a few centimetres, indicating the influence of precise near-field models.

To obtain the mean temperature of the borehole and to calculate the complete soil temperature field $T(r, z, t)$, the heat conduction equation has to be solved as a function of time t, depth z and distance from the borehole r:

$$\frac{1}{\alpha}\frac{\delta T}{\delta t} = \frac{\delta^2 T}{\delta z^2} + \frac{\delta^2 T}{\delta r^2} + \frac{1}{r}\frac{\delta T}{\delta r} \tag{4.2}$$

The height dependence z of the temperature is often ignored and the simplest solution is obtained if a step function heat input is applied at the origin $r = 0$. The solution is known as Kelvin's infinite line source theory and is used to analyse thermal response test data. The temperature at the boundary of the borehole is given by

$$T(r_b, t) - T(t = 0) = \frac{q}{4\pi\lambda}\int_{\frac{r^2}{4at}}^{\infty} \frac{e^{-\beta^2}}{\beta} d\beta = \frac{q}{4\pi\lambda}\left[\ln\left(\frac{4\alpha t}{r_b^2}\right) - \gamma\right] \tag{4.3}$$

The solution is valid for a constant heat flux per metre borehole q, assuming a constant temperature and an infinite length of the borehole. γ is Euler's constant (0.5772).

As the mean fluid temperature is measured from the inlet and outlet temperature during the thermal response experiment, the borehole resistance R_b is used to calculate the fluid temperature as a function of time:

$$T_f(t) - T(t=0) = \frac{q}{4\pi\lambda}\left[\ln\left(\frac{4\alpha t}{r_b^2}\right) - \gamma\right] + q\,R_b \tag{4.4}$$

The measured fluid temperature is then plotted against the logarithm of time to obtain the mean conductivity of the soil from the gradient K:

$$\lambda = \frac{q}{4\pi K}$$

Knowing the thermal conductivity and the heat injected per metre of borehole length, the borehole resistance can also be calculated. For a valid analysis of the rḗsults, a minimum duration for a thermal response test has to be given: $t > 5r_b^2/\alpha$. Measured typical borehole resistances are around 0.1 m KW^{-1}.

If the heat flux is applied at the perimeter of the borehole r_b, the solution to the so-called cylindrical heat source method is given by Carslaw and Jaeger (1947). Both solutions are similar for $\alpha t/r_b^2 > 20$. For a sandy soil with a thermal diffusivity of $4.74 \times 10^{-7}\,m^2\,s^{-1}$ and a radius of 0.1 m, this corresponds to nearly 5 days. However, neither the line source nor the cylindrical source method saturate at large timescales, which is a major drawback. Also, parameters like heat capacity or groundwater flow cannot be quantified.

An extension of the line source approach has been proposed by Eskilson and Cleasson (1988). This includes the height dependence of the temperature field. The integrals in the solution of the temperature field are called g-functions, which have been mostly computed numerically and then used in tabulated form. Recently analytical solutions have been developed for the g-functions by Lamarche and Beauchamp (2007). If the g-function is known, the temperature at the perimeter of the borehole can be easily calculated:

$$T_b(t) - T(t=0) = \frac{q}{2\pi\lambda}g\left(\frac{t}{t_s}, \frac{r_b}{H}\right) \tag{4.5}$$

with $t_s = H^2/9\alpha$ and H as the depth of the borehole; t_s is the time when the transient process ends and the temperature field becomes stationary. The power injected into the borehole then equals the heat given off to the ground. The timescale is very long: for the sandy soil described above and a borehole 100 m deep, it takes 74 years to reach steady-state conditions at constant heat injection rate.

Using the g-function approach, Lamarche and Beauchamp (2007) calculated the time when the height dependence (so-called axial effect) becomes important for a 100 m borehole length as $t \sim t_s/20$. For the wet sandy soil described above, this takes

3.7 years. Thermal interference between two boreholes separated by 5 m becomes relevant for $t > t_s/150$, that is 0.5 years.

Numerical models offer more flexibility in the temperature field calculation for arbitrary geometries and time-varying heat fluxes and inlet temperature levels and have been chosen here for parameter studies. The potentials and limits of direct geothermal energy use for the heating and cooling of buildings is analysed in the following using the results of measurements, and simulation studies carried out by the author.

4.1 Earth Heat Exchanger Performance

4.1.1 Earth to Air Heat Exchanger in a Passive Standard Office Building

In the passive standard office building of the Lamparter Company in Weilheim, Germany, an earth heat exchanger cools the fresh air supplied to the building. It is positioned around the building and consists of two polyethylene pipes with a diameter of 0.35 m and a length of 90 m each. The pipes are laid at a mean depth of 2.80 m at a mutual distance of 0.9 m (Figure 4.2). By ventilating ambient air through the system, the air is cooled in summer and heated in winter.

The earth to air heat exchanger of the office building is mainly designed for winter preheating of ambient air. It reduces the heating demand for fresh air ventilation and prevents freezing of the cross-flow heat exchanger of the heat recovery system. During summer the earth to air heat exchanger helps to meet the cooling loads of the building. A main design goal was to achieve a small pressure loss (Pfafferott, 2003). At the nominal volume flow of 1900 $m^3 h^{-1}$ and a pressure drop of 175 Pa, an air exchange rate of 0.6 h^{-1} is provided.

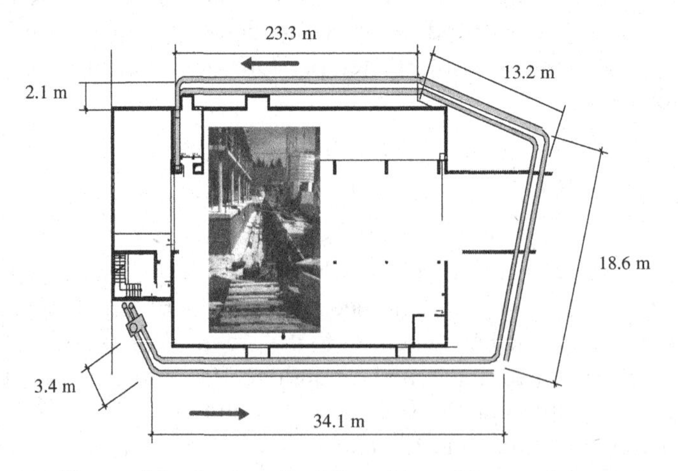

Figure 4.2 Outline of building with earth heat exchanger

The contribution of the earth heat exchanger during daytime operation was investigated both experimentally and theoretically. Inside the earth to air heat exchanger, tubes of temperature sensors are placed at a mutual distance of 9 m. Additionally the humidity of the air is measured at the inlet and the outlet of the tubes. The soil temperature is measured at the front and at the back of the building at various depths and at a range of distances from the tubes of the heat exchanger.

At inlet ambient air temperatures for the earth to air heat exchanger between 9 and 16°C, overlaps occur between heating defined for outside temperature below 15°C and cooling operation fixed at outside temperatures above 15°C. This means that sometimes the air is heated during the cooling period (summer) and vice versa – a control strategy which could certainly be improved. To calculate the coefficient of performance of the earth heat exchanger, it is necessary to determine the additional electrical energy of the fan to overcome the pressure loss of the earth to air heat exchanger. By measuring the pressure loss Δp (Pa) for different volume flows $\dot{V}$ (m^3 h^{-1}), the following empirical equation was obtained for the two tubes of 90 m length:

$$\Delta p = 6 \cdot 10^{-5} \dot{V}^{1.9728} \tag{4.6}$$

With an efficiency of the fan η_f of 57%, the electrical energy P_{el} can then be determined by

$$P_{el} = \frac{\dot{V} \Delta p}{\eta_f} \tag{4.7}$$

The annual coefficients of performance were calculated from the sum of heating and cooling energy divided by the electrical energy consumed. They achieve excellent values of 50, 35 and 38 in the years 2001 to 2003 (Figure 4.3). However, the earth heat exchanger cannot fully remove the daily cooling load: the hygienically required fresh air volume flow limits the cooling power which can be supplied by such systems. From the average internal load of 131 Wh m^{-2} d^{-1} in the south office the earth heat exchanger efficiently provided 24 Wh m^{-2} d^{-1} in the summer period of 2003, that is 18%.

An extensive set of experimental data was used to validate a theoretical model of the earth heat exchanger. A numerical simulation model of the earth to air heat exchanger was implemented, which enables the user to check the performance of the system as well as to test different control strategies (Albers, 1991; Tzaferis *et al.*, 1992; Henne, 1999). It was implemented as a block for the INSEL simulation environment and features a graphical user interface (Schumacher, 2004). In the model the heat exchanger is divided into a number of elements and the analytical steady-state solution of the differential equation of heat transfer is found for each element using a conform transformation. It is possible to account for multiple tubes as well as for the influence of groundwater and nearby buildings.

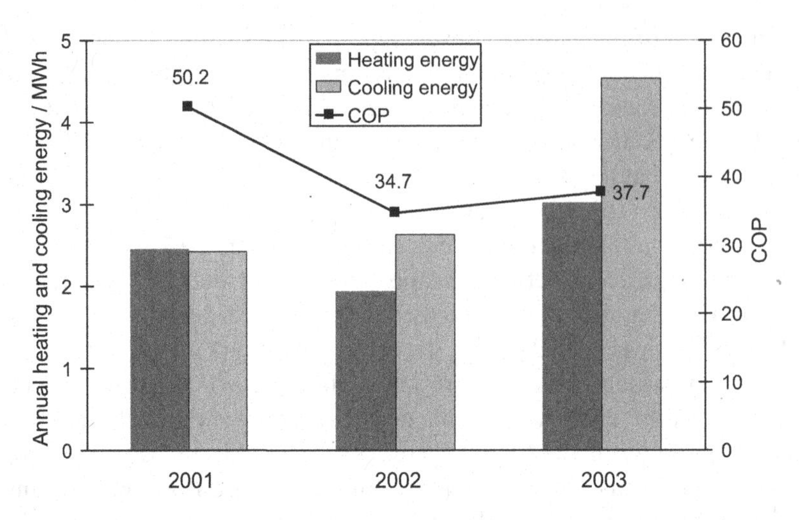

Figure 4.3 Measured cooling and heating energy and COP of earth heat exchanger

The required inputs for the model are the geometrical boundary conditions and the thermal underground parameters, which are considered to be constant. Furthermore, the calculation requires the ambient temperature, undisturbed ground temperature, volume flow and pressure loss for every time step. It is difficult to quantify the ground temperature and even more difficult to simulate it (Dibowski and Rittenhofer, 2000). The ground temperature depends on the history of the ambient temperature and also on the thermal properties of the subsoil. The latter are not exactly known since they strongly depend on the moisture content, which may vary considerably. Pfafferott therefore considers the undisturbed ground temperature a value, which can only be approximated (Pfafferott, 2003). However, Tzaferis *et al.* (1992) state that it must be regarded as one of the key parameters for determining the outlet air temperature. In this work the ground temperature is calculated according to the German guideline VDI 4640, which uses a sinusoidal wave function. All other inputs to the model are measured and are read directly from a data file. A time step of one hour was chosen, which can be regarded as the minimum time step for this kind of model. Outputs of the calculation are the exit air temperature and the thermal power.

The assumption of a sinusoidal soil temperature is obviously greatly simplified. Deviations up to 3 K have been measured over the three years from 1 January 2001 to 31 December 2003 (Figure 4.4). The discontinuity in calculated temperature between the years 2002 and 2003 is due to the fact that the sinusoid is calculated for each year separately using the average temperature and amplitude of that year.

However, despite the simple calculation of the undisturbed ground temperature and the constant thermal underground parameters, the agreement between measured and simulated air exit temperatures for the two moderate years of 2001 and 2002 and the

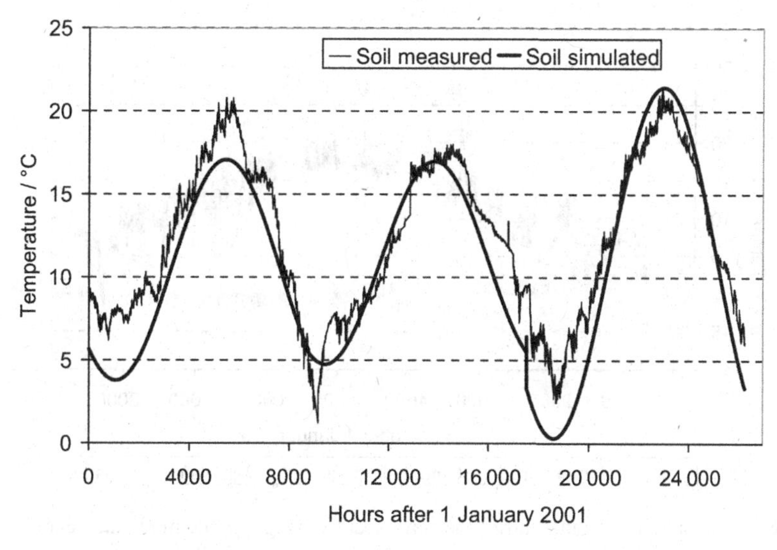

Figure 4.4 Measured and simulated soil temperatures at 2.35 m depth

unusually warm year of 2003 is found to be very satisfactory (see Figures 4.5, 4.6 and 4.7).

Given the limited input effort and the fast calculation process, even dynamic situations can be modelled with acceptable results down to a resolution of one hour. The mean differences between the measured and simulated outlet temperature for the years of 2001 to 2003 are 0.4, 0.7 and 0.4 K respectively (as an example see Figure 4.7).

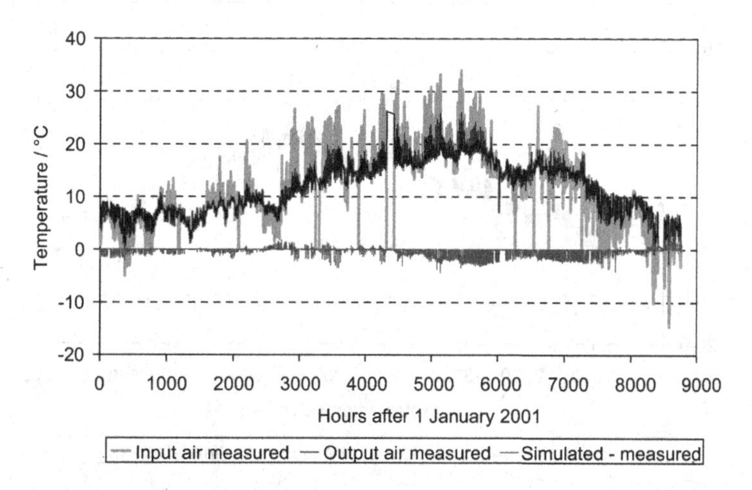

Figure 4.5 Measured ambient and outlet air temperature of the earth heat exchanger in the moderate year of 2001. Also shown is the temperature difference between the simulated and measured value

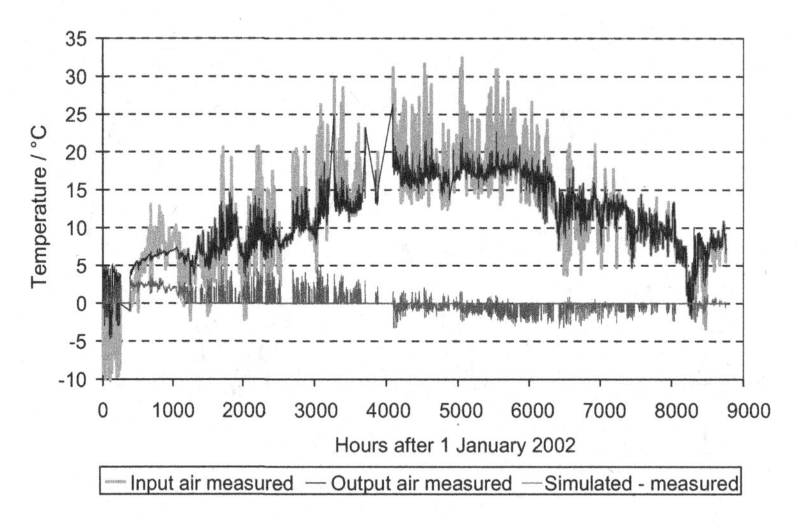

Figure 4.6 Ambient air and outlet air of the earth heat exchanger in the moderate year of 2002. Also shown is the temperature difference between the simulated and measured value

The cumulative frequency distribution of the exit air temperature shows that the agreement between simulation and measurement is excellent at low exit temperatures, that is in winter, and that in summer the simulated temperature is slightly lower than the measured temperature (results from 2001 in Figure 4.8); 95% of all temperatures are below 20°C, which shows the good summer cooling potential, and the temperature is never below 0°C, which is excellent to prevent winter freezing of the heat recovery unit in the mechanical ventilation system.

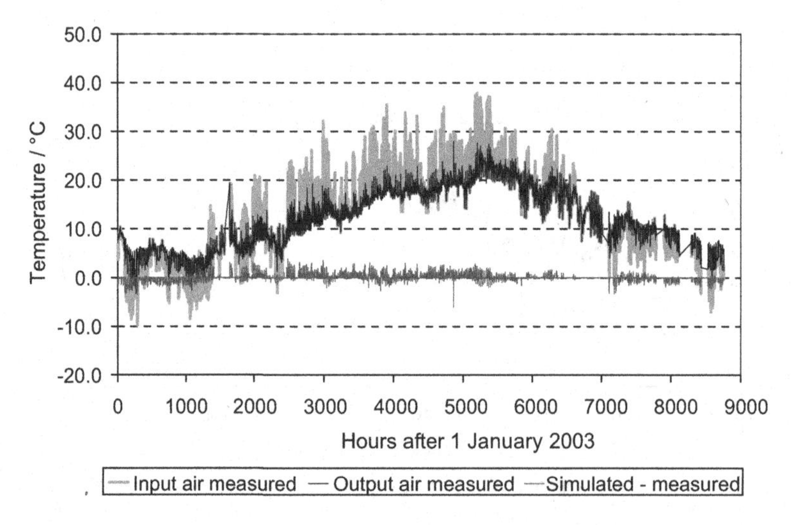

Figure 4.7 Outlet temperatures of the earth heat exchanger in the warm year of 2003. Also shown is the temperature difference between the simulated and measured value

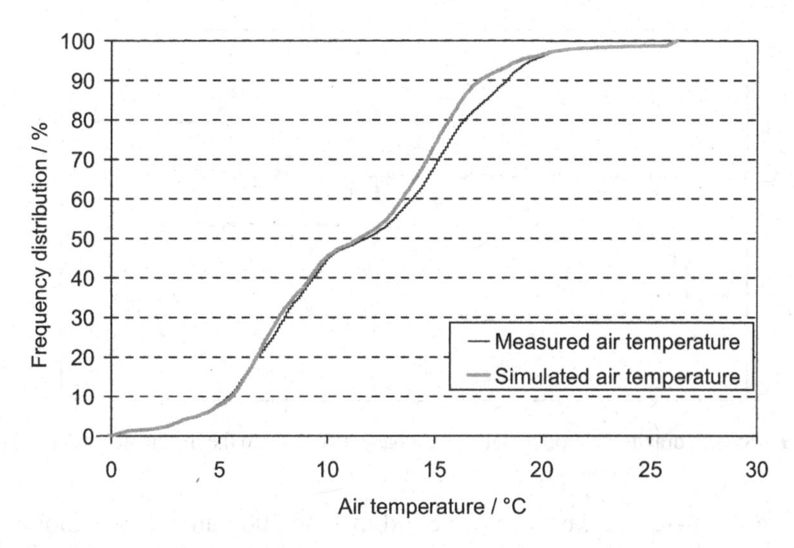

Figure 4.8 Cumulative frequency distribution of measured and simulated exit air temperatures in 2001

4.1.2 Performance of Horizontal Earth Brine to Air Heat Exchanger in the ebök Building

An alternative to the direct-air-based ground heat exchange is the indirect cooling of air by circulating a brine solution through the ground and then using an air to brine heat exchanger. This system has the advantage of avoiding any hygienic problems, which might occur in air-based systems if condensation water is not reliably removed. Furthermore, pressure drops are usually lower in such liquid-based systems. A simple horizontal absorber placed around the perimeter of a rehabilitated office building was experimentally analysed in 2005.

For air preheating and cooling of the ebök building in Tübingen, Germany, five horizontal earth to brine heat exchangers with a length of 100 m each were installed shallowly under the soil surface (about 1.2 m depth). During summer they are used to cool the supply air (see Figure 4.9). The volume flow of air during daytime is about $1750\,m^3\,h^{-1}$.

The earth temperature levels were measured for 2 years directly in between the tubes at 1.2 m depth and a distance of 0.5 m from the building and compared with the undisturbed soil temperatures at 1.2 and 2.0 m depth. The sensors for the undisturbed temperature measurements were placed 3.5 m away from the northern part of the building and 12 m from the next neighbouring building. The time phase shift between the ambient air maximum temperature and soil temperature is about 1.5 months at 2 m depth. When the earth heat exchanger is operating, for example for two months during summer 2006, the daily mean earth temperature increases by about 2–3 K (see Figure 4.10). The regeneration of the soil temperature is very fast in both summer and winter.

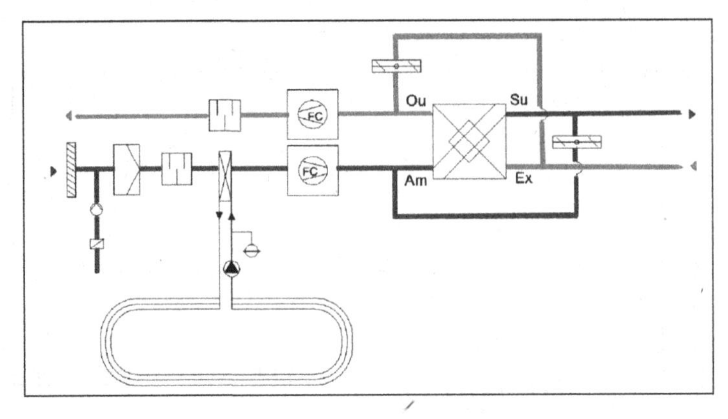

Figure 4.9 System drawing of horizontal heat exchanger coupled to the mechanical ventilation system

During a hot 14-day measurement period in June 2005 an average cooling power of 1.5 kW with a maximum of 4 kW was measured. Due to the close proximity and low depth of the tubes, the maximum heat dissipation per metre of tube is only 8 W.

The pressure drop due to the brine to air heat exchanger amounts to only 12 Pa. The installed fan needs an electrical power of 30 W to overcome this drop, whereas the brine pump consumes about 60 W. This results in maximum COPs of 40 and an average COP of 18.4 (see Figure 4.11). It can be seen that a phase of heating occurred

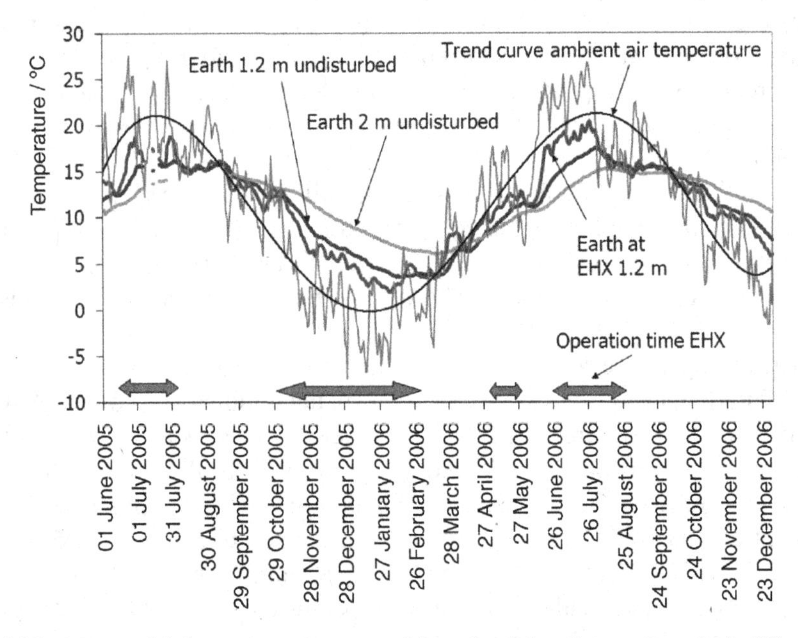

Figure 4.10 Measured daily mean temperatures of the soil at 3.5 m distance from the building ('undisturbed') and directly between the earth heat exchanger (EHX) tubes at 1.2 m depth

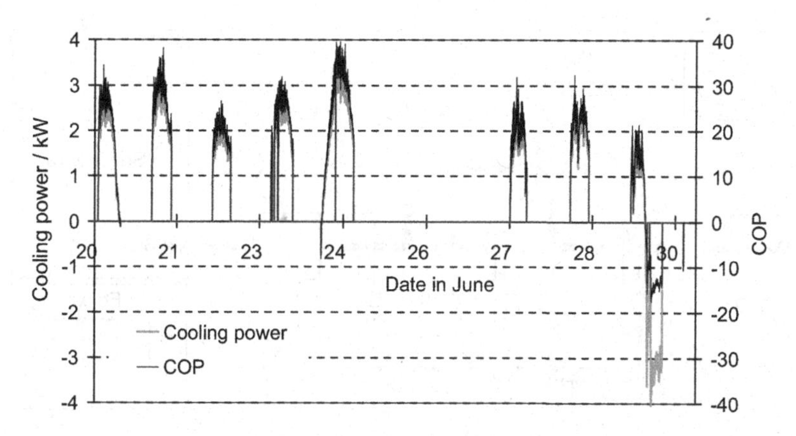

Figure 4.11 Cooling power and COP of horizontal brine heat exchanger in the ebök building during summer 2005

during the last 2 days due to the relatively warm soil temperature. This high soil temperature is mainly due to the shallow depth of the pipes.

The ambient air can be cooled down by as much as 7 K in the heat exchanger. However, supply air to the building is still at 28°C during midday. The soil temperature rises up to nearly 20°C at the end of June (see Figure 4.12). The logarithmic average temperature difference of brine and air across the heat exchanger is about 6.3 K.

4.1.3 Performance of Vertical Earth Brine to Air Heat Exchanger in the SIC Building

In the SIC building, the main cooling strategy is mechanical night ventilation. Only a seminar room with a floor area of 178 m^2 and the lobby area of the building with a

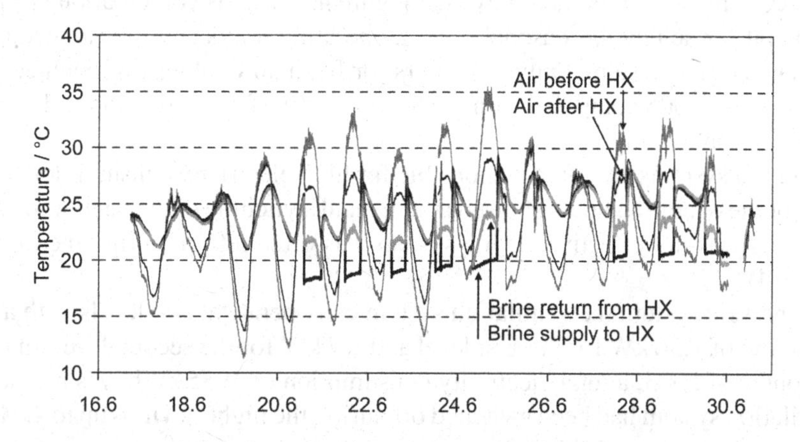

Figure 4.12 Horizontal heat exchanger performance with temperature levels of the brine supply to the heat exchanger (HX) and its return, as well as the ambient air before the HX and after in summer 2005

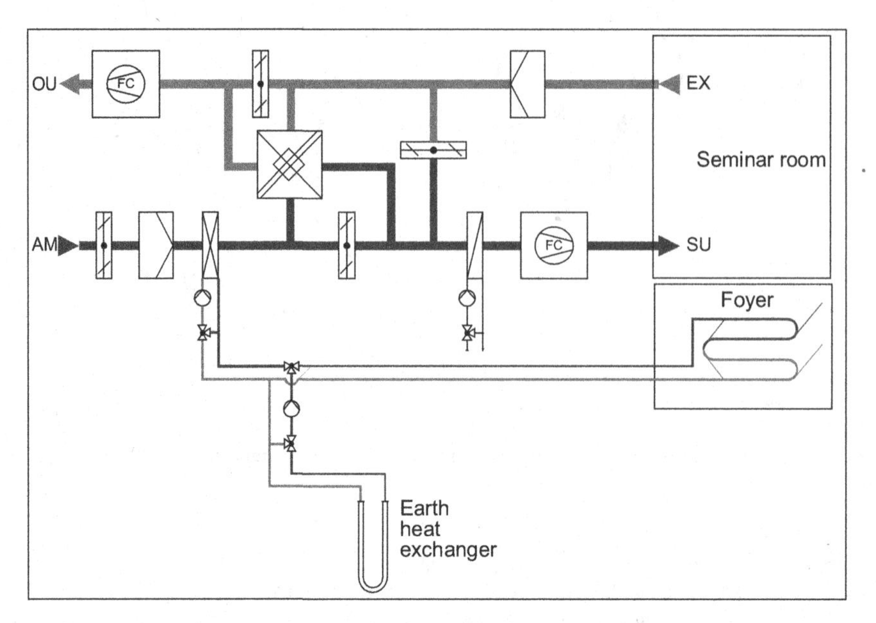

Figure 4.13 System drawing of horizontal heat exchanger coupled to the mechanical ventilation system (AM = ambient air; SU = supply air; EX = exhaust air; OU = outlet air)

157 m^2 activated concrete floor are conditioned by geothermal energy. Five vertical borehole heat exchangers of 80 m each supply cooling and heating energy with an air-based distribution system for the seminar room. The fan is designed for a volume flow of about 5100 $m^3 h^{-1}$. The geothermal cooling system as the cooling supply for the ventilation system is only operated when needed, say if a lecture or seminar takes place. If ventilation air cooling is not required, cooling energy is delivered to the floor cooling system in the lobby (see Figure 4.13). This combination has proven to be crucial for achieving reasonably long operating hours and thus an economical performance of the system. Only 233 hours for fresh air cooling was counted for the seminar room in 2005 compared with 2289 hours for the floor activation and 856 hours for winter air preheating.

Ventilation systems are often responsible for high electricity consumption. A close analysis of the operating hours showed that manual switching of the ventilation system by the users was not effective, as the system was often left on during the night (see Figure 4.14).

In the present system the ventilation system can operate at two levels with a power consumption of 1.36 kW for the first level and 2.7 kW for the second level. In the first three months of 2005, a total electricity consumption of 1985 kWh was measured. If the ventilation system had been switched off during the night, savings up to 42% could have been achieved (see Table 4.1).

In 2006, the ventilation system was still on several times during the night, which was necessary to maintain the room setpoint temperature at 17°C. The main problem

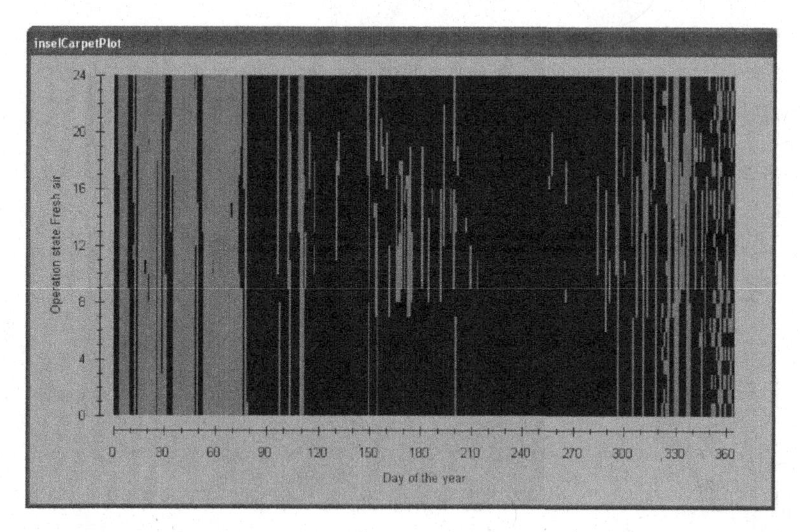

Figure 4.14 Operating hours of ventilation system in SIC building with manual user control (grey = ON, black = OFF) in the year 2005

is an undersized heat exchanger in the ventilation system, which did not allow the supply air temperature to be raised (see Figure 4.15).

Due to the infrequent use of the seminar room ventilation system, the floor cooling system was operating very often for 2289 hours in 2005 and 2922 hours in 2006 (see Figures 4.16 and 4.17).

The measured average soil temperature is close to 16°C during the summer. The brine temperature rises steadily as soon as the pump is turned on, reaching a spread of nearly 3 K during the day. However, the soil recovers quickly during the night when the pump is shut off. Still, a rising trend in the soil temperature is clearly visible from below 15°C to almost 17°C within a week. Supply air temperatures during operation were between 18 and 22°C (see Figure 4.18). Again, the COPs are excellent due to the low pressure drop across the heat exchanger and the low power consumption of the geothermal heat exchanger pump of 170 W. The average COP is 21.6 for cooling and 18.8 for heating.

The heat delivery to or from the ground varies between a maximum of about 15 W per metre of tube in winter and 25 W m^{-1} in summer. The maximum power withdrawal

Table 4.1 Electricity savings for automatic night switch-off of the ventilation system for 10, 8 or 6 hours daily during the months of January to March

Night switch-off	Savings / kWh	Savings / %
21:00–7:00	840	42
22:00–6:00	672	34
23:00–5:00	503	25

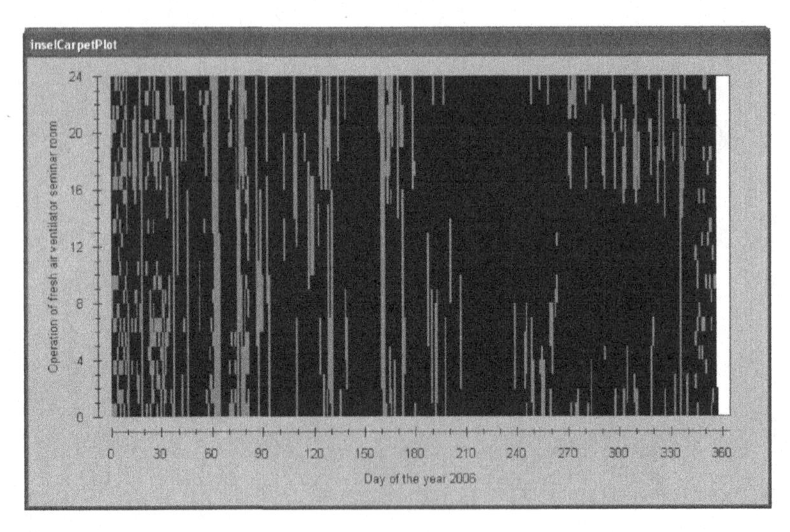

Figure 4.15 Operating hours of ventilation system in SIC building in the year 2006. The night operating times are due to heating demand

is reached in summer, when the ventilation system is switched on. If only the activated floors are used for energy distribution, less heat is taken up from the building and the heat delivery to the ground is reduced (see Figure 4.19).

The measured room and supply air temperatures were also used to validate a dynamic building simulation model of the seminar room. The main unknowns for the simulation are the number of occupants in the room, which were estimated as a function of the measured air volume flow, the shading factor of the external sun

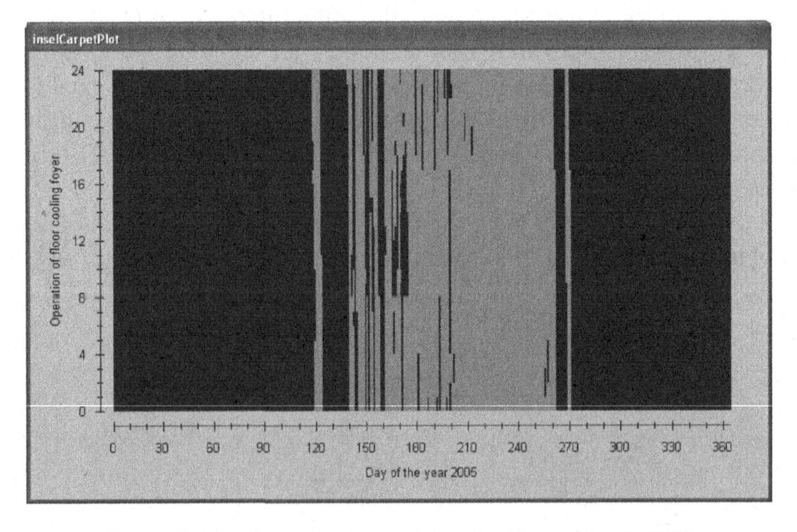

Figure 4.16 Operating hours of floor cooling system in 2005

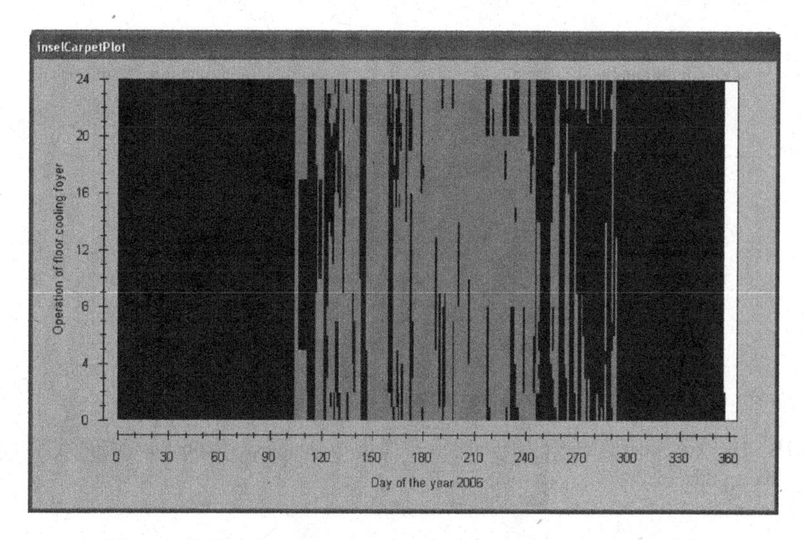

Figure 4.17 Operating hours of floor cooling system in 2006

shades and any natural night ventilation rates through tilted windows. If no external shading were used, the room temperatures would be overestimated by about 5 K. Introducing a low night air exchange of 0.6 per hour reduced the temperature difference by about 1 K. A very good agreement between measured and simulated room temperatures was obtained when the room was shaded on average by 70% during the day and night ventilation was at $0.6\,h^{-1}$ (see Figure 4.20). During occupation of the seminar room, the air temperature could be kept at comfortable levels below 26°C.

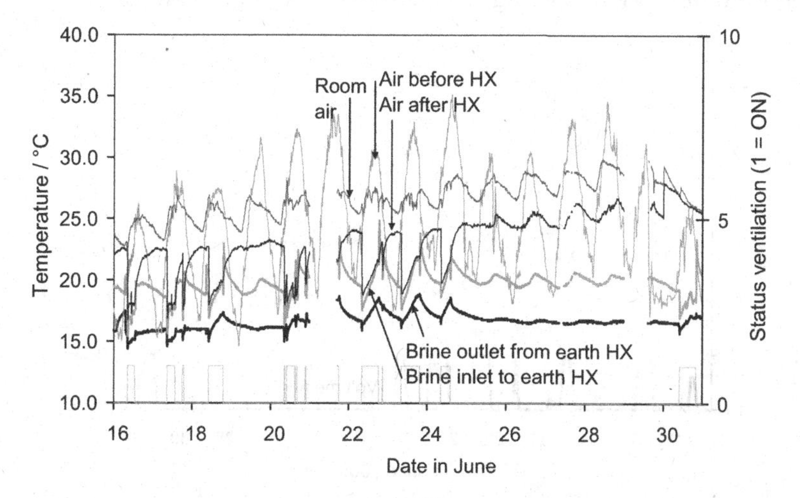

Figure 4.18 Temperature levels of brine solution and air before and after the brine to air heat exchanger. The status of the ventilation system is also shown

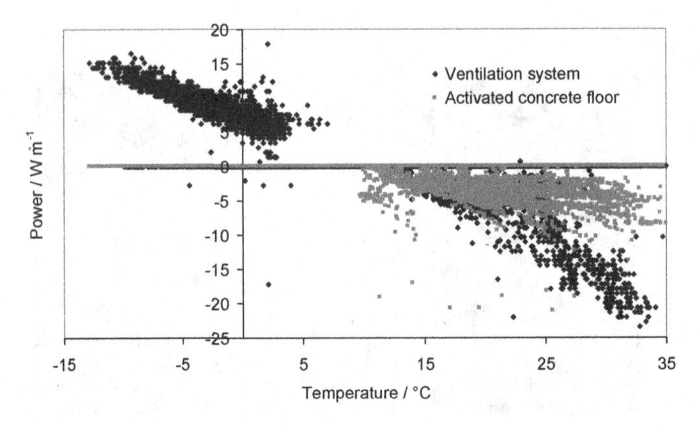

Figure 4.19 Heat transfer of the geothermal vertical heat exchanger as a function of outside temperature for the first six months of 2005

The thermal power dissipated by the heat exchangers is rather low, with a maximum of about 25 W per metre when the ventilation system is operating. Also, during the winter months with ambient temperatures below −10°C the power extracted was less than 20 W per metre (see Figure 4.19). If the floor cooling system is used, the high pressure drop reduces the volume flow rates by a factor of 5 approximately. Thermal power dissipated by the earth heat exchangers drops to 5–10 W m^{-1}. Two factors account for the reduced power: the temperature difference between the floor cooling system and the soil is smaller than for the ventilation system with ambient air temperatures up to 35°C. Furthermore, the heat transfer area between the floor and room air is rather low at 157 m^2. At about 20 W m^{-2} cooling power, only 3 kW cooling load can be taken up by the activated floor. The

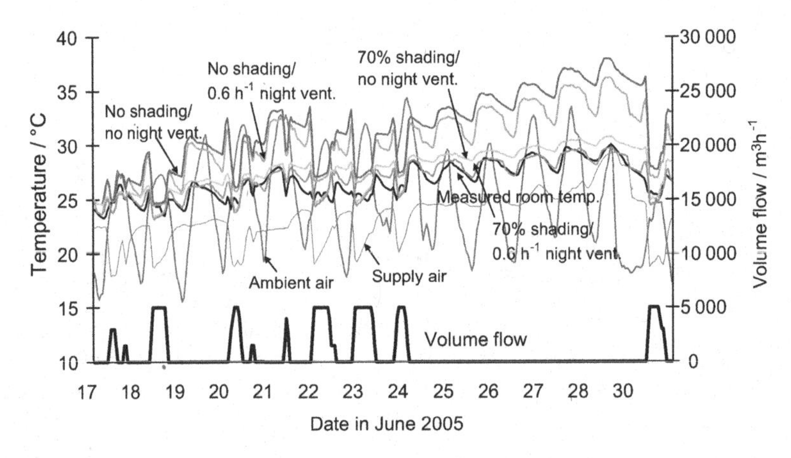

Figure 4.20 Measured and simulated room temperatures during a hot fortnight in summer 2005 with geothermal cooling of the supply air only

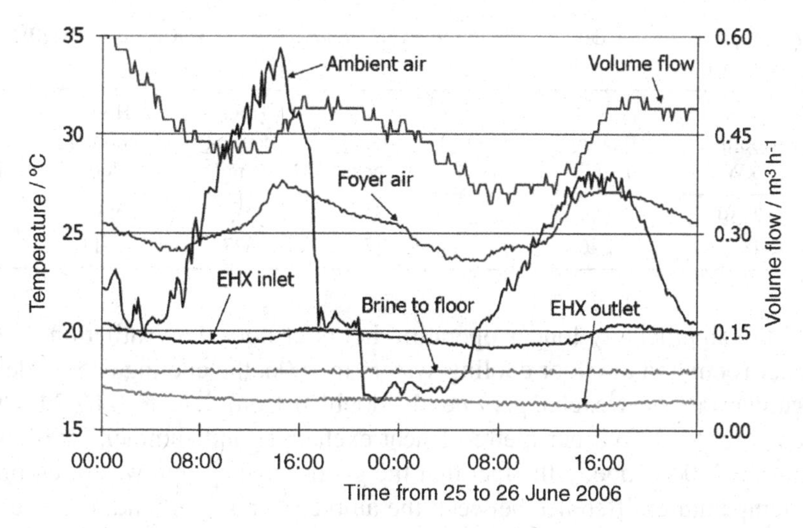

Figure 4.21 Temperature levels and volume flows of activated concrete floor with fixed supply temperature of 18°C to the room

temperature spread between ground heat exchange return and supply is 3 K on average and thus slightly higher than in the case of ventilation use. Increasing the pump power could decrease this temperature spread and thus the surface temperature by 1 or 1.5 K further, but this would not significantly increase the cooling power (see Figure 4.21).

The measurements were repeated during summer 2006 and very similar performance was obtained (see Figure 4.22).

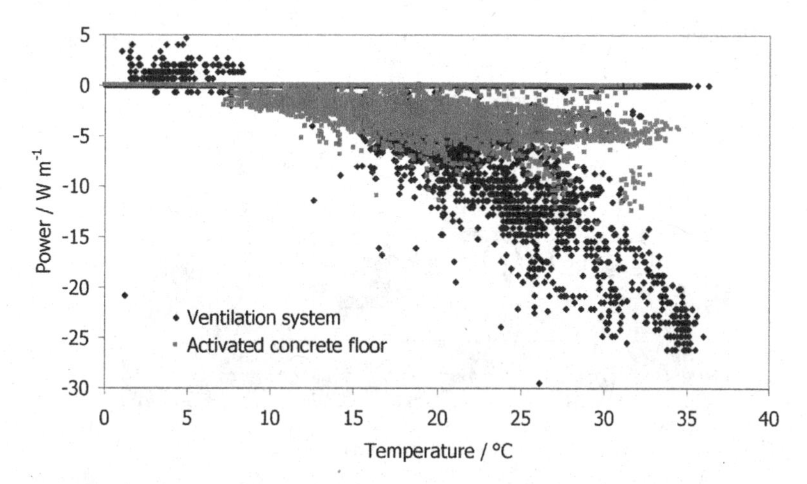

Figure 4.22 Power dissipated per metre of ground heat exchanger for the two operating modes of floor cooling and ventilation system supply in summer 2006

Table 4.2 Measured energy delivered by the five vertical ground heat exchangers to the SIC building in Freiburg, Germany

Year	Cooling / kWh	Heating / kWh	COP cool	COP heat	Hours cooling seminar	Hours cooling floor	Hours heating
2005	2759	2846	20.1	12.8	233	2289	856
2006	4873	246	13.5	5.2	327	2911	289

Due to the significantly longer operating hours of both the ventilation system of the seminar room and the floor cooling system in 2006, the cooling energy delivered by the geothermal heat exchangers nearly doubled from 2.76 to 4.87 MWh. This corresponds to 7–12 kWh per metre of heat exchanger and summer. The low COP for heating in 2006 is due to the fact that the geothermal system was often operated at a low temperature difference between the ambient air and ground heat exchanger. Table 4.2 summarizes the measured results.

4.1.4 Modelling of Geothermal Heat Exchangers

Numerical heat transfer models for geothermal heat exchangers were developed and implemented in the simulation environment INSEL. They are all based on the heat conduction equation with explicit finite difference solutions. The simplest vertical heat exchanger model uses a rectangular discretization geometry. Horizontally, nine zones have been defined and, vertically, 30 elements are calculated for each time step with only one-dimensional heat transfer considered (see Figure 4.23). To determine

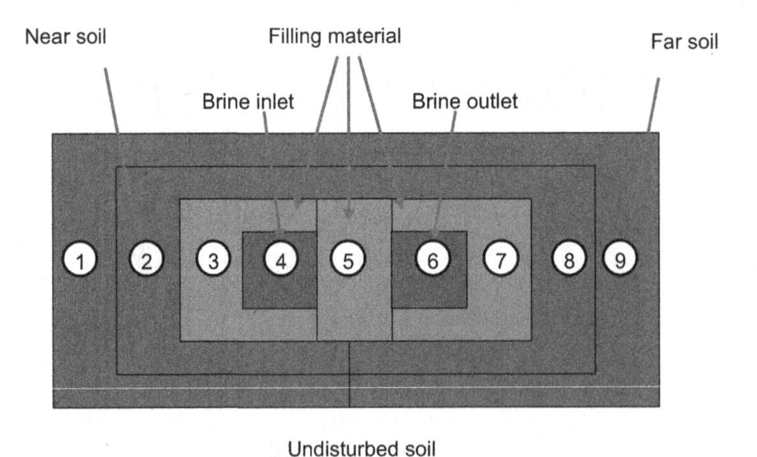

Figure 4.23 Geometry of horizontal discretization of the two-dimensional model for a vertical earth heat exchanger

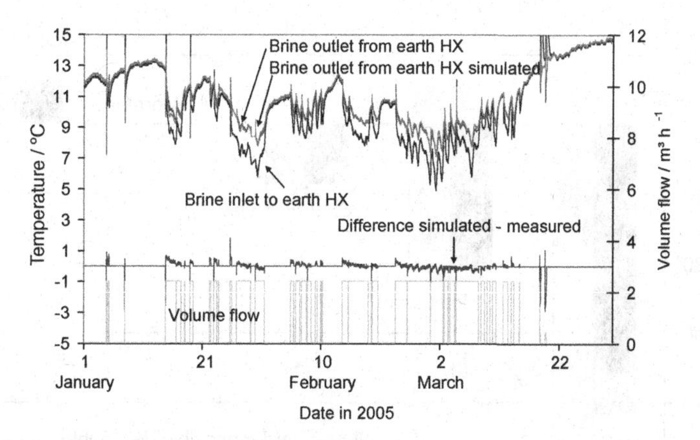

Figure 4.24 Temperature levels of geothermal earth heat exchanger measured and simulated in winter. The volume flows through the tubes are also given

the mutual influence of a field of heat exchangers, a three-dimensional model was developed with a parallelogram geometry for the discretization.

Both models have been validated using the experimental data from the Solar Info Centre project and reproduce the measured temperature and power levels well (see Figure 4.24 for temperature levels in winter using the model for one heat exchanger only).

The differences between simulated and measured temperatures are slightly higher in summer, when the seminar ventilation system is switched off and only low volume flows are used for cooling an activated floor in the foyer (see Figure 4.25).

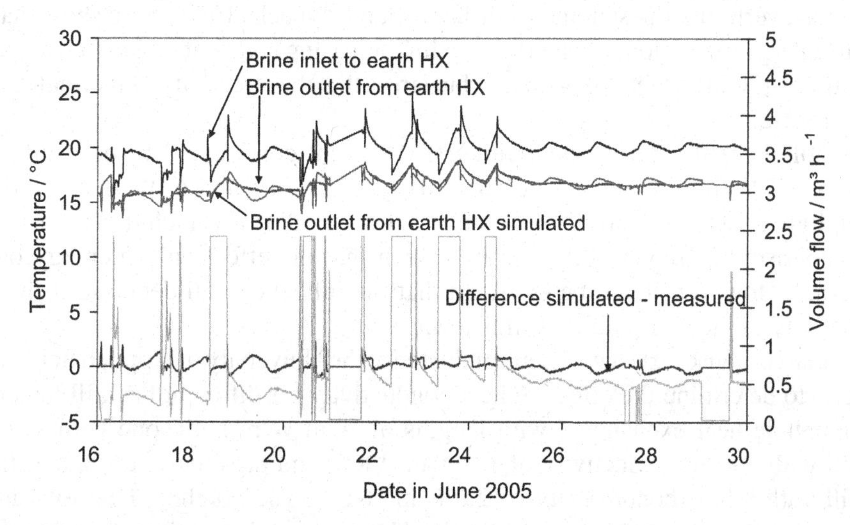

Figure 4.25 Measured and simulated temperature levels for the earth heat exchanger in the SIC building. The main deviations occur at low volume flows

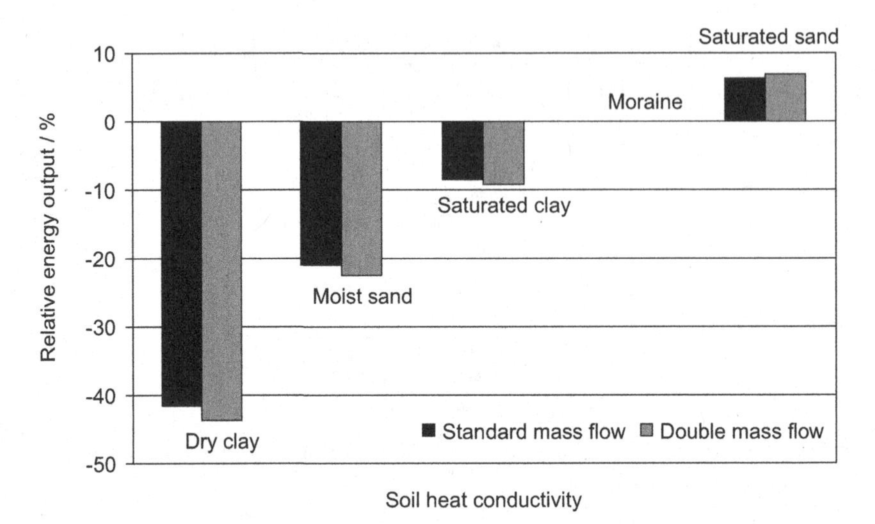

Figure 4.26 Influence of soil heat conductivity on energy output of vertical heat exchangers. The reference value is the moraine soil as in the SIC building context

The three-dimensional validated model was used for parameter studies concerning the backfill and soil heat conductivity, mass flow rates and distance between the geothermal heat exchanger. The boundary conditions are those of the SIC building in Freiburg. The soil heat conductivities range from 0.5 representing dry clay, to 2.5 $W\,m^{-1}\,K^{-1}$ for water-saturated sand. As can be seen from Figure 4.26, the soil heat conductivity has a major influence on the cooling performance. Compared with the standard moraine soil, the energy output is reduced by 39% if the subsoil consists of dry clay, verifying the statements of Sanner and Rybach (1997), who show that the specific energy extraction rate increases with increasing soil heat conductivity. Zhang and Murphy (2003) support this finding but stress that the effectivity of thermal storage may decrease.

Borehole backfill heat conductivities of 0.8 (light concrete), 1.6 (bentonite) and 3.2 $W\,m^{-1}\,K^{-1}$ (high-performance backfill) were investigated in the next step. Although light concrete is rarely used as backfill material, it was included in this study for comparison. In Figure 4.27 it can be seen that the effect on the energy output is relevant. However, it can be expected that the influence will decrease if the heat conductivity of the surrounding soil is low.

Thermal response tests have been conducted at the University of Applied Sciences in Stuttgart to determine the effective heat conductivity for different backfill materials. Two borehole heat exchangers with lengths of 80 m were built, one with standard backfill with a heat conductivity of 1.6 $W\,m^{-1}\,K^{-1}$ and the other with Stuewatherm backfill with a heat conductivity of 2.0 $W\,m^{-1}\,K^{-1}$. The borehole heat exchangers are tested with a thermal power of 3 kW. The heat conductivities of the complete

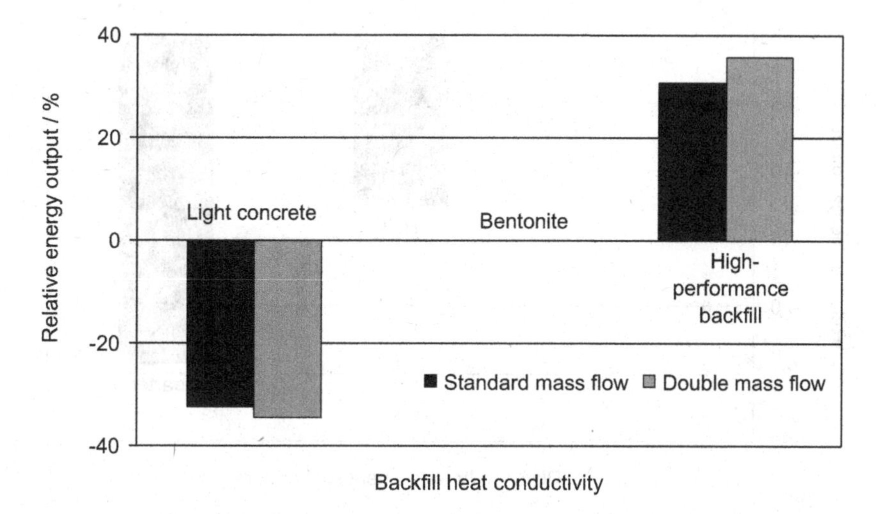

Figure 4.27 Influence of backfill heat conductivity on energy output

system (tube, backfill and surrounding soil), as derived from the first week of thermal response tests, are 1.97 and 2.04 W m^{-1} K^{-1} respectively. Repeated measurements were carried out to determine the measurement errors. The high-performance backfill leads to a 3–7% higher effective heat conductivity. The thermal response tests were also used to validate the above-mentioned numerical model and showed good agreement. Figure 4.28 shows the measured and simulated mean values of input and output fluid temperatures of both heat exchangers. The figure also shows that after 1 week of

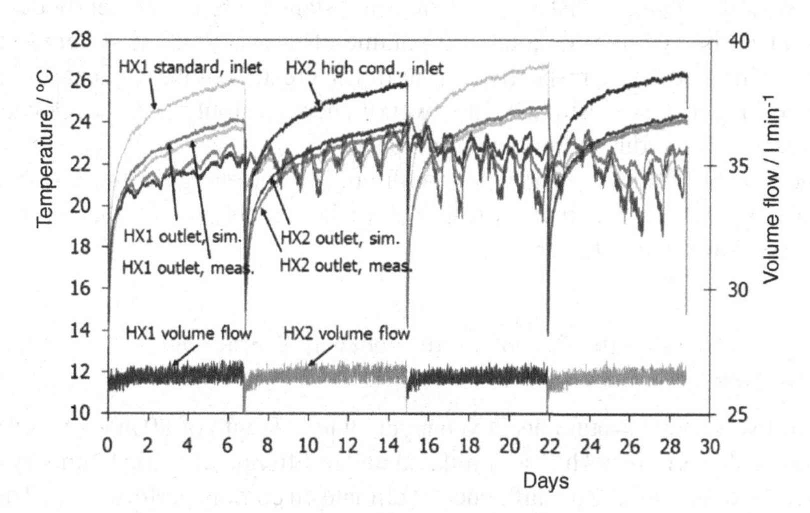

Figure 4.28 Measured inlet and outlet temperatures for individual testing of two different geothermal heat exchangers with standard and higher conductivity backfill material

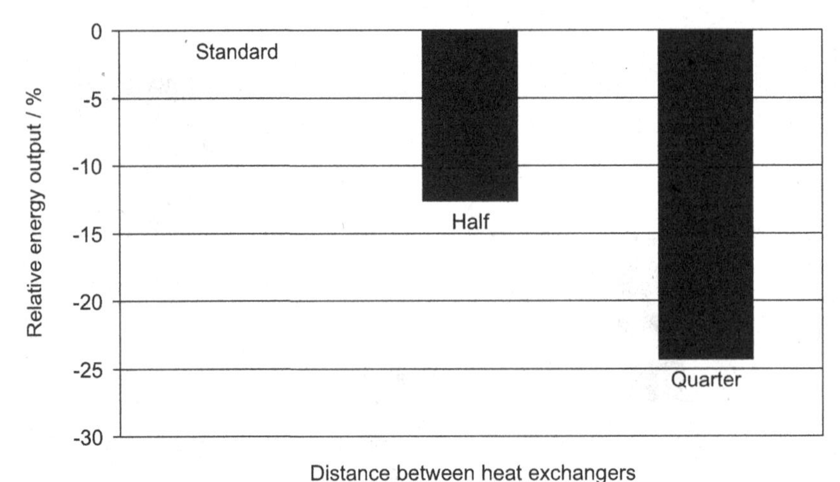

Figure 4.29 Influence of heat exchanger distance on energy output

thermal relaxation the temperature levels are still about 1 K higher than from the undisturbed start.

Regarding the mutual distance of vertical ground heat exchangers, Zhang and Murphy (2003) state that the spacing of the boreholes is of major significance regarding the thermal effectiveness and that the single heat exchangers can be regarded as isolated from each other if the distance is greater than 6 m.

This finding is confirmed by the current numerical parameter study (see Figure 4.29). Starting with the standard distance of 6 m, the distances are subsequently decreased to 3 m and 1.5 m, keeping the total soil volume of the geothermal system constant. This results in a two times, respectively four times, greater number of heat exchangers with decreasing distance. Although the overall energy output rises, that of each heat exchanger is reduced due to the mutual influence.

An increase of mass flow inside the tubes improves the energy withdrawal by up to 9% for high soil heat conductivities. At lower soil heat conductivities the improvement due to mass flow reduces to zero.

Influence of Climate on the Cooling Performance of Geothermal Heat Exchangers

A field of five vertical ground heat exchangers with a length of 80 m each and with a total volume flow of 2.4 $m^3 h^{-1}$ is simulated under different climatic boundary conditions in order to investigate the influence of climate on cooling performance. The average ambient temperatures of several cities in warm climates represent the undisturbed soil temperatures around the heat exchangers. Each system is operated for 4 months,

Table 4.3 Simulation results for different climatic boundary conditions

Location	T_{amb} / °C	T_{max} / °C	P_{mean} / W m^{-1}	P_{min} / W m^{-1}	P_{max} / W m^{-1}	Energy / MWh	Cost / € per kWh
Madrid	13.9	19.4	20.0	17.5	54.0	11.6	0.31
Seville	18.8	21.0	8.0	7.0	21.3	4.6	0.78
Bangkok	28.1	—	—	—	—	—	—
Crete	19.1	21.1	7.3	6.3	19.3	4.1	0.86
Athens	18.3	20.8	9.0	8.0	24.7	5.3	0.68

12 hours every day. An average sand-type soil was assumed as the surrounding soil. The inlet temperature to the ground heat exchangers is set to a constant 22°C, representing the output of a thermally activated concrete slab from the building. Table 4.3 shows the investigated locations, the average ambient temperature (T_{amb}), the maximum outlet temperature of the ground heat exchangers (T_{max}) as well as the mean (P_{mean}), minimum (P_{min}) and maximum (P_{max}) available cooling power. The costs are derived from an annuity of €3570, which corresponds to the typical cost of vertical ground heat exchangers in Germany.

Obviously, the climate in the area of Bangkok is not suitable for direct geothermal cooling, as the inlet temperature from the building is lower than the ground temperature. Active chillers are needed here in order to reach the desired temperatures. However, even in the warm climate of Crete, the outlet temperature of the ground heat exchangers does not exceed 21°C at the end of the summer period. The direct use of geothermal energy via thermally activated concrete slabs is thus possible, although the achievable cooling power is relatively low bearing in mind the high investment cost. At all locations, the ground temperature regenerates to between 0.5 and 1.5° C above the undisturbed temperature at the end of each 12 hours of operating cycle, rising only slightly during the 4-months period. This aspect is important regarding the long-term performance of any geothermal system (Pahud *et al.*, 2002).

In order to determine the influence of the temperature level of the direct cooling system on the performance of the geothermal system, the same geothermal simulation as above was executed with several different inlet temperatures. Crete, with a mean ambient air temperature of 19.1°C, was chosen as the location. Inlet temperatures between 20 and 24°C were chosen to represent the combined operation with a floor cooling system, whereas the inlet temperatures of 35 and 40°C represent the replacement of a standard cooling tower by the borehole heat exchangers.

From Table 4.4 it can be seen that the maximum outlet temperature T_{max}, mean thermal power P_{mean}, minimum thermal power P_{min} as well as maximum specific thermal power P_{max} rise approximately proportional to the inlet temperature. This leads to the conclusion that the ground temperature around the heat exchanger regenerates

Table 4.4 Simulation results for different inlet temperatures for the location of Crete

T_{inlet} / °C	T_{max} / °C	P_{mean} / W m^{-1}	P_{min} / W m^{-1}	P_{max} / W m^{-1}	Energy / MWh	Cost / € per kWh
20.0	19.7	2.3	2.0	6.0	1.3	2.78
21.0	20.4	4.8	4.0	12.7	2.7	1.32
22.0	21.1	7.3	6.3	19.3	4.1	0.86
23.0	21.7	9.8	8.5	26.0	5.6	0.64
24.0	22.4	12.3	10.5	32.7	7.0	0.51
35.0	29.8	39.7	34.5	106.0	22.8	0.16
40.0	33.2	52.2	45.4	139.4	30.0	0.12

almost completely during the downtime of the circulation pump. However, considering the whole system of ground heat exchangers and floor cooling, it is obvious that the total cooling power depends on a momentary equilibrium of energy extracted from the room and energy dissipated into the ground. So the efficiency of the system cannot be increased arbitrarily by raising the inlet temperature. Regarding the operation of the borehole heat exchangers to replace a cooling tower, it can be stated that geothermal systems are an interesting alternative concerning effectiveness and cost.

4.1.5 Conclusions on Geothermal Heat Exchangers for Cooling

In summary, this work presents new experimental results for geothermal heat exchanger performance in some of the best German office buildings today. Earth heat exchangers for precooling and preheating of air have excellent ratios of produced cold compared with electricity used. Even air-based systems are shown to have annual COPs between 35 and 50.

Measured vertical ground heat exchangers also reach high COPs of about 20. Supply air temperatures can be efficiently reduced to temperature levels between 18 and 22°C. The power dissipation measured per metre of heat exchanger was 26 W per metre for 80 m deep vertical heat exchangers. In the case of the vertical heat exchangers, the power was limited mainly by the low cooling energy uptake within the building.

Cheap solutions are available using simple horizontal heat exchangers, which can be placed around the building. However, the heat dissipation measured was only about 8 W m^{-1}, which is mainly due to a low depth and close spacing between the horizontal absorbers. Although temperature levels are then higher than in the case of vertical heat exchangers or deeper horizontal systems, the cooling performance is still good.

Numerical simulation models have been developed and validated against the experimental data and reproduce the measured values well. They can now be used in planning for geothermal heat exchangers and for parameter studies. The simulation results show that performance is improved if the backfill material and obviously the soil have higher conductivity values (e.g. in regions with groundwater flow). The parameter studies showed that the earth heat exchangers can be directly used for building cooling also in warmer Mediterranean climates, although the power dissipation level drops at higher soil temperatures. In a Southern Mediterranean climate, even after a 4-months summer period, the maximum outlet temperature of the heat exchanger was no higher than 21°C, so activated concrete ceilings can still be operated efficiently.

5

Active Thermal Cooling Technologies

If a building cannot be cooled by passive means such as night ventilation or earth heat exchange alone, active cooling technologies have to be employed. Today, the dominant cooling systems are electrically driven compression chillers, which have a world market share of about 90%. The average coefficient of performance (COP) of installed systems is about 3.0 or lower and only the best available equipment can reach a COP above 5.0. To reduce the primary energy consumption of chillers, thermal cooling systems offer interesting alternatives, especially if primary energy neutral heat from solar thermal collectors or waste heat from cogeneration units can be used. The main technologies for thermal cooling are closed-cycle absorption and adsorption machines, which use either liquids or solids for the sorption process of the refrigerant. The useful cold in both cases is produced through the evaporation of the refrigerant in exact analogy to electrical chillers. For air-based cooling systems, desiccant cooling cycles are useful, as they directly condition the inlet air to the building.

Thermal cooling systems are mainly powered by waste heat or fossil fuel sources. Solar cooling systems in Europe have a total capacity of about 6 MW only (Nick-Leptin, 2005). The type of solar thermal collector required to drive the sorption material regeneration depends on the heating temperature level, which in closed systems is a function of both cold water and cooling water temperatures. In open sorption systems, the regeneration heating temperature depends on the required dehumidification rate, which is a function of ambient air conditions.

Low Energy Cooling for Sustainable Buildings Ursula Eicker

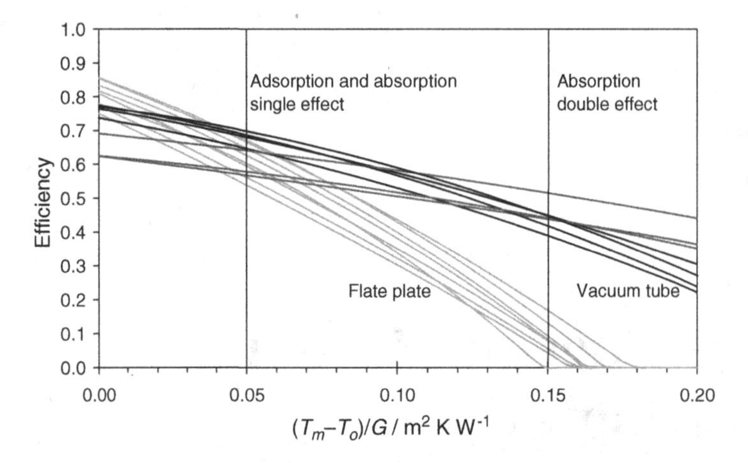

Figure 5.1 Collector efficiency as a function of reduced temperature

Commercial adsorption systems, either open air-based systems or closed adsorption, are designed for heating temperature ranges around 70 °C. Single effect absorption chillers start at operating temperatures of about 70 °C. Commercial machines are often designed for average heating temperatures of 85–90 °C. Double effect systems require 150 °C and more.

An actual database of solar thermal collectors with manufacturer efficiency information was implemented in the software environment INSEL and used to calculate efficiency curves as a function of the difference between mean fluid temperature T_m and ambient air temperature T_o, normalized by the incident irradiance G (the so-called reduced temperature difference). At an irradiance of 800 W m^{-2} and a summer ambient air temperature of 30 °C, the reduced temperature difference for heating temperatures of 70 °C is 0.05 and gets higher at lower irradiance levels. For a mean fluid temperature of 90 °C the reduced temperature is 0.075 and for 150 °C, 0.15. In Figure 5.1 it can be seen that current flat plate collectors available on the market can be well used for driving temperature levels around 70 °C, but are less efficient than vacuum tubes at heating temperature levels of 90 °C. Adsorption or single effect absorption machines can both operate at about 70 °C driving temperatures, if evaporation temperatures are high and cooling water temperatures low. For double effect chillers, only vacuum tubes or parabolic concentrators can provide the required temperature levels.

The main features of the dominating absorption cooling technology and an overview of the historical development will be presented in the following together with some project results from desiccant cooling systems and a study of innovative cooling systems for low-power applications.

5.1 Absorption Cooling

Absorption cooling is a mature technology with the first machine being developed in 1859 by Ferdinand Carré. For the closed-cycle process, a binary working fluid that consists of the refrigerant and an absorbent is necessary. Carré used as the working fluid ammonia/water (NH_3/H_2O). Today the working pair of lithium bromide as the absorbent and water as the refrigerant is most commonly used for building climatization ($H_2O/LiBr$). In contrast to the ammonia/water system with its pressure levels above ambient pressure, the water/lithium bromide absorption cooling machine (ACM) works in a vacuum because of the low vapour pressure of the refrigerant water.

In 1945 the American company Carrier Corp. developed and introduced the first large-scale commercial, single effect ACM using water/lithium bromide with a cooling power of 523 kW. In 1964 the company Kawasaki Heavy Industry Co. from Japan produced the first double effect water/lithium bromide ACM (Hartmann, 1992). The double effect (DE) ACM is equipped with a second generator and condenser to increase the overall COP by reusing the high-temperature input heat also for the lower temperature generator.

Absorption chillers today are available in the range of 5 to 20 000 kW. In the last few years some new developments have been made in the medium-scale cooling range of 10 to 50 kW for water/lithium bromide and ammonia/water absorption chillers (Storkenmaier *et al.*, 2003; Safarik and Weidner, 2004).

While absorption cooling has been common for decades, heat pump applications have only become relevant in recent years, due to the improvement in recent years in the performance figures; small gas-driven absorption heat pumps achieve COPs of approximately 1.5, that is, by using 1 kWh of the primary energy of gas, 1.5 kWh of heat can be produced using environmental energy, which is better than the condensing boilers presently available on the market with maximum COPs of about 1.0. Different manufacturers are producing absorption heat pumps of 10–40 kW output which achieve COPs of about 1.3 at heating temperatures of 50 °C. However, absorption chillers today are mainly used as cooling machines rather than heat pumps.

5.1.1 Absorption Cycles

Absorption cooling machines are categorized either by the number of effects or by the number of lifts: effects refer to the number of times high-temperature input heat is used by the absorption machine. In general, increasing the number of effects is meant to increase the COP using higher driving temperature levels. Lifts refer to the number of generator/absorber pairs to increase successively the refrigerant concentration in the solution and thus to reduce the required heat input temperature level.

The most important restrictions of single effect absorption cooling machines (Figure 5.2) are the limitation of the temperature lifting through the solution field,

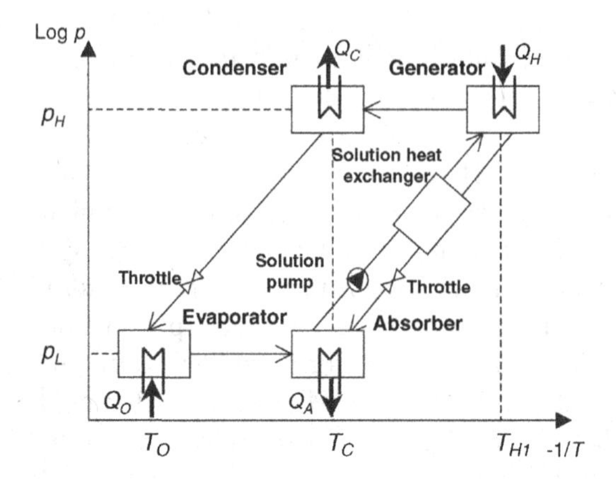

Figure 5.2 Single effect absorption cooling cycle with temperature levels of the evaporator, condenser and absorber and heater

the fixed coupling of the driving temperature with the temperature of the heat source and heat sink and the COP not being larger than 1.0 independent of temperature lifting (Ziegler, 1998). The aim of multi-stage processes is to overcome these restrictions.

Several types of absorption cooling machines are available on the market: single effect (SE) and double effect (DE) ACMs (Figure 5.3) with the working pair

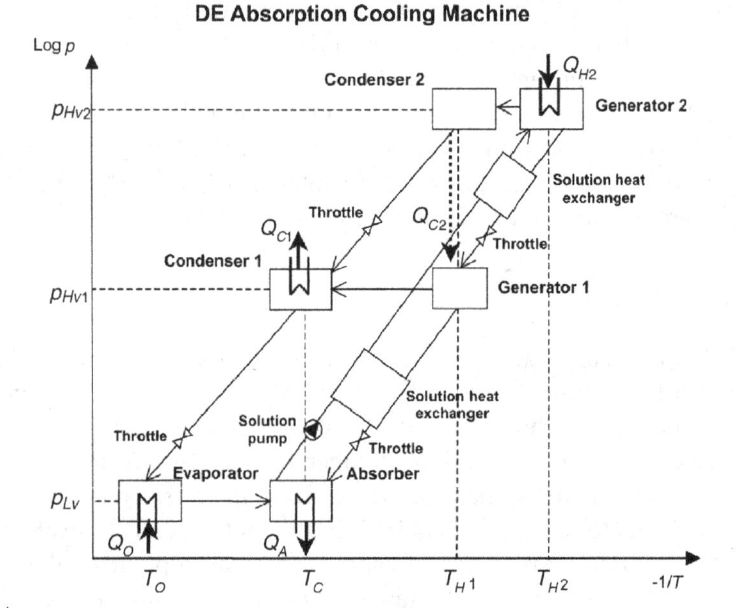

Figure 5.3 Double effect absorption chiller cycle

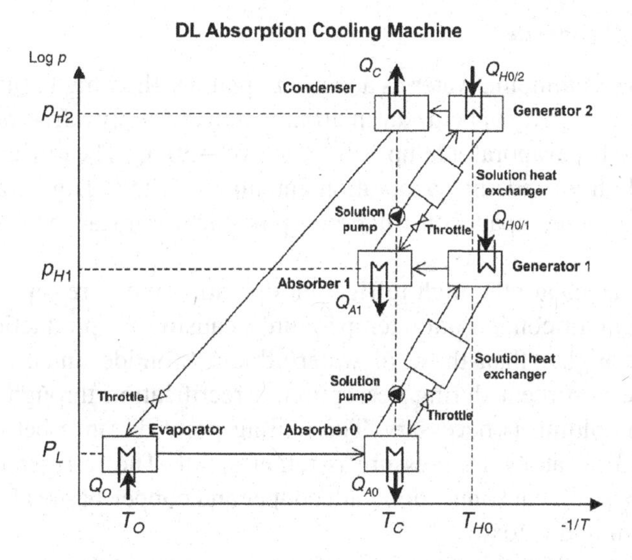

Figure 5.4 Double lift absorption cooling cycle

of water/lithium bromide, as well as single effect (SE) and double lift (DL) ACMs (Figure 5.4) with the working pair of ammonia/water.

The single effect/double lift (SE/DL) cycle with the working pair of water/lithium bromide is a novel technique. Further cycle designs such as triple effect (TE) ACMs, other multi-stage cycles and the use of the described cycles for solar cooling have been investigated by a number of researchers (Kimura, 1992; Lamp and Ziegler, 1997; Höper, 1999; Ziegler, 1999).

Water/Lithium Bromide Properties

The advantages of water/lithium bromide are that water as refrigerant has a very high specific evaporation enthalpy of 2489 kJ kg^{-1} (at +5 °C) and that the lithium bromide solution is not volatile. The working pair itself is odourless and neither toxic nor flammable. The disadvantages are the high freezing point of water and that lithium bromide is not totally soluble in water because of crystallization. Moreover, the vacuum for the machines has to be very tight. Due to the high specific volume of water vapour as refrigerant, the machines are not very compact.

The water/lithium bromide ACMs are usually water cooled by wet cooling towers to avoid the possibility of crystallization of the lithium bromide solvent. As water is used as refrigerant, the external evaporator temperature is restricted to temperatures above the freezing point of at least +5 °C. Therefore the possible temperature range of such machines is limited to pure air-conditioning and cold water generation.

Ammonia/Water Properties

The advantages of ammonia/water as a working pair are their high affinity, their high stability and their property of being an environmentally friendly refrigerant. There is no vacuum required for evaporator temperatures above −30 °C. The refrigerant ammonia has a reasonably high specific evaporation enthalpy of 1258 kJ kg^{-1} (at +5 °C), it is lighter than air and a compact construction is possible, for instance through plate heat exchangers.

The disadvantages are the high refrigerant operating pressure (up to 25×10^5 Pa) at typical ambient air condensation temperatures, causing the production costs of the machines to be higher than those of water/lithium bromide units. Because of the volatility of the absorbent during desorption, a rectification through a dephlegmator/rectification column is necessary. The boiling point distance between ammonia and water of 133 K at 10^5 Pa pressure is rather small. The refrigerant ammonia is corrosive, for example, in connection with copper and copper-based alloys, has a very unpleasant odour and is toxic.

The thermodynamic properties of the refrigerant ammonia determine the operating temperature range of the ammonia/water ACM. The critical temperature of the refrigerant ammonia is 132.4° C. Ammonia has a low freezing point at −77.7 °C and the boiling point temperature is −33.3 °C at 10^5 Pa pressure. Ammonia/water ACMs can be air cooled for air-conditioning applications but usually they are also water cooled (Chinnappa, 1992). By using ammonia as a refrigerant, the evaporator temperature can go down even to −60 °C. Thus, the temperature range of the machines is suitable for air-conditioning and for industrial refrigeration processes, for example in the chemical industry.

Performance

Typical performance characteristics for the closed ACM cycles described above are given in Table 5.1 for water/lithium bromide and in Table 5.2 for ammonia/water.

In contrast to the ideal Carnot process, the COP stays nearly constant as soon as a certain minimum driving temperature level is reached. The COP increases from approximately 0.7 for single effect (SE) processes to 1.3 for double effect (DE) absorption chillers and 1.7 for the triple effect (TE) as shown in Figure 5.5 (Grossmann, 2002).

For ammonia/water absorption chillers, the COP is approximately 0.6 for single effect (SE) cycles and reduces to 0.4 for double lift (DL) cycles because of the reduced

Table 5.1 Water/lithium bromide absorption chiller characteristics

Cycle type	SE	DE	SE/DL
Cold temperature / °C	6–20	6–20	6–20
Heating temperature / °C	70–110	130–160	80–100
Cooling water temperature / °C	30–35	30–35	30–35
COP	0.5–0.7	1.1–1.3	0.4–0.7

Table 5.2 Performance of ammonia/water absorption chillers and diffusion–absorption machines with auxiliary gas

Cycle type	SE	DL	SE
Auxiliary gas	—	—	hydrogen/helium
Cold temperature / °C	−30–20	−50–20	−20–20
Heating temperature / °C	90–160	55–65	100–140
Cooling water temperature / °C	30–50	30–35	30–50
COP	0.4–0.6	0.4	0.2–0.5

driving heat temperatures. The COP of those chillers with auxiliary gas is even lower, between 0.2 and 0.5.

5.1.2 Solar Cooling with Absorption Chillers

In the 1970s, the American company Arkla Industries Inc., (now owned by Robur SpA, Italy) developed the first commercial, indirectly driven, single effect H_2O/LiBr ACM for solar cooling with two different nominal cooling capacities. One was the 3-TR Solaire or Solaire 36 unit with a 10 kW cooling capacity and the other Solaire unit had a 75 kW cooling capacity. The driving heat temperatures were in the range of 90 °C and the cooling water temperature was 29 °C for 7 °C cold water temperature. The machine was installed in more than 100 demonstration projects in the USA (Loewer, 1978; Lamp and Ziegler., 1997; Grossmann, 2002). Arkla and also Carrier Corp., USA, then developed a small-size single effect H_2O/LiBr ACM that could work with air cooling. There was no market success, mainly because of the high investment costs for solar cooling. Carrier Corp. further decreased the driving temperature of a water-cooled single effect H_2O/LiBr ACM by using a falling film generator with a large surface

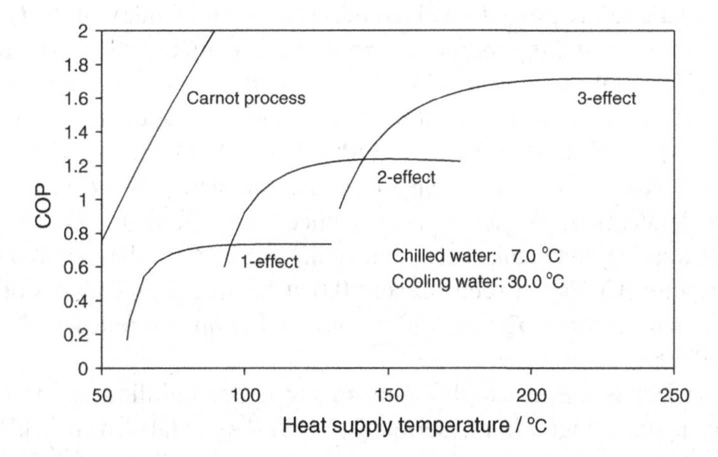

Figure 5.5 COP as a function of the heat input temperature for different water/lithium bromide absorption cycles (Grossmann, 2002)

area. The driving heat temperature was 82 °C and the cooling water temperature was 28 °C for 7 °C cold water temperature (Lamp and Ziegler, 1997). The production of these ACMs was stopped and the technology's licence was given to the Japanese company Yazaki. Up to the beginning of the 1990s, Yazaki offered H_2O/LiBr ACMs with 5–10 kW cooling power (such as the WFC-600 with 7 kW), which were used for solar cooling projects. Due to falling demand, the production was eventually stopped.

At the beginning of the 1980s, Arkla developed a double effect H_2O/LiBr ACM in which the lower temperature generator was supplied with solar energy, while in fossil mode the double effect generator was fired using the higher COP. Due to the lack of demand for solar cooling, the production of this cooling machine was stopped and the technology was also licensed to Yazaki. The company sold the machines for several years, but they are no longer available today.

In the following, selected past and current international projects are described which consist of partly solar-assisted or completely solar-powered water- or air-cooled single effect water/lithium bromide ACM. A general view of some old H_2O/LiBr ACM projects before 1979 is extensively presented in Loewer (1978). As there is not much experience available with solar cooling installations, the main focus of the study is to summarize the dimensioning of the solar collector field and storage tank in relation to the cooling machine capacity, operating experiences and temperature levels used and the associated costs. A simulation-based method for solar cooling system design is then presented in Chapter 6.

Projects with Single Effect ACMs

In the medium-sized performance range, most project experience of solar cooling relates to the single effect water/lithium bromide ACM WFC-10 from Yazaki, with a cooling power range of 35–46 kW. There are only a few published results concerning COPs, but the operation is generally rated as dependable and unproblematic.

An early solar cooling project involved the cooling of a winery of the Groupement Interproducteurs du Cru Banyuls in southern France, set up in 1991. The wine cellar is operated by a Yazaki WFC-15 with 52 kW cooling capacity and 130 m^2 of vacuum tube collectors (VTCs) with only 1.0 m^3 of hot water storage. The operating heating input temperature T_h is 80 °C and the reached COPs 0.57–0.58 (Quinette and Albers, 2002).

The Hotel Belroy in Benidorm, Spain, is also an early project where a 125 kW, single effect H_2O/LiBr ACM has been used since 1992. There are 344.5 m^2 of VTCs and three hot water storage tanks each with 12 m^3 installed for the required heat input. The ACM reaches COPs between 0.5 and 0.6 at heating temperatures of 96/86 °C, cooling water temperatures T_c of 29/36 °C and cold temperature levels T_o of 14/9 °C (Hansen, 1993a).

Another project is the air-conditioning of the office building of the J. Wolferts GmbH company in Cologne-Porz, Germany (1995). Two solar-driven 46 kW WFC-10 ACMs are installed for an air-conditioned surface over four floors of 1628 m^2 and are powered by 176 m^2 of VTCs at T_h 95/88 °C, T_c 29/35 °C and T_o 22/16 °C. A 5.0 m^3 cold water storage tank was also installed as a buffer (Ohn, 1995; Karbach *et al.*, 1997; Karbach 1998).

The office building of the Ott & Spieß company in Langenau, Germany (started in 1998), is equipped with a 35 kW ACM which is driven by 30 m^2 of VTCs. A 2.0 m^2 hot water storage tank and a 1.0 m^3 cold water buffer tank are also installed. The operating parameters are T_h 76/71 °C, T_c 24/29 °C and T_o 17/13 °C at a COP of 0.73 (Mößle, 2000; Henning, 2001).

Two further projects are located in Berlin, Germany, each with two solar-assisted ACMs (2000). In the administrative building of the German Federal Information Ministry (BPA) there are two 44 kW WFC-10 ACMs installed with a COP of 0.73, powered by 230 m^2 of VTCs and 1.6 m^3 hot water storage tanks with the operating parameters T_h 88/82 °C, T_c 26/32 °C and T_o 16/10 °C. The second administrative building is the German Federal Ministry of Traffic, Building and Housing (BMVBW). Here there are also two solar-assisted, single effect WFC-10 ACMs installed, each with a 42 kW cooling capacity and a COP of 0.73. The solar heat is produced by 210 m^2 of high-performance flat plate collectors (FPCs) and stored in 6.0 m^3 hot water storage tanks. There is also a 0.5 m^3 cold water storage tank for the chilled water. The operating parameters are T_h 88/82 °C, T_c 27/35 °C and T_o 18/12 °C (Albers, 2002).

The M + W Zander Holding AG company built an office and production building in Weilimdorf, Germany, and installed a single effect ACM with a nominal cooling capacity of 650 kW and 360 m^2 of VTCs for a solar-assisted cooling system which has been in operating since 2000. The operating temperature ranges are for T_h from 65 to 95 °C, for T_c from 22/27 to 27/32 °C and for T_o from 12/8 to 21/15 °C (Wolkenhauer and Albers, 2001).

Further commercial applications since 2001 include the Hotel Rethymno Village in Greece where a 105 kW H_2O/LiBR ACM is installed with 500 m^2 of FPCs and also seven hot water storage tanks with a total of 10.0 m^3 used for solar heating. An additional project is the Hotel Olympic II in Greece where a 105 kW ACM has also been running since 2001. There are 450 m^2 of FPCs and five hot water storage tanks with a total of 5.0 m^3 installed (Balares, 2003).

In Spain the office building of the Viessmann company in Madrid has a single effect Yazaki WFC-30 ACM with 105 kW cooling capacity and 105 m^2 of FPCs. There are no published operational results yet. A further project is a reconstructed office building (EAR Tower) in Pristina, Kosovo, where two WFC-10 ACMs each producing 45 kW of cooling capacity have been in operation since 2003. They are solar driven by 227 m^2 of FPCs at operating temperatures of T_h 87/77 °C, T_c 25/31 °C and T_o 13.5/8.5 °C and a COP of 0.7. A 4.0 m^3 hot water and a 1.0 m^3 cold water storage tank are used (Holter and Meiβner, 2003).

The project in an office building at Basse Terre, Guadeloupe, with a 30 kW ACM and 61.2 m^2 of VTCs has also been in operation since 2003. The heating input temperature T_h is 85°/95 °C at a cooling input temperature T_c of 24 °C and T_o of 12/7 °C at a COP of

0.7. A 570 m^2 floor area is cooled by the solar-powered system which also consists of a small hot water buffer tank (less than 0.1 m^3) and a cold water storage tank (Henning, 2004b).

Further projects are a 35 kW SE ACM with 70 m^2 of VTCs in Perpignan, France, and an SE ACM with a 35 kW cooling capacity and 60 m^2 of collectors at the CSTB in Valbonne, France.

Low-Power SE ACM Projects

After the experiences in the 1970s with the low-power systems from Arkla, some projects were recently carried out once again with small-scale ACMs. Since 1999 there has been a 7 kW H_2O/LiBr ACM Yazaki WFC-600 installed for laboratory and office cooling at ZAE Bayern in Garching, Germany. The driving heat is produced by 19.6 m^2 of VTCs and stored in two 0.6 m^3 hot water storage tanks. The operating characteristics of the ACM are T_h 88 °C and T_o 16/14 °C with a COP of 0.6.

A newly developed SE H_2O/LiBr ACM from the German company Sonnen Klima with a 10 kW cooling capacity was installed for its first field test at an office building in Berlin Treptow, Germany, in August 2003, providing 152 m^2 of cooled office space. The ACM is driven alternately with 42 m^2 of FPCs or 25 m^2 VTCs and a 0.75 m^3 hot water buffer tank is also installed. The nominal operating temperatures are T_h 72/62 °C, T_c 27/35 °C and T_o 18/15 °C and the COP ranges between 0.76 and 0.82 (Storkenmaier *et al.*, 2003; Kohlenbach *et al.*, 2004).

Further field tests, each based on a 15 kW H_2O/LiBr ACM (Wegracal SE 15 from the EAW company, Germany), have been carried out since 2003. The first is at a sales store in Neumarkt, Italy, where 55 m^2 of FPCs and a 0.75 m^3 hot water storage tank are used as heat sources for the ACM. The second field test is being performed at the technology centre TZ Köthen, Germany, for office cooling with 77 m^2 of VTCs. The determined operating parameters are T_h 90/80 °C, T_c 32/38 °C, T_o 21/15 °C and an average COP of 0.5. Within the system, there is also a paraffin latent heat storage with a melting temperature of 90 °C, a 0.5 m^3 conventional hot water storage tank and a 0.2 m^3 cold water buffer tank (Safarik and Weidner, 2004). A further EAW H_2O/LiBr ACM has been used since 2003 for a laboratory test stand at the Technical University of Ilmenau, Germany. This test cooling machine, however, has a reduced cooling capacity of 10 kW. The ACM is driven by 10 m^2 of FPCs and a 2 m^3 hot water buffer tank. The nominal operating temperatures are T_h 85/75 °C, T_c 30/35 °C and T_o 14/8 °C. The reach COPs about 0.4 (Ajib *et al.*, 2004).

There are only a few laboratory prototypes of air-cooled, SE H_2O/LiBr ACMs. A prototype of an air-cooled H_2O/LiBr absorption chiller can be found at the Polytechnic University of Catalunya in Terrassa, Spain. The unit has measured cooling capacities between 1.2 and 1.8 kW and COPs from 0.3 to 0.5 for an input temperature T_h of 75–95 °C, input cooling air T_c of 32 °C and input chilled water T_o of 20 °C (Castro, 2003).

Double Effect ACMs

Tanaka introduced a combined single/double effect H_2O/LiBr ACM that works in double effect mode by using fuel at a higher COP and in SE mode using solar energy. Two ACMs of 20–30 kW cooling capacity and COPs of 1.1 (DE)/0.6 (SE) have been installed in the office building of the Energy Engineering Department of Oita University in Kyushu, Japan, and at the Kabe office building of the Chugoku Electric Company in Hiroshima, Japan (Kimura, 1992).

The American company Solargenix Energy, LCC (formerly Duke Solar Energy, LCC) has had a commercial building project in Sacramento, California, since 1997, where a DE McQuay ACM with a 70 kW cooling capacity and a COP of 1.1 to 1.2 is used, which is driven by 200 m^2 of integrated compound parabolic concentrator (ICPC) VTCs at 165 °C heating temperature. These special VTCs are internal mirror coated (Duff *et al.*, 2003). Solargenix Energy also developed a roof integrated system with a fixed parabolic reflector and tracking evacuated-tube receiver and has been running a modified hot-water-driven Yazaki 176 kW DE ACM with a 930 m^2 Power Roof combined with a 7.5 m^3 storage tank at a large commercial office building in Raleigh, North Carolina, since July 2002. The chiller has a COP of 1.23 (Gee *et al.*, 2003; Guiney and Henkel, 2003). Another project of Solargenix Energy (since 2003) is the municipal utility building at the Sand Hill Power Plant control centre in Austin, Texas. The solar system consist of 360 m^2 of non-evacuated CPC solar thermal collectors and a 9.5 m^3 pressurized thermal storage tank which drives a 95 kW Yazaki hot water SE ACM. Further detailed operating data for the ACM is not available (Sklar, 2004).

A further solar cooling project of Solitem GmbH is the Hotel Iberotel Sarigermepark in Dalamen, southern Turkey. A steam-fired, DE Broad ACM with a 140 kW cooling capacity has been running since April 2004 on superheated steam which is produced by 180 m^2 of newly developed parabolic trough collectors (PTCs). The operating conditions are a pressure level of 4×10^5 Pa at a temperature T_h of 144 °C for the saturated steam, and a cooling water inlet temperature of 27 °C. A COP of 1.3 has been reached (Lokurlu *et al.*, 2002; Krüger *et al.*, 2002). Further projects are under construction or planned, for example at the Hotel Grand Kaptan in Alanya, Turkey, and at the Hotel Iberotel in Ortaca, Turkey.

Projects with NH_3/H_2O SE ACMs

An early project was the 12 kW NH_3/H_2O SE ACM with FPCs for a laboratory test stand at the Gradevinski Institute of the University of Split, Croatia, which was installed in 1979. The unit was in operation for 200 hours only in 1979. The installed ACM reached a COP of 0.55 at temperatures T_h 80/76 °C, T_c 25/33 °C and T_o 12/6 °C (Podesser, 1982).

At the TZ technology centre in Köthen, Germany, a SE NH_3/H_2O ACM from ABB with a 15 kW cooling capacity has been operated since 1998 for air-conditioning a

260 m^2 office space. The ACM is driven by 77.3 m^2 of VTCs at working conditions of T_h 95/88 °C, T_c 27/31 °C, T_o 14/8 °C and COPs between 0.4 and 0.5 (Safarik *et al.*, 2002).

A further small-scale ACM project is the 20 kW NH_3/H_2O ABB ACM, which was installed in 2001 at the Innovation Centre Wiesenbusch (IWG) in Gladbeck, Germany. The produced refrigeration is used to demonstrate comfort cooling, refrigeration for food preservation and also for 1.0 m^3 of ice storage. The driving heat T_h of 100/90 °C is produced by 72 m^2 of VTCs. Evaporator temperatures T_o down to −6 °C are reached at a COP of 0.63 (Albring, 2001; Braun and Hess, 2002).

A field test has been being carried out on a 10–17 kW NH_3/H_2O ACM developed by the Joanneum Research-Institut für Energieforschung and S.O.L.I.D. GmbH (both from Austria), which has been installed at the Peitler winery in Schloßberg, Austria (2003). The 100 m^2 of FPCs, 4 m^3 hot water storage tanks and a 0.5 m^3 sole storage tank are parts of the solar cooling system which operates at T_h 100/85 °C, T_c 25/35 °C and T_o 8/3 °C and at higher driving heat temperatures at T_o −10 °C down to −15 °C (Meißner *et al.*, 2004).

Only a few laboratory prototypes of air-cooled SE NH_3/H_2O ACMs and components have been developed and set up so far.

A prototype of an air-cooled, SE, NH_3/H_2O ACM from AoSol, Portugal, and INETI/IST, Portugal, is driven by 14.3 m^2 of CPC VTCs or gas. The ACM has a cooling capacity of 5–6 kW at COPs between 0.54 and 0.62. The heating temperature T_h is between 105 and 110 °C and the evaporator temperature T_o is 10 °C (Afonso *et al.*, 2003).

The Technical University TU Delft, in the Netherlands, developed a prototype of an air-cooled solar-driven absorption NH_3/H_2O ACM with a cooling capacity of 10 kW. The required heating temperature T_h is around 92 °C and the cooling water temperature T_c is 30 °C (Kim *et al.*, 2003).

An air-cooled Solar-GAX absorption prototype (GAX stands for heat exchange between absorber and generator) with the working pair of NH_3/H_2O and a 10.6 kW cooling capacity has been developed at the University Autónoma de Baja California, Mexico, and the University Autónoma de México. The unit is powered in a hybrid manner by natural gas and solar energy. The determined COP of the ACM is 0.86 (Velázquez *et al.*, 2002).

The ITW institute of the University of Stuttgart, Germany, set up a test and demonstration prototype of a solar-driven absorption NH_3/H_2O ACM with a cooling capacity of 10 kW and which is mainly based on plate heat exchangers. The cooling machine has been in operation since September 2004 and is exclusively cooled by a closed water cycle with ambient air. The heating temperatures T_h vary for different operating parameters between 70/65 °C and 127/117 °C and, with that, the cooling water temperatures T_c for the condenser are 22.5/24.5 °C or 38/41 °C and for the absorber, T_a 24/27 °C or 38.5/43.5 °C. The reached evaporator temperatures T_o are between 13 and 5 °C at COPs of 0.6 and 0.77 (Brendel *et al.*, 2004).

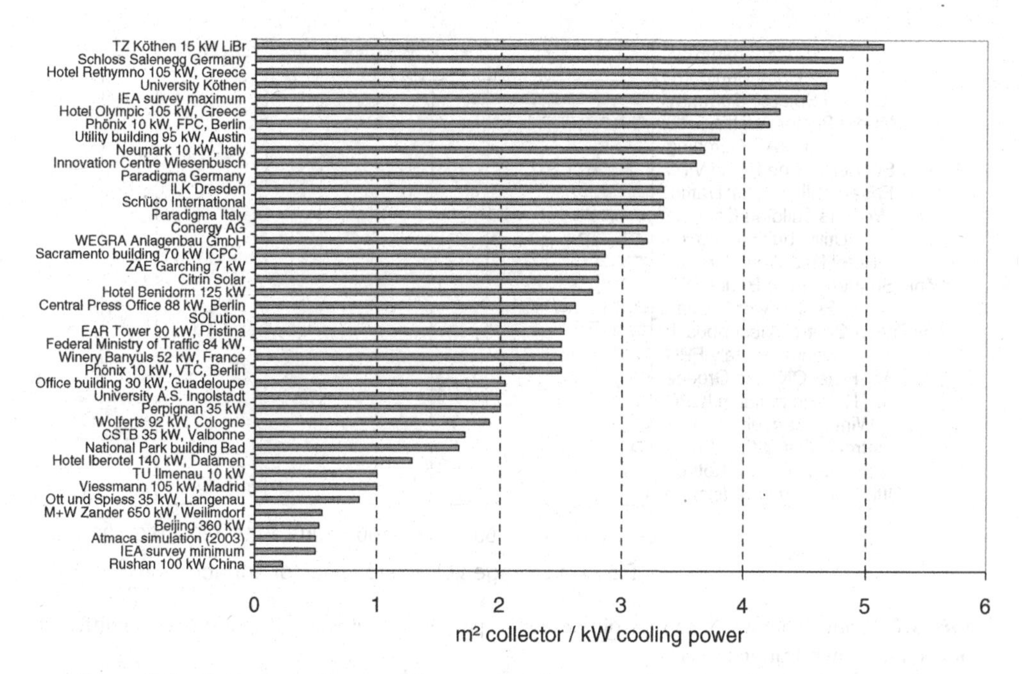

Figure 5.6 Collector surface per kilowatt of cooling power in various demonstration projects

This overview shows that dimensioning of solar cooling systems is a complex issue, where planners often do not have adequate tools to determine the energy yield and solar fraction. The ratios between solar collector surface area and cooling power or storage volume and collector surface in the various demonstration projects vary strongly (see Figure 5.6). Under comparable climatic conditions – in Austria and Germany, for example – less than 1 and more than 5 square metres of collectors have been installed per kilowatt of cooling power.

Also the ratio of storage volume (in litre water volume) to installed collector surface varies by more than a factor of 20 in the different demonstration projects. While warm water solar thermal systems or heating support systems have typical storage volumes between 50 and 100 litres per square metre of collector surface, in the solar cooling projects storage volumes are often much lower with less than 30 litres per square metre of collector. However, there are also some projects with significant storage volumes over 100 litres per square metre (see Figure 5.7).

The ratio of installed collector surface to the cooled building surface covers a wide range of values from below 10% up to 30% of the building surface area (see Figure 5.8). It is clear that the solar contribution to the total cooling demand must vary significantly. Unfortunately, very few published results of measured solar fractions are available today. These facts have led to the development of simulation models for a combined solar cooling plant and building performance simulation, which will be described in the next chapter.

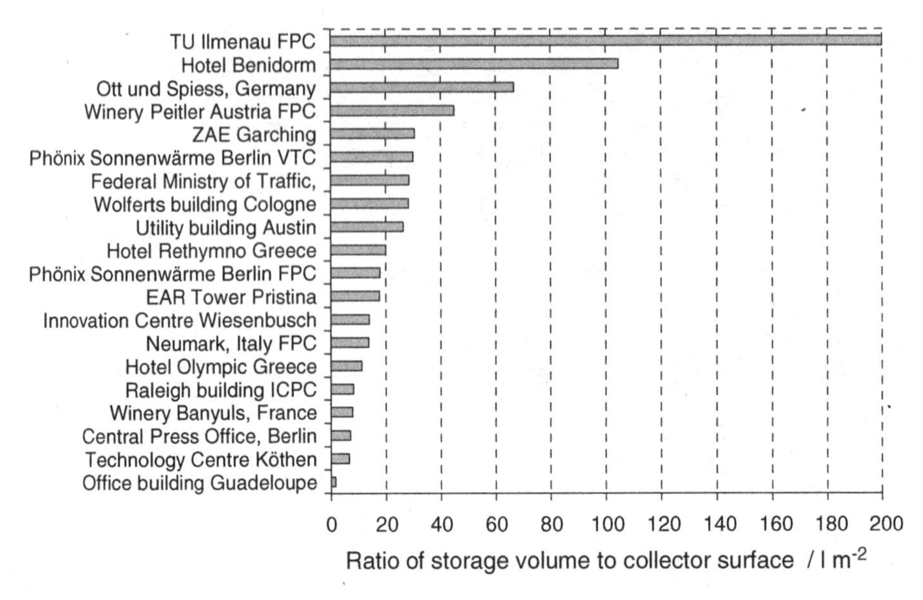

Figure 5.7 Ratio of hot water storage volume in litres per square metre of collector surface for different solar cooling demonstration projects

For some of the solar cooling systems described above, total investment costs are available. Depending on system size and technology chosen, the total investment costs vary between € 1900 and 6000 per kilowatt of cooling power installed (see Figure 5.9).

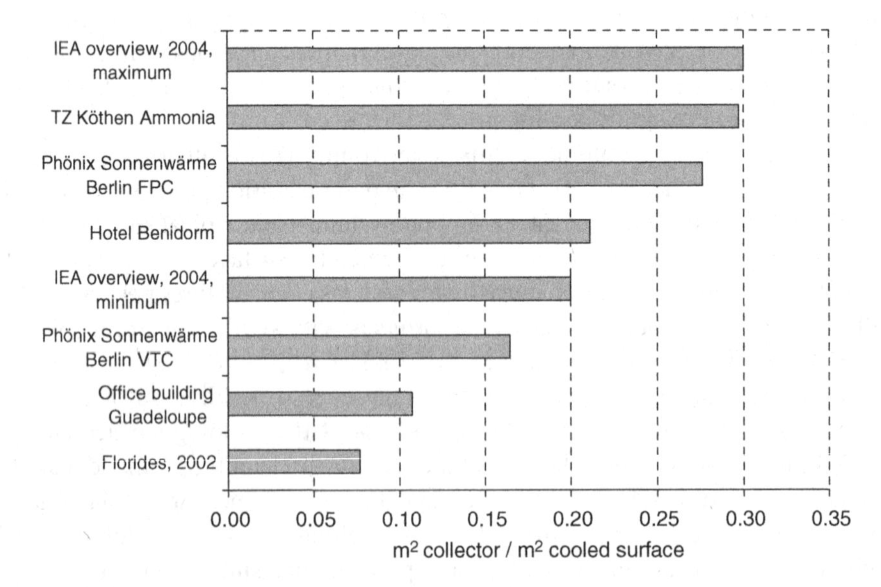

Figure 5.8 Ratio of installed collector surface area to building surface area in various demonstration projects

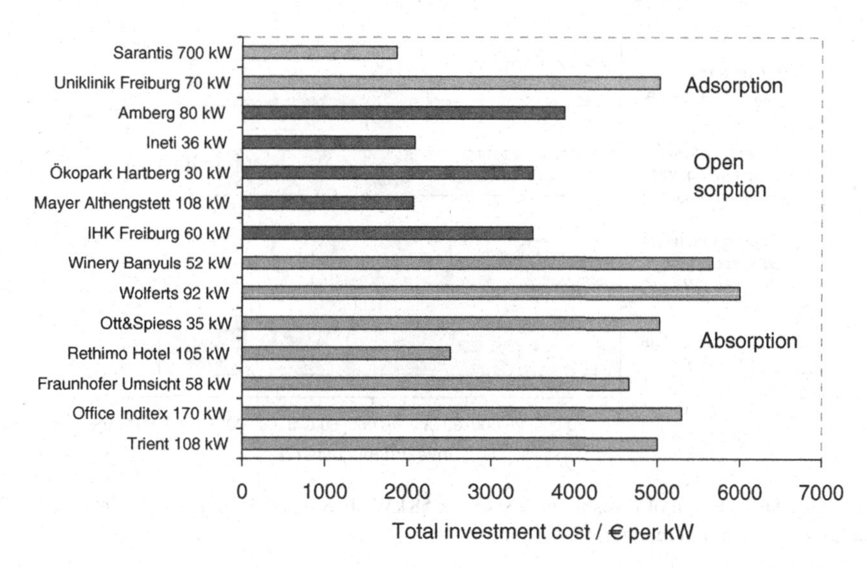

Figure 5.9 Total investment costs for different solar cooling systems implemented in the last decade in Europe

For some plants, details of the cost distribution are given. As an example, an absorption cooling system with a Yazaki WFC 10 installed at the Fraunhofer Institut Umsicht in Oberhausen, Germany, is analysed. The system consists of 108 m^2 of Soleado Lux 2000 VTCs, which are operated at design temperatures of 105/98 °C at a nominal efficiency of 69.6%. The cooling tower has 134 kW_{th} at operating temperatures of 24/31 °C, which are lowered with decreasing wet bulb temperature. The system also has a heat storage of 6.6 m^3 and a cold storage of 1.5 m^3. The total connected electrical power including cold distribution is 10 kW, but usually does not exceed 3.5 kW (Noeres, 2004). The distribution of costs is shown in Figure 5.10.

A more detailed analysis of investment and operating cost as a function of the chosen solar cooling system design is carried out in Chapter 6. Some results from the author's own projects on desiccant cooling systems will be presented in the next section.

5.2 Desiccant Cooling

Desiccant cooling systems are an interesting technology for sustainable building climatization, as the main required energy is low-temperature heat, which can be supplied by solar thermal energy or waste heat. Desiccant processes in ventilation mode use only fresh air, which is dried, precooled and humidified to provide inlet air at temperature levels between 16 and 19 °C. The complete process is shown in Figure 5.11 with the fresh air side (dark heavy lines) and the exhaust air side (light heavy lines). Outside air (1) is dried in the sorption wheel (2), precooled in the heat

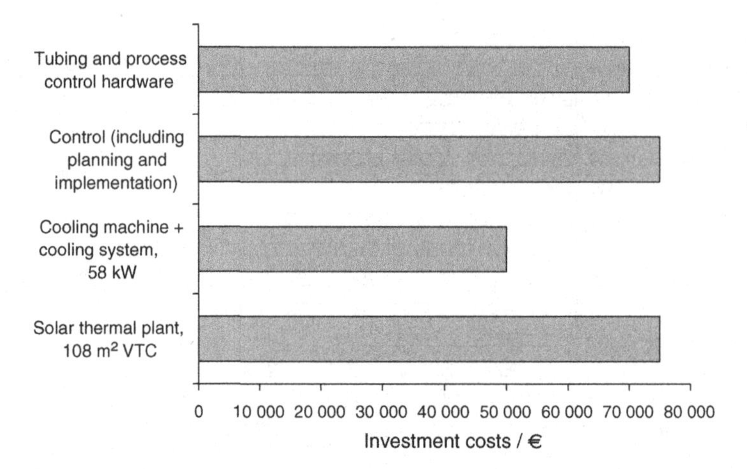

Figure 5.10 Distribution of investment costs for a 58 kW absorption cooling system installed in 2001 at Fraunhofer Umsicht in Oberhausen, Germany

recovery device with the additionally humidified cool space exhaust air (3) and afterwards brought to the desired supply air status by evaporative cooling (4). The space exhaust air (5) is maximally humidified by evaporative cooling (6) and warmed in the heat recovery device by the dry supply air (7). In the regeneration air heater the

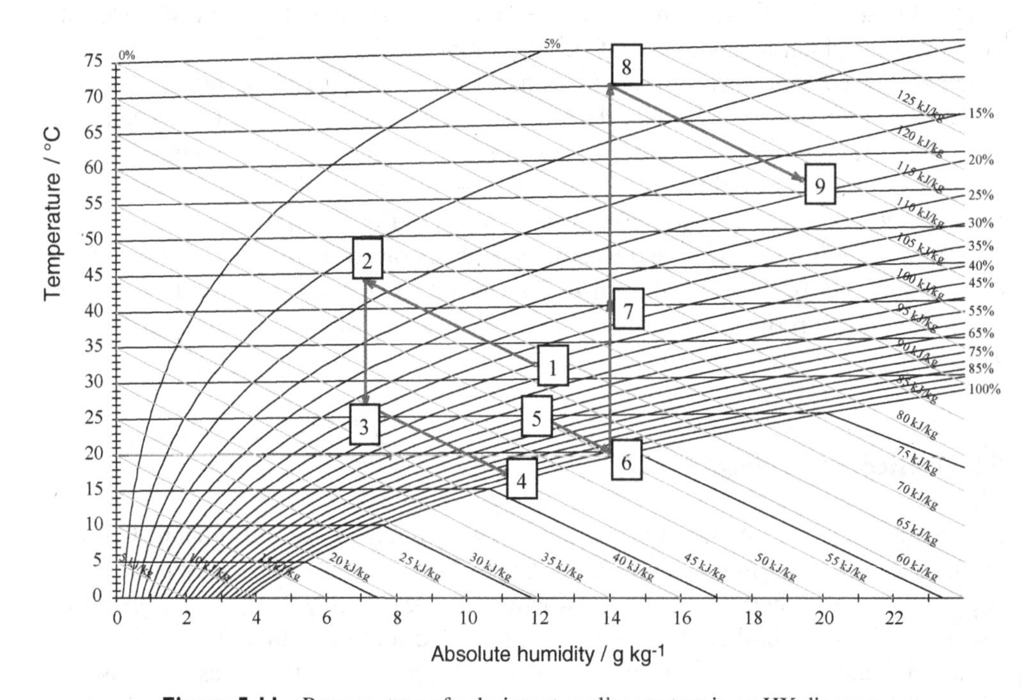

Figure 5.11 Process steps of a desiccant cooling system in an HX diagram

exhaust air is brought to the necessary regeneration temperature (8), takes up the water adsorbed on the supply air side in the sorption wheel, and is expelled as warm, humid exhaust air (9). If room air is recirculated, the desiccant wheel is used to dry the room exhaust air, which is then precooled using the rotating heat exchanger and humidified to provide the cooling effect. Regeneration of the desiccant wheel and precooling of the dried recirculation air is done by ambient air, which is first humidified, then passes the rotating heat exchanger, is heated to the necessary regeneration temperature and finally used to regenerate the desiccant wheel.

The concept of desiccant cooling was developed in the 1930s and early attempts to commercialize the system were unsuccessful. Pennington patented the first desiccant cooling cycle (Pennington, 1955), which was then improved by Munters in the 1960s (Munters, 1960). Good technology overviews are given by Mei *et al.*, (1992) Lavan *et al.*, (1982) or Davanagere *et al.*, (1999). The most widely used desiccants are silica gel, lithium chloride or molecular sieves, for example zeolites. Solid desiccants such as silica gel adsorb water in their highly porous structure. Lithium chloride solution is used to impregnate for example a cellulose matrix or simpler cloth-based constructions and can then be used to absorb water vapour from the air stream (Hamed *et al.*, 2005).

The thermal COP is defined by the cold produced divided by the regeneration heat required. For the hygienically needed fresh air supply the enthalpy difference between ambient air and room supply air can be considered as useful cooling energy. If the building has higher cooling loads than can be covered by the required fresh air supply, then the useful cooling energy has to be calculated from the enthalpy difference between room exhaust and supply air, which is mostly lower. The thermal COP is obtained from the ratio of enthalpy differences (state points are given in brackets):

$$COP_{thermal} = \frac{q_{cool}}{q_{heat}} = \frac{h_{amb(1)} - h_{supply(4)}}{h_{waste(9)} - h_{reg(8)}} \tag{5.1}$$

Related to ambient air, COPs can be near to 1.0 if regeneration temperatures are kept low, and reduce to 0.5 if the ambient air has to be significantly dehumidified. COPs obtained from room exhaust to supply air are lower, between 0.35 and 0.55 (Eicker, 2003). The maximum COP of any heat-driven cooling cycle is given for a process in which the heat is transferred to a Carnot engine and the work output from the Carnot engine is supplied to a Carnot refrigerator. For driving temperatures of 70 °C, ambient air temperatures of 32 °C and room temperatures of 26 °C, the Carnot COP is 5.5:

$$COP_{Carnot} = \left(1 - \frac{T_{ambient}}{T_{heat}}\right)\left(\frac{T_{room}}{T_{ambient} - T_{room}}\right) \tag{5.2}$$

However, as the desiccant cycle is an open cycle with mass transfer of air and water, several authors have suggested the use of a reversible COP as the upper limit of the desiccant cycle, which is calculated from Carnot temperatures given by the enthalpy differences divided by entropy differences (Lavan *et al.*, 1982). This allows the calculation of equivalent temperatures for the heat source (before (7) and after the regenerator heater (8)), evaporator (room exhaust (6) minus supply (4)) and condenser (waste air (9) minus ambient(1))

$$T_{equivalent} = \frac{\sum m_{in}h_{in} - \sum m_{out}h_{out}}{\sum m_{in}s_{in} - \sum m_{out}s_{out}} \tag{5.3}$$

Under conditions of 35 °C ambient air temperature and 40% relative humidity, defined by the American Air-conditioning and Refrigeration Institute (so-called ARI conditions), reversible COPs of 2.6 and 3.0 were calculated for the ventilation and recirculation modes (Kanoğlus *et al.*, 2007). The real process, however, has highly irreversible features such as adiabatic humidification. Furthermore, the specific heat capacity of the desiccant rotor increases the heat input required.

Crucial for the process is an effective heat exchange between the dried fresh air (state 2) and the humidified exhaust air (state 6), as the outside air is dried at best in an isenthalpic process and is warmed up by the heat of adsorption. For a rather high heat exchanger efficiency of 85%, high humidification efficiencies of 95% and a dehumidification efficiency of 80%, the inlet air can be cooled from design conditions of 32 °C and 40% relative humidity to below 16 °C, which is already slightly below the comfort limit (see Figure 5.12).

Simple models have been used to estimate the working range of desiccant cooling systems, for example to provide room conditions not just for one setpoint, but for a range of acceptable comfort conditions (Panaras *et al.*, 2007). The performance of the desiccant rotor itself can be evaluated by complex heat and mass transfer models

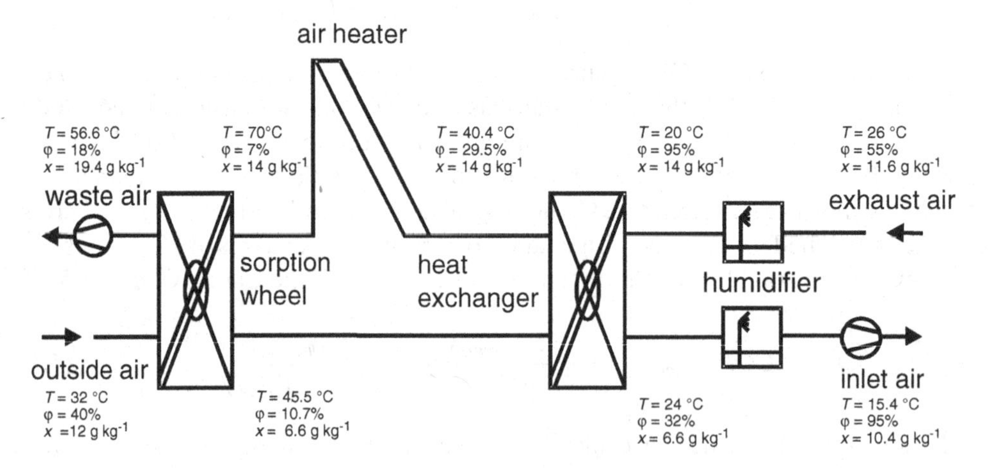

Figure 5.12 Temperature and humidity levels in a high-performance desiccant cooling process

based on the Navier–Stokes equations (Maclaine-Cross, 1988; Gao *et al.*, 2005). This allows the evaluation of the influence of flow channel geometry, sorption material thickness, heat capacity, rotational speed, fluid velocity, etc. They are mostly too time consuming to be used in full system simulations including solar thermal collectors, where mostly simpler models are available, based either on empirical fits to measured data (Beccali *et al.*, 2003) or on models of dehumidification efficiency.

Different control strategies have been compared by Ginestet *et al.*, (2003) to study the influence of air volume flow and regeneration temperature. As the increase in regeneration temperature does not linearly lower the supply air temperature, the study concluded that increased air flow rates are preferable to increased thermal input, if the cooling demand is high. Mean calculated COPs for the climatic conditions of Nice, France, were between 0.3 and 0.4. Henning and others also remarked that increasing the air flow is useful in desiccant cooling mode, but that the minimum acceptable flow rate should be used in adiabatic cooling or free ventilation mode to reduce electricity consumption (Henning et al, 1999). When thermal collectors with liquid heat carriers are used in combination with a buffer storage, Bourdoukan and others suggested operation of the cooling system only in adiabatic cooling mode during the morning and then allowing desiccant operation in the afternoon, using heat from the buffer storage (Bourdoukan *et al.*, 2007). However, in many applications, dehumidification is already required during the morning hours. Also, if cheaper air collectors are used, heat storage is not possible.

In Europe there are only a few implemented demonstration plants which are powered by solar energy, so operational experience from such plants is very scarce. To introduce the technology onto the market, information about the energetic performance (efficiency, heat and electricity consumption), water consumption and maintenance issues needs to be provided. In the following, experimental results from two large European installations are presented together with laboratory measurements on desiccant rotor systems. Special emphasis is placed on the performance analysis of each individual component as well as the optimization of the overall control strategy.

5.2.1 Desiccant Cooling System in the Mataró Public Library

A desiccant cooling plant with a process air volume flow of 12 000 $m^3\ h^{-1}$ was installed in the public library building in Mataró (Spain) with a 3500 m^2 cooled surface area. The building was equipped with four conventional air-conditioning units, where the cooling energy was provided by heat exchangers from a central electrical compressor chiller. One of the air-conditioning units for the children's reading and multimedia room with a 510 m^2 surface area was replaced by a desiccant cooling plant and provides cool fresh air via 15 ceiling air outlets. The building has a ventilated photovoltaic façade of 244 m^2 and 330 m^2 of shed roofs with a total electrical power of 55 kW_p. The heat produced by the photovoltaic modules is transferred into on air gap 14 cm wide, which is exhausted by the desiccant cooling regeneration fan. Two additional

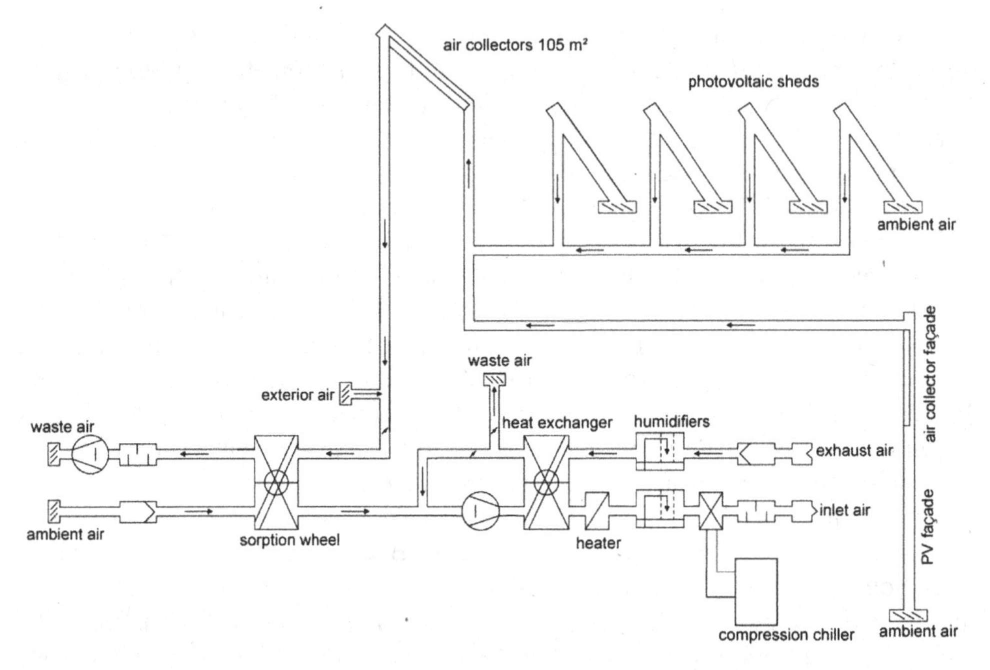

Figure 5.13 System concept of the desiccant cooling plant in Mataró with ventilated photovoltaics and series-connected solar air collectors

air collector fields in the façade (50 m^2) and roof (105 m^2 at 34° tilt angle) increase the temperature level to the required regeneration air temperature (see Figure 5.13). The common regeneration ventilator is volume flow controlled to provide a regeneration temperature between 50 and 70 °C. With a yearly irradiance of 1020 $kWh\,m^{-2}\,a^{-1}$ on the vertical south façade and 1570 $kWh\,m^{-2}\,a^{-1}$ on the shed roofs the combined solar thermal energy system is calculated to produce nearly 70 000 kWh of useful thermal energy from April to October, covering 93% of the cooling demand of 44 000 kWh. Exterior air can be added to the regeneration air just before the sorption wheel, so that temperature peaks from the solar collectors after stagnation can be avoided. The sorption wheel is a silica gel rotor with a nominal rotation speed of 15 h^{-1}. Auxiliary cooling energy is supplied from the existing compression chiller via a heat exchanger after the fresh air humidifier.

The thermal efficiency of the ventilated photovoltaic system is rather low at 12–15%, because flow velocities in the many parallel large air gaps reach only 0.3 $m\,s^{-1}$. The maximum air temperature level increase is between 10–15 K. The complete volume flow through the ventilated PV system is between 3000 and 9000 $m^3\,h^{-1}$ and is fed into three parallel air collector fields. Flow velocities in the 9.5 cm air channels are between 3 and 9 $m\,s^{-1}$ and efficiencies are in the range of 50%.

An important result from the design study was that only 9% of the total cooling energy is provided with a full volume flow of 12 000 $m^3\,h^{-1}$ in desiccant cooling mode; 24% of the time the volume flow is between 6000 and 12000 $m^3 h^{-1}$ and 27% at 6000 $m^3 h^{-1}$. The rest of the cooling energy comes from free cooling operation. Variable volume flow is thus extremely important for reducing the rather high electrical energy consumption of the ventilators. The calculated COPs are 0.52 on average over the whole operation period, with higher values of 0.65–0.73 in July and August and decreasing values of 0.2–0.53 in April and October (Mei and Infield, 2002).

If the regeneration air flow is kept at a constant value of 9000 $m^3 h^{-1}$ the solar fraction to the total cooling demand is 84%. The monthly distribution of total cooling energy, desiccant cooling contribution and auxiliary compression cooling as calculated with the simulation system INSEL is shown in Figure 5.14.

Control Strategy

If the room air exceeds the setpoint temperature, a supply temperature cascade is executed by the control system. Process and room exhaust air are set to 50% of the maximum volume flow, that is to 6000 $m^3 h^{-1}$. The exhaust air humidifier is then switched on together with the rotating heat exchanger. In the following, the supply air humidifier is switched on with each of the three stages of the contact matrix evaporator. Finally the sorption rotor is started and regeneration takes place. If the cooling load can still not be met, the volume flows are gradually increased up to 12 000 $m^3 h^{-1}$. The regeneration air volume flow is reduced from the maximum value of 9000 $m^3 h^{-1}$ until the air collector exit temperature reaches 70 °C. The room exhaust air humidity is limited to a maximum allowed value and the controller switches off the supply air humidifier as soon as the maximum admissible humidity is reached.

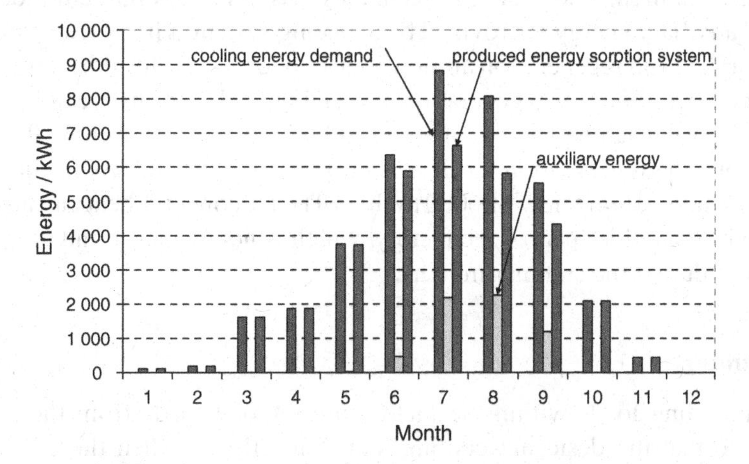

Figure 5.14 Monthly cooling energy demand of the public library rooms and the contributions of the sorption system and the auxiliary cooling source

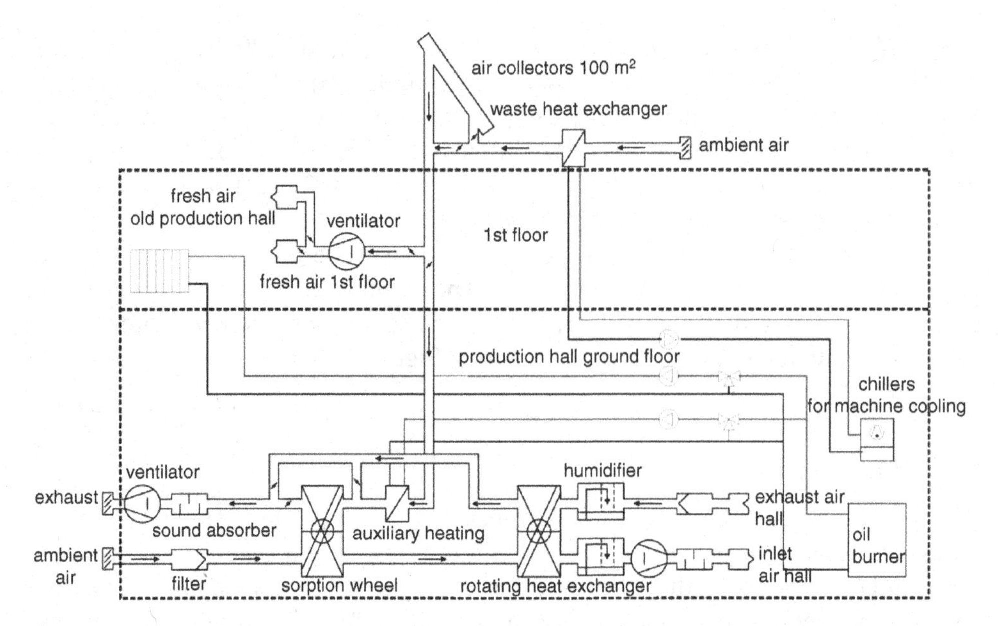

Figure 5.15 Schematic of the desiccant cooling unit in the factory in Althengstett, Germany

5.2.2 Desiccant Cooling System in the Althengstett Factory

A desiccant cooling unit with a similar-sized air collector field of 100 m^2 was built for a plastics factory climatization in Althengstett, Germany (see Figure 5.15). The air collector field consists of two parallel strings each with 50 m^2 of series connected collectors, orientated towards the south-west with an azimuth angle of 232° and a tilt angle of 30°. The air collector type is the same as the Spanish collector with flow channels 9.5 cm high. The desiccant technology used is a lithium chloride sorption wheel and auxiliary energy is provided by a heating system. The process volume flow is 18 000 m^3 h^{-1} and regeneration air flow is at 11 000 m^3 h^{-1}. In addition to the rather small air collector field for regeneration air heating, 45 kW of heat can be recovered from the plastics machine coolers. In order to limit the specific volume flow through the collectors, the regeneration air flow is only partially heated by the air collectors with a maximum volume flow of 6000 m^3 h^{-1}. The air collector heat can also be used directly to heat an older part of the factory, which is not insulated and often requires heating even during the summer months.

Control Strategy

The main cooling loads within the factory are internal loads from the production machines so that the desiccant cooling unit is mostly on when there is a production run. The process and exhaust air volume flows are constant at 18 000 m^3 h^{-1}, and the regeneration volume flow is set to 11 000 m^3 h^{-1}. If the room temperature setpoint is exceeded, a supply air temperature cascade takes place. First the supply air humidifier is switched on, then the exhaust air humidifier together with the rotating heat exchanger and finally the sorption wheel. The minimum supply air temperature is limited to 17 °C. The regeneration air stream is preheated by waste heat from the factory machines. A flap to the air collectors opens if the temperature difference between the absorber and entry temperature exceeds 6 K with a maximum volume flow through the 100 m^2 collector field of 6000 $m^3 h^{-1}$. The total regeneration air flow can be heated by an additional auxiliary heater with a maximum allowed regeneration temperature of 72 °C, as a cellulose rotor with LiCl solution is used. Again the room exhaust humidity is limited, here to a maximum value of 11.7 $g\,kg^{-1}$.

5.2.3 Monitoring Results in Mataró

The monitoring results obtained during the commissioning period in 2002 confirm the correct functioning of the whole desiccant cooling unit: during full desiccant cooling operation the cooling power reaches 55 kW (enthalpy difference between room air and exterior air) or 35 kW when referred to the enthalpy difference between room exhaust and exterior air. When the machine starts up for afternoon operation, the heating power is initially very high as the solar collector field is at standstill temperature. The COP is correspondingly low and then stabilizes around 0.6 (see Figure 5.16). The main factor limiting the COP is the rather low heat recovery efficiency of the rotating heat exchanger. On average the measured heat recovery efficiency is 68% for volume flow rates of 12 000 $m^3 h^{-1}$. Both fresh and exhaust air humidifiers worked satisfactorily at 85% humidification efficiency.

The 12 000 $m^3 h^{-1}$ of fresh supply air volume flow is cooled down from 31°C outside air temperature to 17°C supply air at regeneration temperatures of about 70°C (see Figure 5.17). The regeneration air flow measured is 8000 $m^3 h^{-1}$.

Higher COPs close to 1.0 were obtained at lower regeneration temperatures around 50 °C at similar regeneration volume flows of 8000 $m^3 h^{-1}$ (see Figures 5.18 and 5.19).

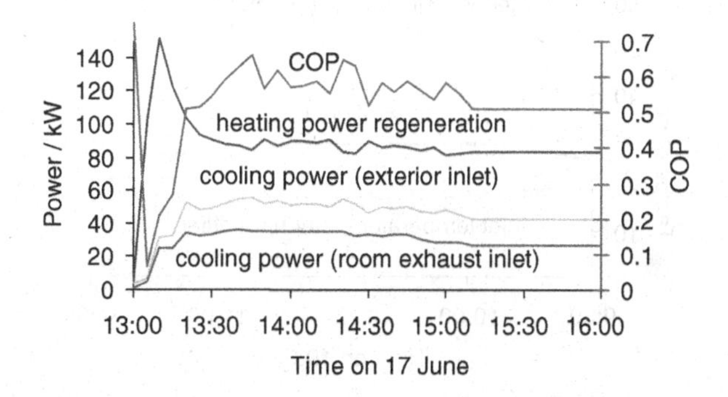

Figure 5.16 Heating and cooling power measured during plant commissioning in Mataró. The COP relates the ambient air cooling power to the heating power

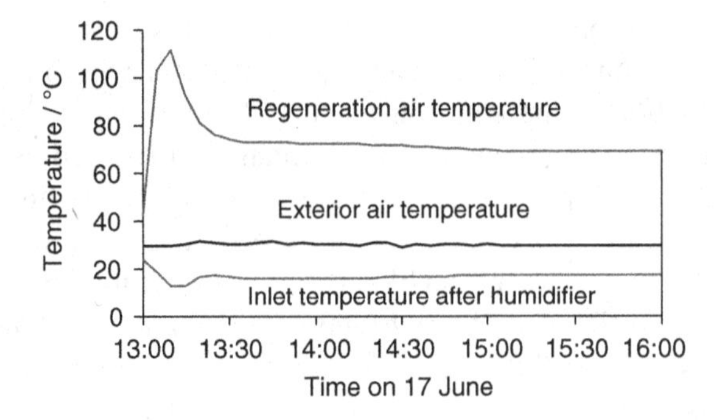

Figure 5.17 Temperature levels under full desiccant cooling operation

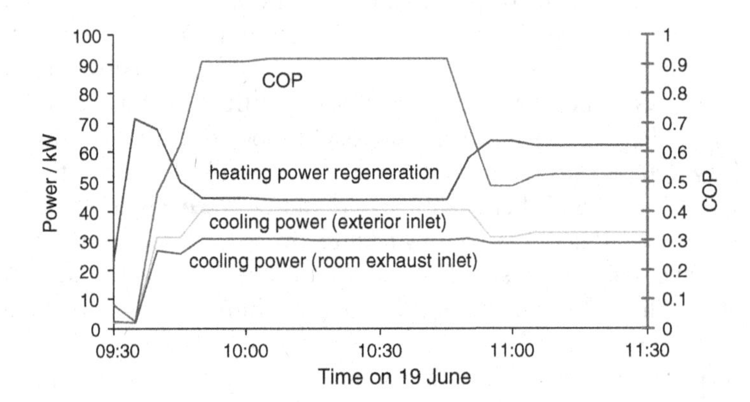

Figure 5.18 Power and COP at low regeneration temperature

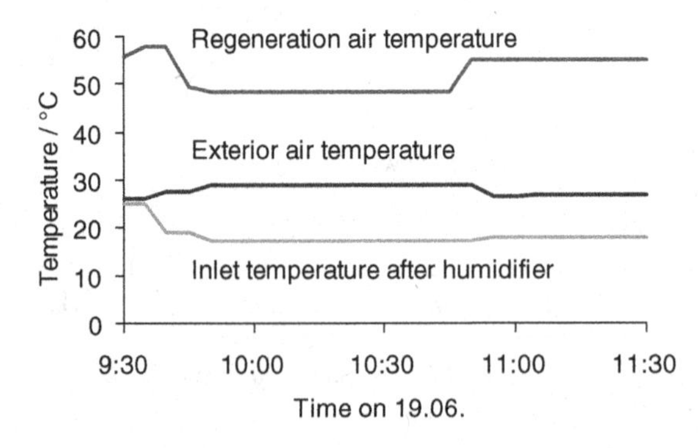

Figure 5.19 Temperature levels during commissioning with regeneration air temperatures around 50°C

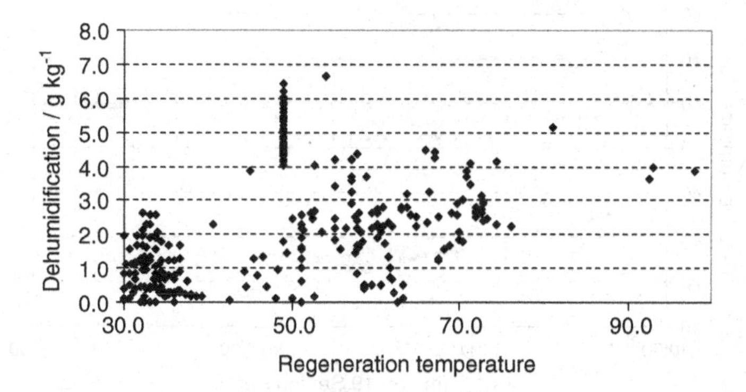

Figure 5.20 Dehumidification of the process air measured during the commissioning procedure

If the dehumidification rate is sufficient at such low regeneration temperatures, this low-temperature regeneration mode is preferable in terms of energy efficiency.

The dehumidification rates obtained from measured temperature levels and relative humidities rise with increasing regeneration temperature, but fluctuate strongly (see Figure 5.20).

To determine the cause of the fluctuations in absolute humidity change, more detailed measurements were done on the 2.1 m diameter wheel. Four temperature and humidity sensors (type ebro EBI-2TH-611) were placed in the process air stream behind the sorption wheel for a detailed comparison with the averaging temperature sensor and single point humidity sensor, which are connected to the building management system (from the Sauter company). The single point temperature measurements of the exterior air before the sorption wheel correspond very well (Figure 5.21).

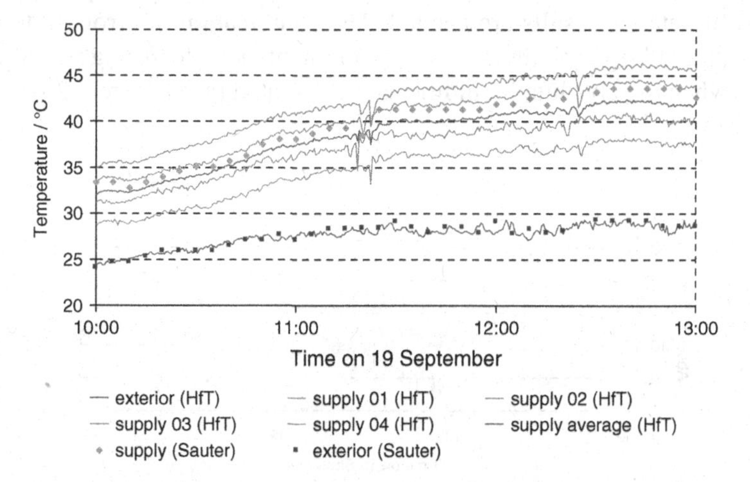

Figure 5.21 Temperature levels before and after the sorption wheel with one building management system sensor and four distributed sensors. The sensors were provided either by the building management company Sauter or by the University of Applied Sciences in Stuttgart

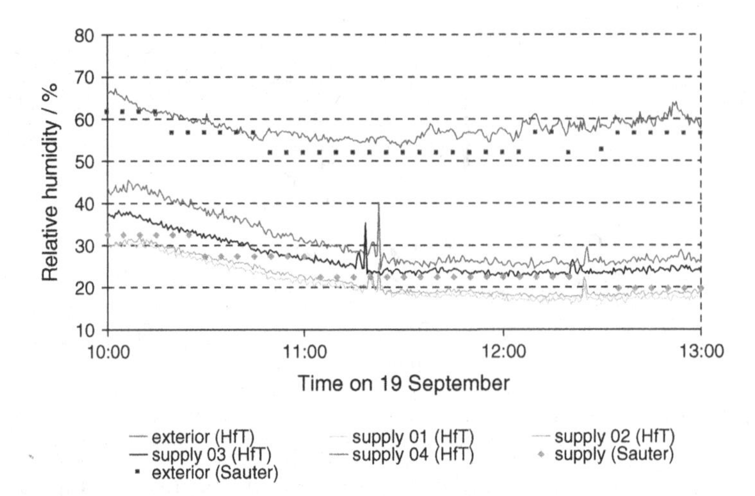

Figure 5.22 Relative humidity measurements on the sorption wheel in Mataró

The temperature after the drying process is highest on the side where the dry sorption material enters the process air stream and adsorbs most of the moisture. Here the temperature level is up to 8 K higher than at the other end of the sorption wheel, where the humidity has been adsorbed during the rotation half period. The averaged temperature signals deviate by 1–2 K from the building management sensor. In the measurement interval the solar supplied regeneration air temperatures steadily increased from 51 °C at 10:00 to 62 °C at 13:00.

The relative humidity measurements as recorded by the building management system show a stepwise change for every 5% change in humidity (see Figure 5.22). This is purely due to the recording strategy of the building management system and should be adjusted if detailed results are needed. The combination of errors in temperature measurement and imprecise humidity measurements lead to strong deviations of up to $1\ \mathrm{g\,kg^{-1}}$ when the absolute humidities are calculated (see Figure 5.23).

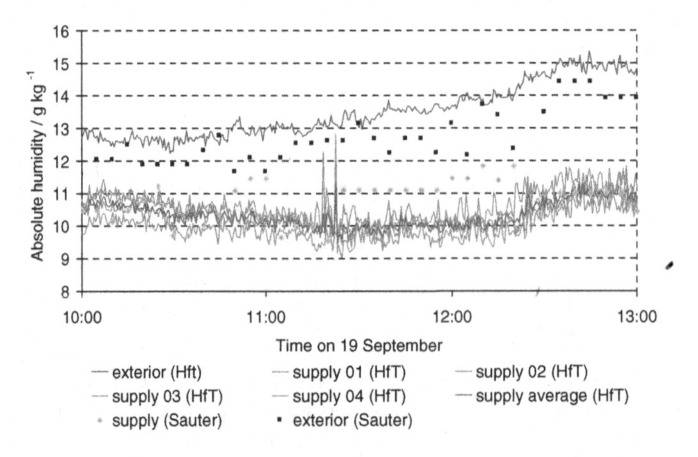

Figure 5.23 Absolute humidities calculated from measured relative humidities and temperatures

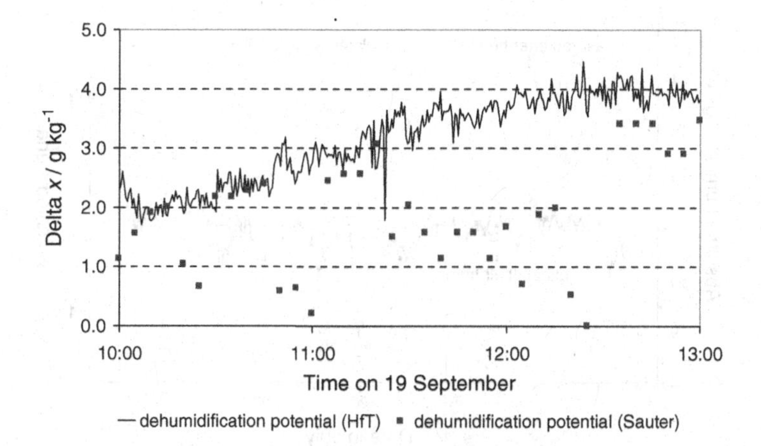

Figure 5.24 Measured dehumidification from the building management system (Sauter) and averaged individual sensors (HfT)

If the difference between the exterior air absolute humidity and the supply humidity after the sorption wheel is now calculated, it can be seen that the values taken from the building management system are not usable (see Figure 5.24). As similar experiences were found in the commercial desiccant cooling unit in Germany, more measurements were done on the laboratory test side in Stuttgart in order to establish the enthalpy changes during dehumidification. If the enthalpy change is known, the dehumidification can then be calculated from the temperature change alone.

5.2.4 Monitoring Results in Althengstett

Component and Control Strategy Analysis

Monitoring results are available for the whole summer period from March to September 2002. First of all, efficiencies were determined for all components from the desiccant cooling plant. The heat recovery efficiency of the rotating heat exchanger was only 62% at rotation rates of 600 turns per hour. At a measured mass flow ratio of supply to exhaust air of 1.16 the manufacturer's given value was 73%. The contact evaporators reached 85% humidification efficiency, compared with the 92.4% given by the manufacturer. Using this constant humidification efficiency, the humidification rates can be simulated based on the room exhaust air temperature and relative humidity and correspond well with the measured data (see Figure 5.25). The steady-state model, however, does not consider the dynamics of the humidification process, so the calculated absolute humidity after the humidifier immediately takes on the exhaust air humidity, whereas the measured humidity stays high (see Figure 5.25 with the switching signal set to 1 for humidifier operation).

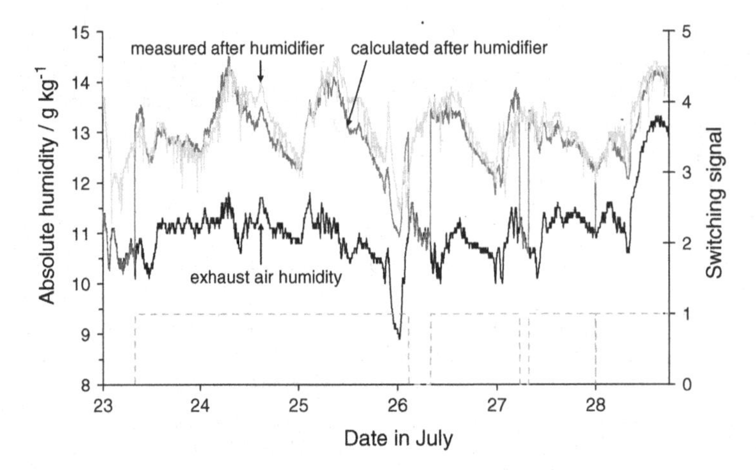

Figure 5.25 Measured and calculated absolute humidities of the exhaust air with 85% humidification efficiency

Each component has specific time delays between the control signal (release) and showing the effect in operation. The humidifiers, for example, need about 5–10 minutes after release to reduce the air temperature at the outlet significantly and the cooling effect does not stop until 45 minutes after the release is switched off (see Figure 5.26). So if the control signals are used for the evaluation of the humidifiers, in each period of operation, about 40 minutes of real cooling operation is not considered. In the detailed analysis of one month's performance (July), the different energy performance based on the control signal evaluation and the real evaporative cooling were compared.

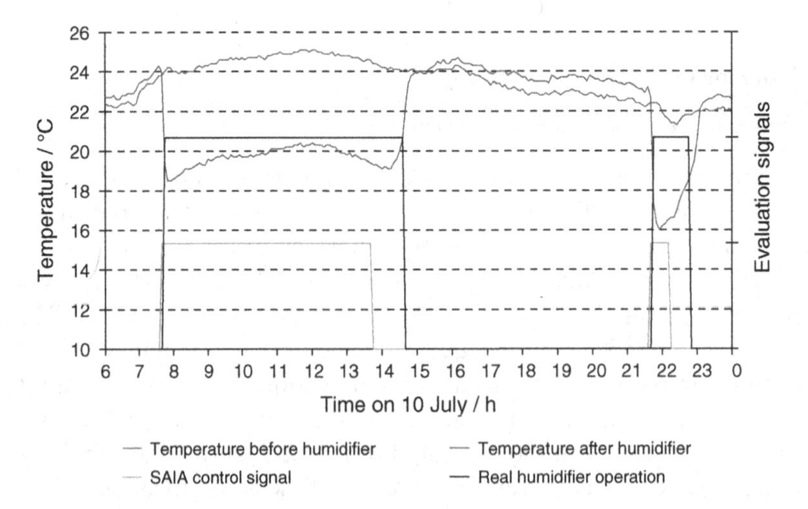

Figure 5.26 Time delay between the control signal and reaction of the humidifier together with the signal representing real operation of the humidifier, when the temperature difference is at least 3 K

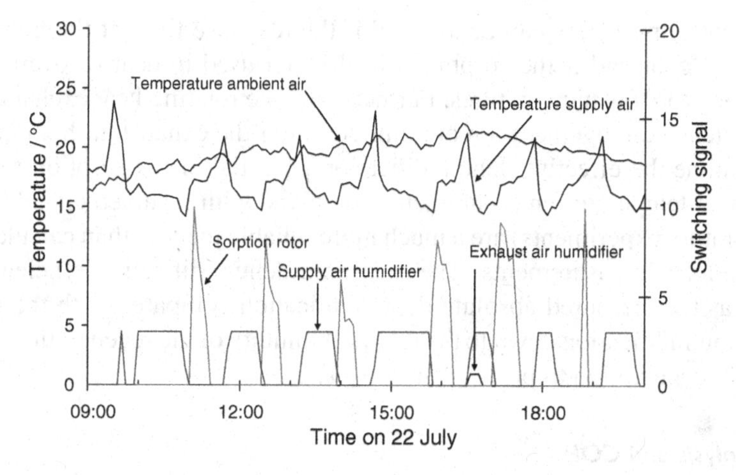

Figure 5.27 Fluctuations of supply air temperature due to control strategy

This dynamic and partially slow response of the components is not adequately considered in the control strategy of the plant. For example, the measured supply air temperature often fluctuates by about 6 K when supply air humidifiers and heat recovery strategies work against each other. At very moderate ambient air temperatures of 22 °C the combination of exhaust air humidification and full humidification of the three-stage supply air humidifier leads to a temperature decrease in the supply air below the setpoint of 17 °C in less than 30 minutes (see Figure 5.27).

The humidifiers are consequently switched off and the rotational speed of the sorption rotor is increased to raise the temperature level in heat recovery mode. Furthermore, the rotating heat exchanger is switched on early (about 10 minutes before the exhaust air humidifier), so that supply air temperatures increase well above 20 °C and then drop again below 15 °C (see Figure 5.28).

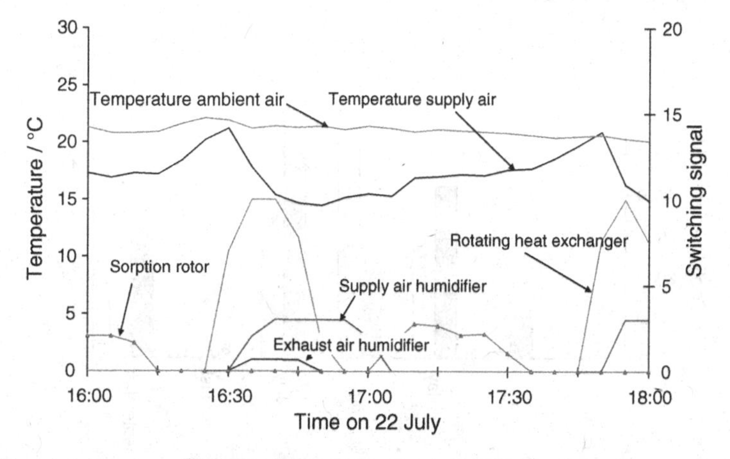

Figure 5.28 Details of switching signals and resulting supply air temperature

These control problems can be avoided if the response time of the humidifiers is taken into account and if the sorption wheel is not used in heat recovery mode for compensating too low temperatures. Furthermore, the rotating heat exchanger should only be switched on together with or 5 minutes after the exhaust air humidifier.

To determine the effective dehumidification potential of the sorption rotor, it can be shown that temperature measurements combined with assumed enthalpy changes (from laboratory experiments) are a much more reliable method than calculation from relative humidity measurements. The measured dehumidification efficiency, which is defined as the measured absolute dehumidification compared with the maximum possible dehumidification down to the relative humidity of the regeneration air, is 80% at a ratio of regeneration to process air of 66%.

Energy Analysis and COP

During the seven months of measurements, 34 710 kWh of cooling energy was produced. For the calculation of all energy balances only sensible cooling energy was calculated (ambient–inlet air temperature difference), as the enthalpy measurements are too unreliable. At an average COP of 0.95 the heating energy required was 36 460 kWh. A main feature of the desiccant process is that heating energy is only required for full desiccant operation and the humidification processes alone do not require additional energy. If the heating energy is related only to those hours with full desiccant operation, the COP is 0.5.

Full desiccant cooling operation is mainly required in the summer months, when the humidification process alone is not sufficient for exterior air cooling. The average COPs are highest when the process does not require heating energy and reach average monthly values up to 4.8 (see Figure 5.29).

The low COP in June was to a large extent due to the malfunction of the exhaust air humidifier, when the supply air temperatures could not be reached and the regeneration

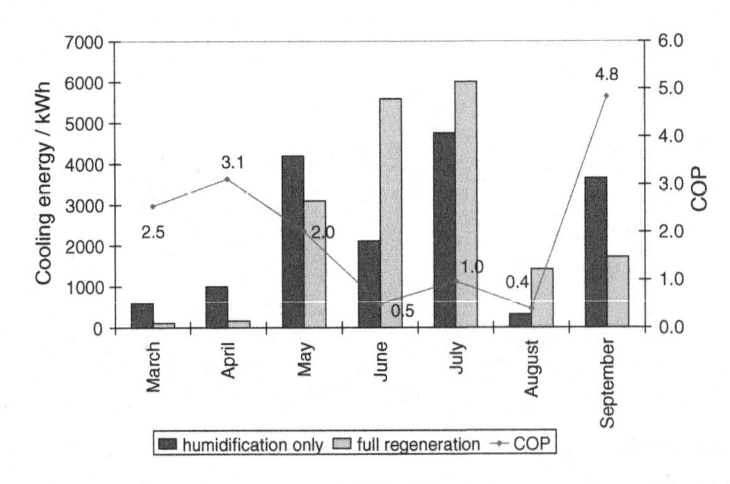

Figure 5.29 Monthly cooling energy and COP of the desiccant cooling unit in Althengstett

air was heated to a maximum and consumed a high amount of energy. In July, the problems were solved and the system was working properly. In August, the operating time of the system was low due to summer vacations. In September, the auxiliary heating system was disconnected, so that pure solar and waste heat operation took place. Due to the limited regeneration air temperatures, the average COP was very high. Even when the heating energy is only related to full desiccant cooling operation, the COP is 1.5. This demonstrates the benefit on the performance results of allowing low-temperature regeneration.

During the whole operation period more than half of the cooling energy was obtained with full desiccant operation; for the rest of the operating time the humidification mode was sufficient. From the total required heating energy of 36 460 kWh only 6996 kWh was covered by the solar air collector field, and a further 9557 kWh was provided by the preheater from machine waste heat. More than half of the required heating energy came from an auxiliary heater. However, the solar air collector heating energy was used to a large extent for heating the old production hall, which was required throughout the summer, so that the total collector thermal energy yield from march to September was 24 637 kWh (see Figure 5.30).

It is interesting that about 30% of the auxiliary energy is spent when the outside temperatures are rather low (2975 kWh for outside temperatures below 20 °C and 8353 kWh for temperatures between 20 and 25 °C). Although the temperature reduction between the outside air and inlet air is small for these conditions, the dehumidification process requires a high amount of energy. Auxiliary cooling would be preferable under such conditions. If this is not possible, the controller must ensure that the regeneration air temperature is kept at its minimum and not set to fixed values.

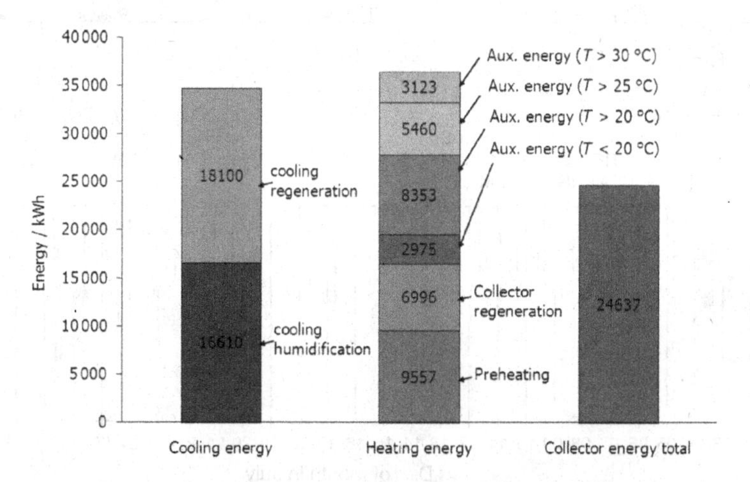

Figure 5.30 Complete energy balance for the Althengstett desiccant cooling unit. The auxiliary energy is split into different ambient air temperature ranges. The cooling energy is separated into full regeneration mode and humidification mode only

The total electrical power consumption for the fans and wheel motors is 17.2 kW. A conventional air-conditioning system without an exhaust air humidifier, sorption wheel and air collector field would have about 4 kW less electrical power demand. For 1167 operating hours in cooling mode the additional electrical energy consumption for the desiccant unit is 4668 kWh producing a total of 34 710 kWh. If the total electrical energy consumption is considered, about 20 000 kWh of electrical energy is needed. This is a consequence of an air-based cooling distribution system. The situation can only be improved if variable volume flows are possible at partial load conditions. However, as the pollutant levels are very high in the plastics factory, the air volume flows had to be kept high to maintain a reasonable air quality. In addition to the 1167 h in active cooling mode, 2473 h of operating time was in free cooling or free ventilation mode, also requiring electrical energy for the fan operation.

Air Collector and System Performance

The low heating energy contribution of the solar thermal air collector of about 70 kWh m^{-2} a^{-1} was analysed in more detail for the month of July, where data availability was highest at 83% and there were no more problems with operation the humidifier. From a total of 421 operating hours, 83 hours needed full regeneration mode, but 90 hours pure evaporative cooling was sufficient. The dominant operating mode even in July was free cooling with 178 operating hours (see Figure 5.31).

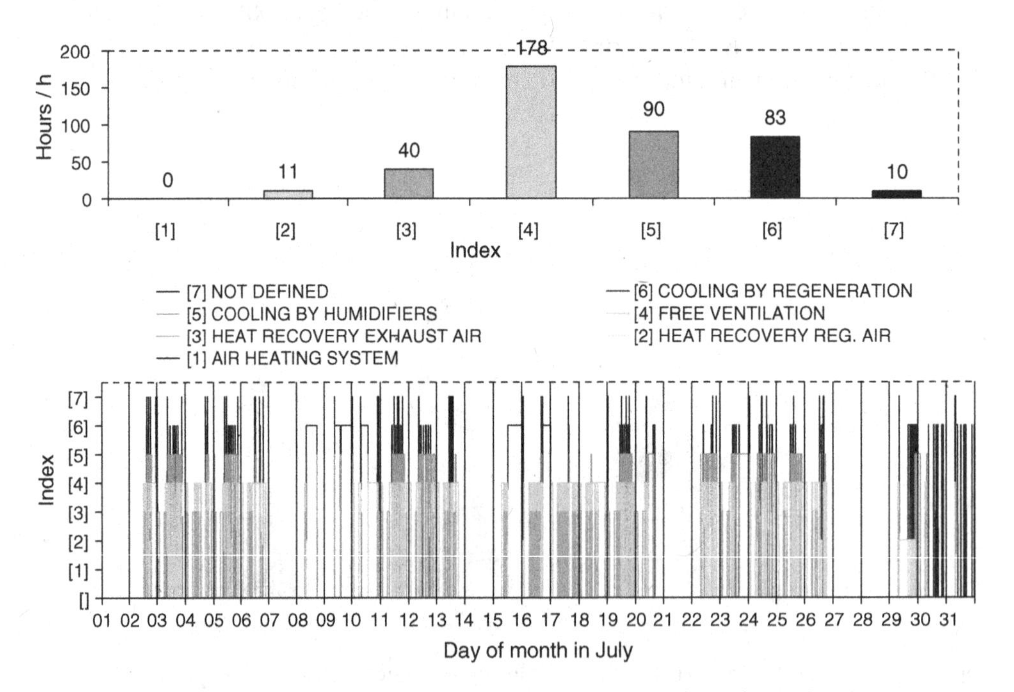

Figure 5.31 Operating hours of different desiccant cooling modes for all days in July

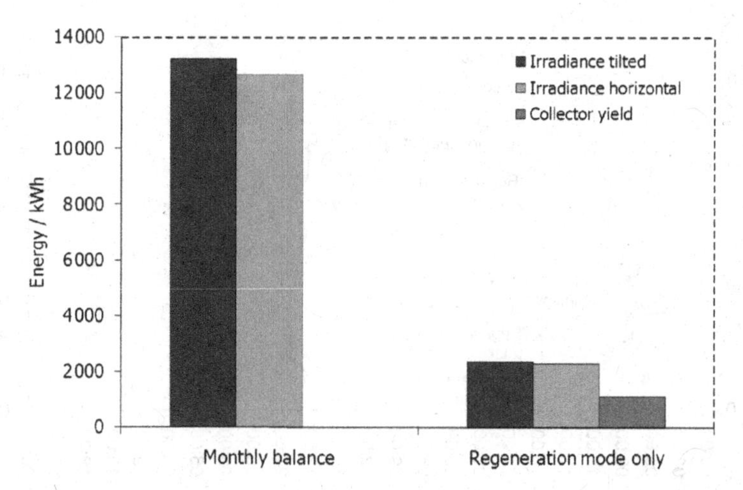

Figure 5.32 Total irradiance on collector surface and horizontal plane during July and energy yield during regeneration hours only

From the 83 hours of required regeneration, the collector provided energy for 53 hours, that is during 64% of the time, and produced a total of 11 kWh m^{-2} and a month which is about the same amount that the waste heat delivered from the factory. The other days were too cloudy to obtain the minimum temperature difference of 6 K. The mean collector efficiency during the operating hours was 48%, which is a good performance value (see Figure 5.32).

Looking in more detail at the control strategy and the timing of the regeneration mode, the temporal sequences and phase shifts between regeneration operation and solar thermal energy contribution can be clearly seen (Figure 5.33). During regeneration, the temperature levels vary typically between 60 and 70 °C. The two parallel-connected air collector fields are capable of delivering this temperature level for nearly 7 hours during a sunny day. As the air collector field only heats 55% of the total regeneration air flow due to its small size, and the waste heat from the factory only delivers air temperatures of 40 °C, the complete regeneration volume flow of 11 000 m^3 h^{-1} reaches maximum temperature levels of 60 °C and auxiliary heating was always switched on. This could be avoided by increasing the collector field size, which is not economical for the low number of regeneration hours. Also, the complete system simulation showed that the auxiliary heating was not even necessary to reach the desired inlet air temperatures and that the energy consumption of the auxiliary heating could be largely avoided.

The fixed high-temperature levels for regeneration also create a problem for the early start-up of the desiccant unit, when solar thermal energy is not yet available at sufficient temperature levels (see Figure 5.33). With air-based systems without thermal storage this is difficult to avoid. It would be advisable to operate the desiccant mode at low solar regeneration temperatures and admit lower dehumidification rates.

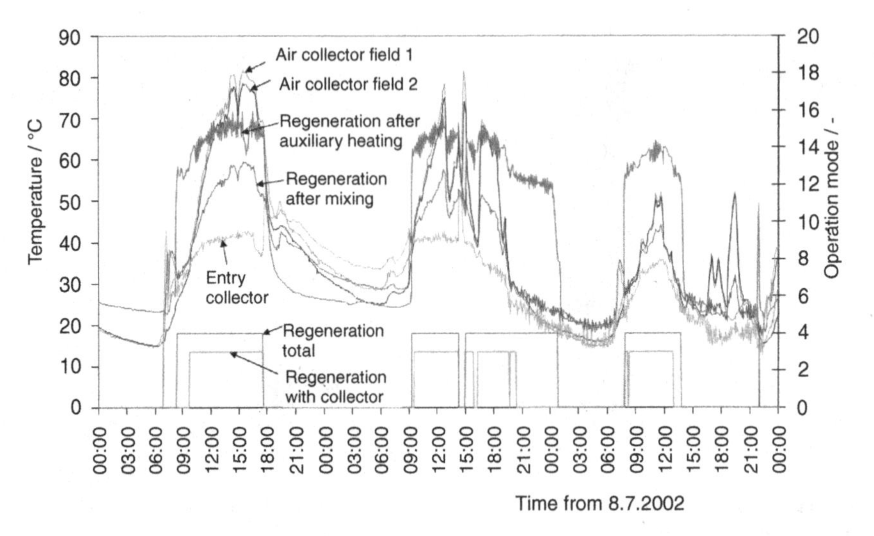

Figure 5.33 Temperature levels in the regeneration air stream after preheating with waste heat, air collector heating, mixing of the two volume flows and auxiliary heating. The total operating hours of regeneration and regeneration with collector heating are also shown

Temporary increases of room humidity could then be reduced when solar energy is available. To use the solar cooling energy also for long operating hours until midnight would only be possible if the storage mass of the building itself were used.

The analysis showed that the additional measured auxiliary heating energy demand mainly occurred on days with rather low ambient air temperatures, when the auxiliary heating system was on especially in the mornings and evenings. In several cases, free cooling with a higher supply air volume flow would have been sufficient to cover the cooling load. In Figure 5.34 it can be seen that the sorption rotor was often switched on for short time intervals (operating mode = 4), which can mainly be attributed to a heat recovery mode, when the humidifiers decreased the supply air temperature too quickly.

The total solar fraction for the month is low at 18%. The COP is 33% for all full regeneration hours, if the cooling energy is calculated from the enthalpy difference between ambient air entering the machine and room supply air. In the Althengstett system, a large discrepancy in temperature levels and as a result in COPs was noted between the ambient air temperature measured with a shaded sensor in the supply air entry channel on the building roof and the temperature measured in the entry channel of the machine placed on the ground floor. The machine entry temperature was up to 6 K lower than the roof air temperature, so that the net cooling effect of the DEC system was much lower than calculated using ambient air temperature. If the roof temperature is taken for the COP calculations, the value for full regeneration doubles. However, this is not due to the DEC system, but rather due to precooling of the ambient air within the building, indicating that better thermal insulation should be used.

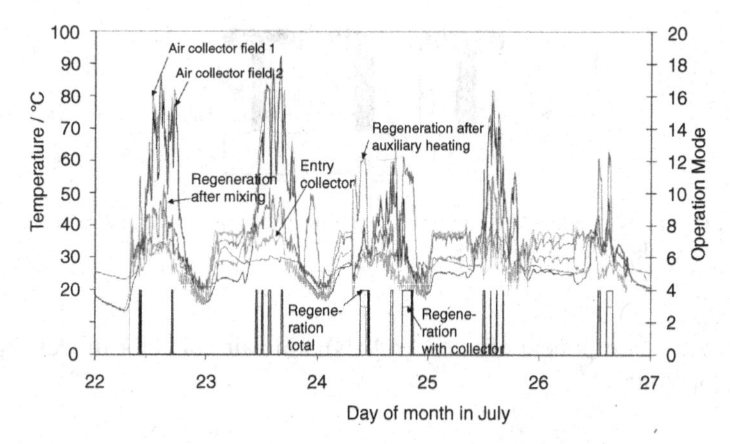

Figure 5.34 Regeneration air temperatures and regeneration operation on days with low irradiance

In contrast to typical design conditions, where outside air enthalpy is higher than that inside, the measured enthalpy difference between room exhaust air and supply air is higher on average. The heating energy related to the total produced room cooling energy results in COPs of 1.13 (see Figure 5.35).

5.2.5 Simulation of Solar-Powered Desiccant Cooling Systems

A complete desiccant cooling simulation tool was developed in the simulation environment INSEL to improve the control strategy and to analyse the influence of solar

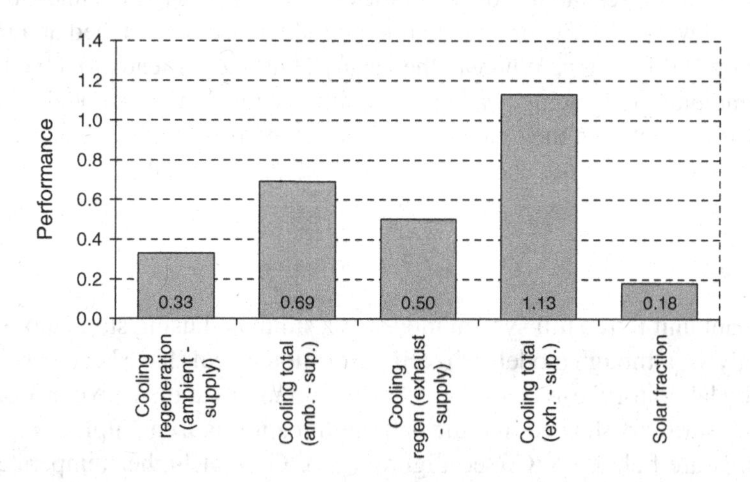

Figure 5.35 COPs for regeneration mode only and total cooling energy calculated relative to ambient air enthalpy and room exhaust enthalpy. The solar fraction of the total heating demand is also given

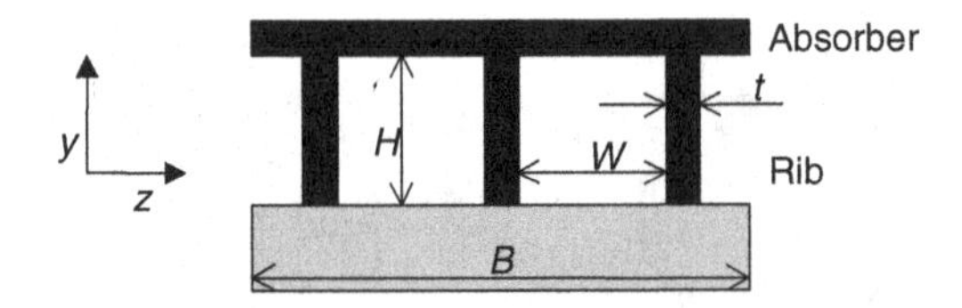

Figure 5.36 Geometry of a finned air collector with flow channels of height H and width W and ribs of thickness t to calculate the rib efficiency

collector size and component efficiency. A dynamic air collector model is part of the model development.

Air Collector Model

The air collector model uses Nusselt correlations for laminar or turbulent flow conditions and a rib efficiency for a channel of height H and width W derived by Altfeld (1985) (see Figure 5.36). In addition, heat capacities were attributed to the three temperatures nodes of the absorber, the air flow in the gap and the back insulation surface. The energy balance for the three nodes was solved with the known air entry conditions and then subsequent elements were calculated in the flow direction. The long-wave radiative exchange between the absorber and back surface and the heat loss from the absorber to the environment are calculated using the temperature values from the last time step.

The accuracy of the dynamic model was studied on a sunny day with fast-moving clouds. The collector volume flow through each of the 50 m^2 collector fields is 3000 m^3 h^{-1}, which results in flow velocities in the 9.5 cm high channels of 9 m s^{-1}. Temperature levels of 80 °C or 47 K temperature increase are reached at a maximum irradiance of 1000 W m^{-2}. Whereas the steady-state model leads to very high temperature gradients at changing irradiance conditions, the dynamic model is capable of very precisely simulating the dynamic collector response (see Figure 5.37). The time constant for the temperature change is approximately 15 minutes.

System Model

The desiccant unit in the full system model was simulated using static models for the energy analysis, although models of rotating sorption wheels have been developed and validated by laboratory results. Laboratory measurements at the University of Applied Sciences in Stuffgart showed that the dehumidification is isenthalpic if regeneration temperatures are below 75 °C (see Figure 5.38). Only at higher temperature levels does heat transfer from the regeneration side lead to an enthalpy increase during the dehumidification.

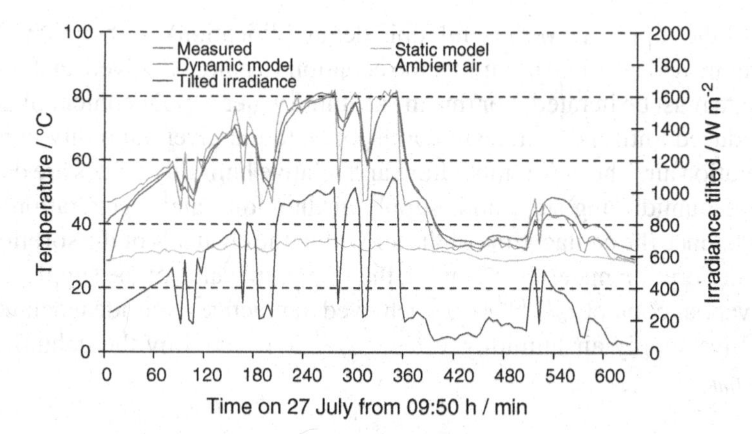

Figure 5.37 Measured and simulated air collector temperatures on a day with fast-moving clouds and high irradiance gradients

The main assumption of a simple dehumidification model is that the sorption isotherms, represented as a function of the relative humidity, are independent of temperature and coincide. The minimum water load of the sorption material is given by the relative humidity of the regeneration air.

The supply air humidity can then be reduced to a minimum, which is given by the regeneration air relative humidity. Exact knowledge of the functional form of the sorption isotherms is not necessary. For the final state of the dried air the relative humidity, namely the regeneration humidity, is known, as well as the enthalpy

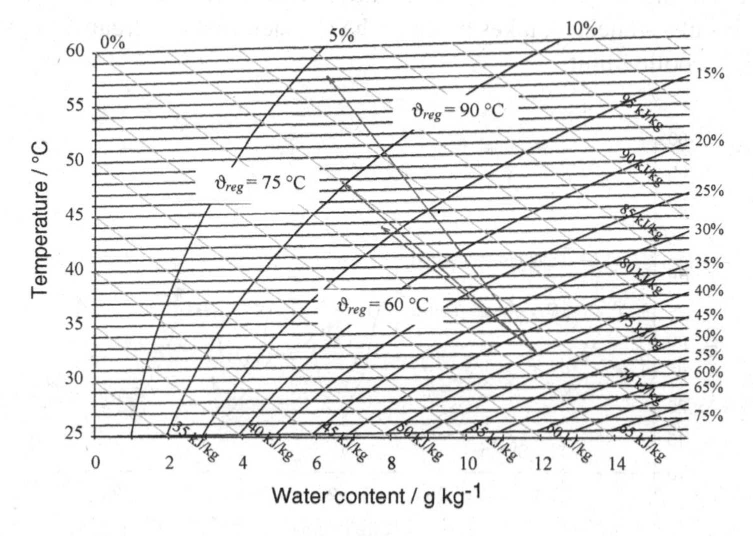

Figure 5.38 Measured dehumidification process of a silica gel sorption rotor at different regeneration temperatures

(enthalpy of the supply air with isenthalpic dehumidification). Although both depend on temperature and absolute humidity, the equations cannot be solved analytically and the final state must be iterated. For this the absolute water vapour content of the supply air x_o is reduced until the final state corresponds to the given humidity condition of the regeneration air. The new temperature and relative humidity values are determined after each dehumidifying step, until supply air humidity and regeneration humidity correspond. Since the contact time of the air within the channels of the sorption rotor is short and the sorption material is limited, the minimum value of the supply air humidity with a vapour content ${x_{dry}}^{ideal}$ is not achieved in practice. Non-ideal dehumidifying at an effective supply air humidity value ${x_{dry}}^{eff}$ is covered by the dehumidification efficiency η_{dh}

$$\eta_{dh} = \frac{x_o - x_{dry}^{eff}}{x_o - x_{dry}^{ideal}}$$

$$x_{dry}^{eff} = x_o - \eta_{dh}\left(x_o - x_{dry}^{ideal}\right) \tag{5.4}$$

The best fit to the experimental data was obtained for a dehumidifcation efficiency of 80%. Measured and simulated temperature levels after the sorption wheel (supply side) coincide very well, which shows the suitability of the simple model (see Figure 5.39).

The component parameters from the experimental analysis of the two desiccant cooling units in Spain and Germany are summarized in Table 5.3.

The system model includes the preheating of the regeneration air using the factory waste heat. Mixing of the maximum collector volume flow of 6000 m^3 h^{-1} and the air, which is only preheated, takes place in the regeneration air stream before entering the desiccant cooling unit.

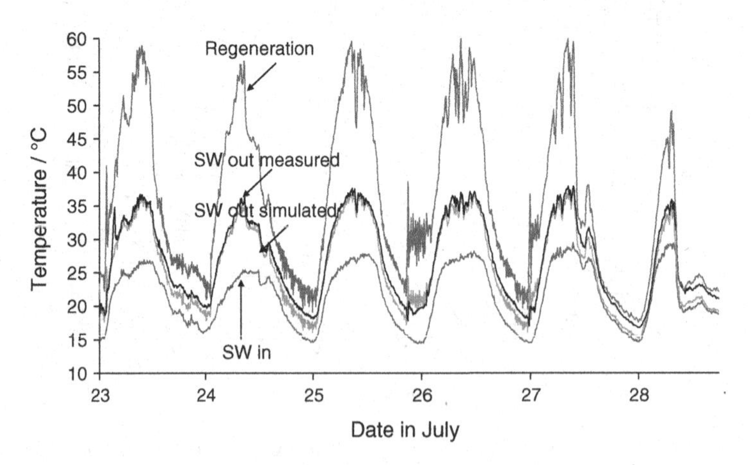

Figure 5.39 Measured and simulated temperature levels after dehumidification by the sorption wheel (SW) on the supply air side together with inlet and regeneration temperatures

Table 5.3 Summary of measured efficiencies of desiccant cooling units

	Mataró	Althengstett
Dehumidification efficiency	80	80
Humidification efficiency	86	85
Heat recovery efficiency	68	62

Control Strategy Modelling

The goal of the simulation was to determine the influence of the control strategy on the system performance and to establish whether different sizing or orientation of the collector field could improve the solar fraction and reduce the auxiliary energy demand. The model can work with building cooling loads $\dot{Q}_{cool}$ calculated using dynamic simulation tools or from measured loads obtained from the supplied cooling energy (calculated using the supply air volume and the measured enthalpy difference between room exhaust and supply air). For each time step, the programmed system controller calculates the required supply air temperature to cover the cooling demand of the room for a given supply volume flow. To save electrical power, the supply air volume flow is first set to 50% of its nominal value. If the resulting supply air temperature is too low or if the cooling demand cannot be covered by this reduced volume flow, the flow rates are gradually increased up to the nominal flow:

$$T_{required} = T_{exhaust} - \frac{\dot{Q}_{cool}}{c_p \dot{V}_{supply}} \tag{5.5}$$

The controller then switches on each of the components subsequently (first either the exhaust air humidifier plus rotating heat exchanger for the Mataró control strategy or the supply air humidifier for the Althengstett strategy) and calculates the supply air temperature. If this temperature is still higher than the required temperature, the next component is switched on, that is the next humidifier. Finally the sorption wheel starts operating and if temperature levels are still not sufficient, auxiliary heating or cooling takes place, depending on the system configuration.

Simulation Results

Using measured July data from the Althengstett system with a time resolution of 5 minutes, the model can be validated. Inputs to the model are the global horizontal irradiance, ambient temperature and humidity. Room and supply air temperature and the supply volume flow were used for the cooling load calculation, which the controller needs as an input.

The component switch-on times of the simulation model and measured results in general correspond well (see Figure 5.40). The real component operating times

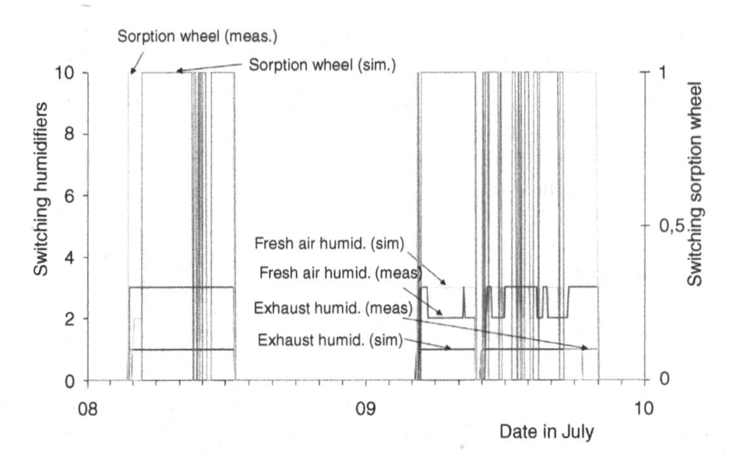

Figure 5.40 Measured and simulated switching times of the humidifiers and the sorption rotor for 2 days in July

are slightly longer than the model calculations. The time discrepancy is especially noticeable in the morning and evening, when the sorption wheel starts about 1 hour earlier and stops up to 3 hours later than the simulated controller. During this time period, free cooling is sufficient to provide the required cooling power.

The main difference between real operation and simulated operation was the additional auxiliary heating demand: in real operation, regeneration operation nearly always meant auxiliary power consumption, whereas in the simulated process, the temperature levels of the collectors and preheated air were sufficient to cover the cooling load. For example, on 3 days from 8 to 10 July the auxiliary heating was on during the complete day period, whereas in the simulation model, only about 1 hour of auxiliary heating was necessary (see Figure 5.41). The problem in the implemented control strategy is that regeneration always means a fixed and rather high temperature level, whereas often lower temperature levels would be sufficient.

As a consequence, the simulation model applied to the month of July resulted in 90% less auxiliary heating energy demand! Some 42 kWh auxiliary heating was obtained from the simulation, whereas the measured auxiliary heating consumption was 3900 kWh.

Using the full simulation model, some parameter studies could be carried out. Doubling the collector size to 200 m^2 increases the collector thermal energy production by 70% and reduces the auxiliary energy production by 25%. The specific collector yield drops slightly by 10%. As the specific energy production of the solar thermal collector is quite low anyhow, the amount of auxiliary energy reduction would never justify the additional investment costs for the collector field. Only significantly longer operating times for the regeneration mode, for example for air drying or other heating purposes, could economically justify higher collector surface areas.

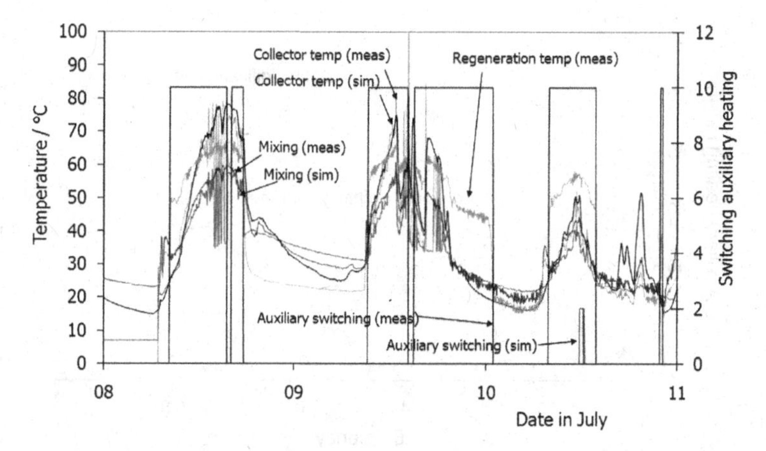

Figure 5.41 Measured and simulated temperature levels on the regeneration side together with measured and simulated switching times for the auxiliary heating

Also, the influence of the desiccant system component efficiencies was studied. A major influence on auxiliary heating demand is given by the humidifier efficiencies. Decreasing the humidification efficiency from 85% to 75% nearly doubles the auxiliary heating demand. A dramatic rise in auxiliary heating occurs for humidification efficiencies below 70% (see Figure 5.42).

At 80% humidification efficiency and 62% heat recovery efficiency, the auxiliary demand more than doubles if the sorption wheel dehumidification reduces to 70%. The auxiliary heating energy can be reduced almost to zero if the efficiency of the heat exchanger is improved from the measured 62% to 75% (see Figure 5.43).

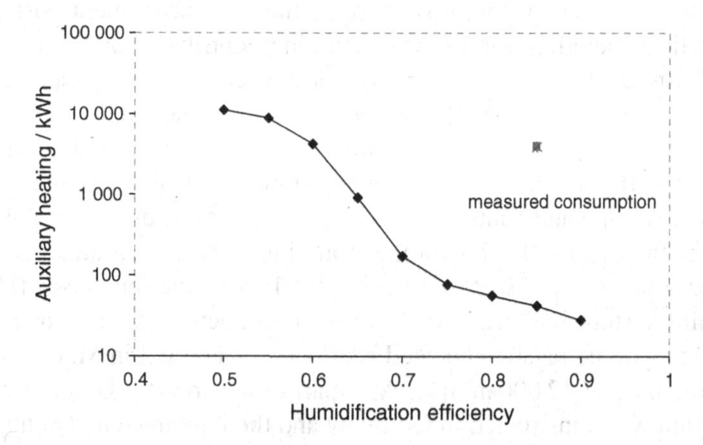

Figure 5.42 Simulated auxiliary heating energy demand as a function of the efficiency of both humidifiers

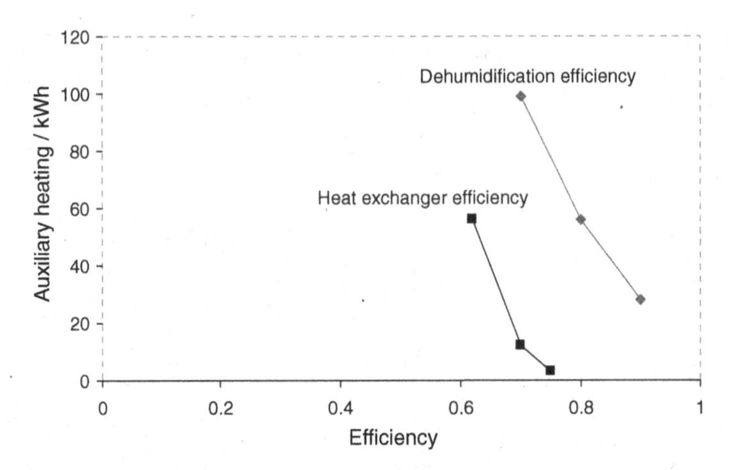

Figure 5.43 Auxiliary energy demand for varying heat recovery efficiencies (at constant dehumidification and humidifier efficiency both of 80%) and for varying sorption wheel dehumidification efficiencies (at constant heat recovery efficiency of 62% and humidification efficiency of 80%)

5.2.6 Cost Analysis

The costs for desiccant cooling units are often full system costs, as the desiccant unit already includes parts of the conventional distribution system such as fans, fresh air channels, exhaust air channels, filters, etc. Furthermore, the machine itself contains humidifiers and heat exchangers, which are also available when a conventional air handling unit is installed.

As an example the total capital, consumption and operation-related costs were analysed for the Althengstett demonstration system with a volume flow of 18 000 $m^3 h^{-1}$. The investment costs are dominated by the desiccant unit itself and its system integration, which together comprise 72% of the total investment costs. The solar air collector field including tubing and mounting contributes only 14% of the total investment costs at 300 euros per square metre of collector area (see Figure 5.44). The total investment costs are € 12.4 per $m^3 h^{-1}$. For another well-analysed system of the IHK in Freiburg, Germany, with a smaller volume flow of 10 000 $m^3 h^{-1}$ the costs were higher at € 16.6 per $m^3 h^{-1}$. Also in this system, the DEC unit together with the air channels, mounting and control comprised about two-thirds of the total investment costs (46% for the unit itself, 17% for the mounting and channels and 6% for control). The solar air collector field was responsible for 10% of the total costs (Hindenburg, 2002). A third system with 105 m^2 of solar air collectors connected in series to a ventilated photovoltaic façade and shed roof was constructed in Mataró, Spain, with a total volume flow of 12 000 $m^3 h^{-1}$. As a part of a European Union demonstration project, the unit was constructed in Germany and then mounted and connected to the existing building management system by the Spanish project partners. Here the DCS unit comprised 33% of the total costs of €179 300, and a high amount of 23% was used

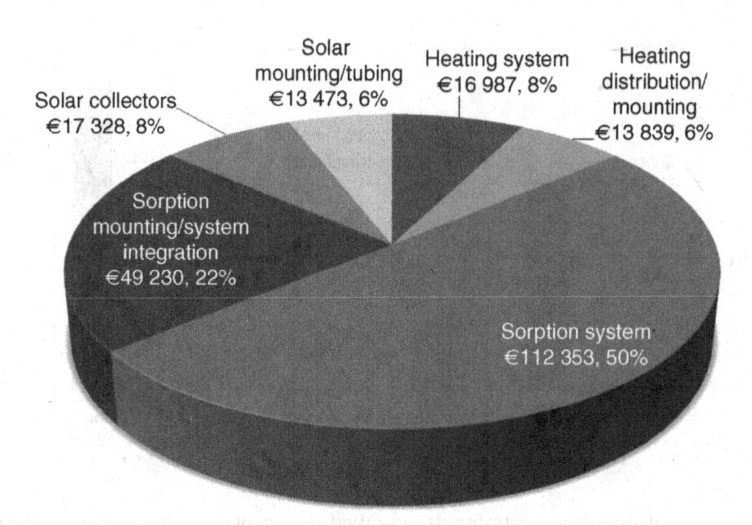

Figure 5.44 Total investment costs for hardware, tubing and system integration for the solar-powered desiccant cooling plant in Althengstett with 18 000 $m^3 h^{-1}$ volume flow

for the unit control and its connection to the existing building management system. The collectors were responsible for 12% of the total investment costs, but mounting and system integration were expensive at 15%. In total the price per cubic metre and hour of air flow was similar to the German system at € 15 per $m^3 h^{-1}$.

From the total investment costs in the Maier factory project, an annuity of € 26 070 capital-related costs results. If the funding for the DEC investment costs of € 100 000 is taken into account, the annuity reduces to € 14 122. Consumption-related costs for heat, electricity and water occur together with the demand charge to provide a given electrical power. In total the annual consumption costs for the desiccant cooling system were calculated at € 3147, about 40% less than for a conventional air-conditioning system (see Figure 5.45). In the Spanish Mataró project, the savings calculated from electrical peak power cost reduction were € 4200 per year.

In addition, operation-related costs for maintenance and repair arise. Repair costs are usually between 1 and 3% of investment costs, so for the calculations, 2% was chosen. Maintenance costs are in a similar range: 76% of the total annual costs are capital costs, the rest are operation and consumption costs (see Figure 5.46).

The annual heating energy saving through the solar thermal collector field of about € 1500 can be subtracted from the total annual costs. The remaining cooling costs for the investigated year with 34 710 kWh of cold production result in a specific cold price of € 0.94 per kWh without funding and € 0.6 per kWh including the investment funding. By comparison, the costs for a conventional air handling unit with humidification and an electrical compression chiller were calculated in this project at € 0.65 per kWh. The high price per kilowatt hour is largely due to the low total cooling demand in the building. At a nominal power of the system of approximately 100 kW, the cooling

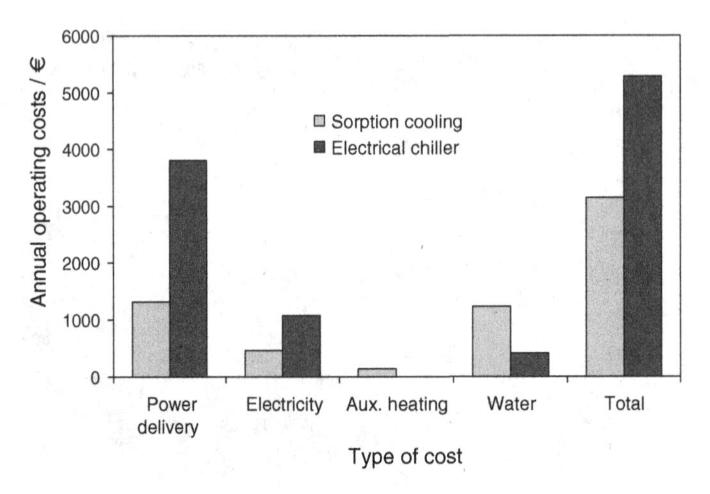

Figure 5.45 Annual operating costs for the installed desiccant cooling system compared with an electrical chiller

energy corresponds to only about 350 full load hours. In climates with higher cooling demand and 1000 load hours, the price could then go down to € 0.3 per kWh. An Austrian research team compared the costs of a district-heating-powered small DEC system (6000 m^3 h^{-1}) including an air handling unit with an electrical compression chiller. For 960 full load hours, the team obtained cooling costs for the DEC unit of about € 0.55 per kWh compared with € 0.51 to 0.56 per kWh for the electrical cooling system (depending on the tariff structure). Also, here the capital costs are about two-thirds of the total annual costs of the system (Simader and Rakos, 2005).

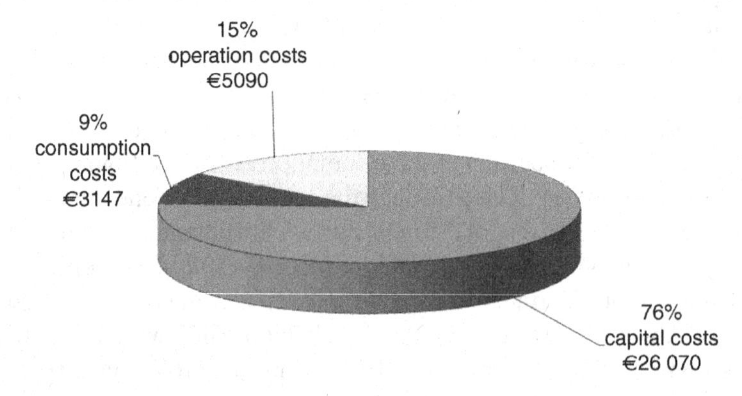

Figure 5.46 Distribution of total annual costs for the installed system including subsidy on the capital costs

5.2.7 Summary of Desiccant Cooling Plant Performance

Two large desiccant cooling systems with solar air collector energy supply were conceptually designed and monitored by the University of Applied Sciences in Stuttgart. Both systems work well with average coefficients of performance around 1.0, when all operating modes are included. If only full desiccant operation is considered, COPs are between 0.5 and 0.6. A detailed measurement analysis in the Althengstett system showed that COPs under full regeneration were around 0.37 if the air temperature measurement was taken directly at the DEC machine entry rather than on the roof, especially as the air channels are long and the entry air is precooled in the building. This reduces the total system efficiency and care should be taken to insulate the air channels well.

Using the full system simulation with validated component models, it can be shown that the cooling energy demand could be covered with very little auxiliary energy consumption. The main problem in real machine operation was that the regeneration air temperature was set to a fixed value. As the deviation of room temperature from the setpoint was often high, all components including full auxiliary heating were switched on, although pure solar operation with lower regeneration temperatures would have given the same supply air temperatures. Adjustable regeneration temperatures are absolutely necessary to reduce auxiliary energy consumption. The solar thermal air collectors have a relatively low energy yield during summer, although the efficiency is good at near 50% during full regeneration mode. The reason for the low thermal yield is that the number of regeneration hours is quite low (e.g. only 20% of the total operating hours or 83 hours during July in Germany), from which the collector can supply energy for 64% of the time. During the remaining hours solar irradiance is too low for a significant collector contribution. Increasing the collector surface area obviously increases the solar fraction during those days with high irradiance, but is a very costly measure for just a few hours of regeneration mode. To achieve higher solar fractions, the control strategy of the desiccant system must be adapted more to the solar air collector system, allowing full regeneration operation, whenever solar irradiance is available and using the building's heat and humidity storage capacity to dry and cool it down more than required by static setpoints.

5.3 New Developments in Low-Power Chillers

Whereas thermal cooling technologies are available on the market in the medium- and high-power range, there are hardly any systems below 10 kW of cooling power. However, these small air-conditioning units dominate today's market, which is almost exclusively covered by electrical compression systems. In order to supply this market with sustainable technologies the development of cost-effective solar or waste-heat-driven cooling systems is necessary.

The author's research team has therefore developed a thermally driven diffusion–absorption cooling machine (DACM) with ammonia/water and pressure-compensated auxiliary gas. For a residential building in Germany the annual primary energy-saving potential by using a solar thermal supported absorption cooling/heat pump system is over 50% compared with an electrical heat pump/chiller, at estimated operating hours of 700 h a^{-1} and 1800 h a^{-1} for cooling and heating operation, respectively.

Also, a new desiccant cooling cycle to be integrated in residential mechanical ventilation systems has been developed by the author. The process shifts the air treatment completely to the return air side, so that the supply air can be sensibly cooled by a heat exchanger. Purely sensible cooling is an essential requirement for residential buildings with no maintenance guarantee for supply air humidifiers. As the cooling power is generated on the exhaust air side, the dehumidification process needs to be highly efficient to provide low supply air temperatures.

5.3.1 Development of a Diffusion–Absorption Chiller

In 1899 H. Geppert was the first to propose the introduction of an auxiliary inert gas (non-condensable gas) into the Carré absorption cooling cycle resulting in a third cycle, the so-called auxiliary gas cycle. By heat input to the generator, the gas is first displaced out of the generator and then out of the condenser. It collects inside the evaporator and the absorber, where its pressure compensates for the refrigerant partial pressure difference between the generator and absorber and between the condenser and evaporator. This special kind of absorption cooling machine is called a diffusion–absorption cooling machine (DACM). The usual throttle for the pressure compensation is removed and the solution pump can be a simple bubble pump because the whole unit now has the same total pressure p_{total}. Geppert suggested using air as the inert gas, but the COP that he reached was too low for commercial use. This was due to two reasons; the first one is that the diffusion of ammonia into air is very slow so that only a very small cooling capacity can be achieved. The second and most important reason is that the gas mixture of air and ammonia is less dense than air alone and because of this it was not possible to get a thermosyphon circulation of the gas mixture. For these reasons, Geppert had to use a ventilator for the circulation of the gas cycle.

In 1922 the two Swedish engineers von Platen and Munters developed the idea of a DACM with pressure-compensated auxiliary gas and no mechanically moving parts independently of Geppert's proposal. They suggested hydrogen as the auxiliary gas and at first they also obtained low COPs. Hydrogen alone is lighter than the gas mixture of ammonia and hydrogen, and with this mixture a thermosyphon circulation is possible. With the use of a gas heat exchanger between the evaporator and absorber, von Platen and Munters were able to reduce the cooling losses in the auxiliary gas cycle and thus reached higher COP values. Consequently, this diffusion–absorption cooling technology became interesting for commercial use for example, for absorption refrigerators. Later, helium was also used as the auxiliary gas.

Conventional gas- or electrically driven diffusion–absorption refrigerators (DARs) with their directly powered generator/gas bubble pump were theoretically and experimentally investigated in numerous research projects for the operative range of refrigeration as well as air-conditioning (Watts and Gulland, 1958; Stierlin, 1964; Narayankhedkar and Maiya, 1985; Kouremenos *et al.*, 1994; Chen *et al.*, 1996; Smirnov *et al.*, 1996; Srikhirin and Aphornaratana, 2002). The Electrolux AB company, Sweden (today Dometic AB), brought out the first refrigerator in 1925 (Niebergall, 1981). The cooling power of these diffusion–absorption refrigerators is between 40 and 200 W. Another company, Servel Inc. in Evansville, USA (today Robur SpA, Italy), produced refrigerators with small capacity under licence starting in the 1930s and later, according to their own design, Servel developed and produced a gas-driven heat pump in 1937. This pump was water cooled and the auxiliary gas used was hydrogen. The unit had a COP of 0.192 with a cooling capacity of 1.75 kW and a heating capacity of 9.12 kW (Plank and Kuprianoff, 1960).

In the 1990s, these domestic DARs were modified and improved for use as directly heated, gas-driven diffusion–absorption heat pumps (DAHPs). Values of the COP for heating applications, COP_{heat}, were between 1.2 and 1.35 for a heating capacity of 80 to 205 W and with a cooling capacity between 25 and 51 W (Herold, 1996). Another group of researchers developed a directly gas-heated DAHP with a heating capacity between 3.0 and 3.5 kW at heating temperatures of 150 °C and evaporator temperatures ranging from −15 to +5 °C (Schirp, 1990; Stierlin and Ferguson, 1990). The COP for heating was between 1.4 and 1.5. The industrial conversion of this directly heated, gas-powered DAHP was carried out by BBT Thermotechnik GmbH, formerly Buderus Heiztechnik GmbH, in combination with a condensing boiler for a near-market unit, but is not yet commercially available. The output heating capacity of the DAHP is approximately 3.6 kW at COP_{heat} of 1.5 and it requires a 1.2 kW power input out of environmental heat through the evaporator (e.g. by a solar air collector) and a 2.4 kW heating capacity through the gas burner/generator (Schwarz and Lotz, 2001). Another industrial conversion of the DAHP has been done by Entex Energy Ltd. The company realized DAHPs with a 2.6 kW up to 8.0 kW heating capacity with COP_{heat} of about 1.5. The company also realized a gas-driven DACM with a 1.0 to 3.5 kW cooling capacity (Entex, 2004).

Current thermally driven DACMs with ammonia/water (NH_3/H_2O) and a pressure-compensated auxiliary gas circuit (helium or hydrogen) are only commercially used in the smallest power range up to 100 W. The priority operating criterion is the absolute noiselessness (hotel mini-bars) and the autonomous power supply (camping gas refrigerators).

Some prototypes of commercial absorption refrigerators with indirectly solar-powered generators and hydrogen or helium as the inert gas have also been experimentally and theoretically investigated. In these studies, COPs of 0.2 to 0.3 and cooling capacities between 16 and 62 W were reached at heating temperatures between 160 and 230 °C and evaporator temperatures of −6 down to −18 °C (Keizer,

1979; Bourseau et al, 1987; Gutierrez, 1988; Ajib and Schultheis, 1998). One research group used the DAHP of the BBT company and modified the DAHP by substituting the direct gas-fired generator by an indirectly heated one (Braun and Hess, 2002; Stürzebecher et al, 2004). The solar thermal heating capacity of 1.8 kW is provided by VTCs. The cooling capacity is approximately 1 kW and the COP is 0.59 at a heating temperature of 175 °C and an evaporator temperature of 2 °C.

Considering the circumstances that no suitable indirectly driven ACMs with small-scale cooling performance (1 to 5 kW) are available on the market, the University of Applied Sciences in Stuttgart has developed a solar-heated SE ammonia/water DACM with a design cooling capacity of 2.5 kW and helium as the inert gas (Jakob *et al.*, 2003). A detailed investigation of the performance potential has been carried out as well as the experimental characterization and optimization of a first prototype and a second improved one, especially of the newly developed, indirectly heated generator/bubble pump. A third prototype is currently in the testing phase. Moreover, the performance of the optimized DACM pilot plant is theoretically analysed and evaluated using a detailed expanded component model of the DACM process which was developed based on the characteristic equation of sorption chillers.

Design of the Pilot Plants

An indirectly powered generator/gas bubble pump is used, which has the function of a solution pump and a thermal compressor. The core components of a DACM are the generator, condenser, evaporator and absorber as shown in Figure 5.47.

The solution heat exchanger (SHX), the gas heat exchanger (GHX) in the auxiliary gas circuit and the dephlegmator for the condensation of the evaporated solvent are vertical tubular, plate or coaxial heat exchangers, which are welded hermetically tight.

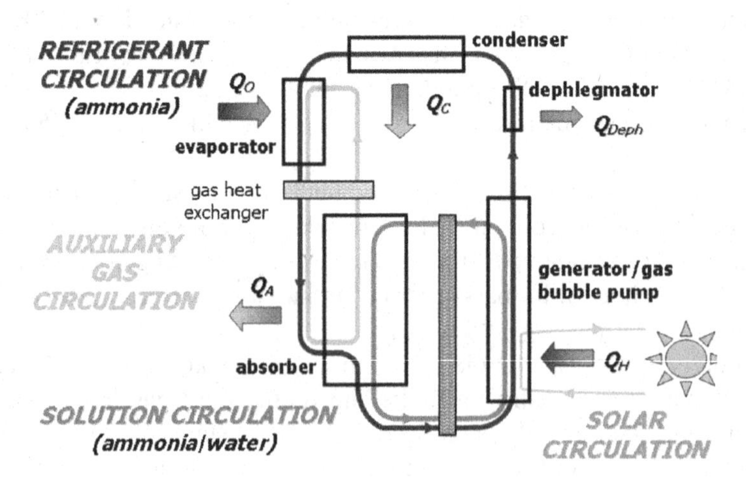

Figure 5.47 Configuration of a diffusion–absorption chiller

Table 5.4 Design conditions for a diffusion–absorption chiller

Generator	Heating capacity	5.2 kW
	Heating water in/out	130/120 °C
Dephlegmator	Recooling capacity	0.9 kW
	Recooling water in/out	42/47 °C
Condenser	Recooling capacity	2.8 kW
	Recooling water in/out	35/37 °C
Evaporator	Refrigerating capacity	2.5 kW
	Cold brine in/out	13/10 °C
Absorber	Recooling capacity	4.0 kW
	Recooling water in/out	30/35 °C
COP		0.48

The used working pair for the solution circulation is an ammonia/water mixture. The inert auxiliary gas used for the pressure compensation between the high and low ammonia partial pressure level and for the gas circulation is helium.

The prototypes of the single effect solar-driven DACMs are designed for the application area of air-conditioning as water chillers with an evaporator temperature of 6–12 °C as well as for the operation of cooled ceilings with an evaporator temperature of 15–18 °C. Design temperatures and power levels are summarized in Table 5.4.

Gas Bubble Pump Performance

The directly driven bubble pump of DARs usually consists of a single lifting tube where the heat input is restricted to a small heating zone by a heating cartridge or the flame of a gas burner with a high heat flux density. The indirectly driven DACM consists, on the other hand, of a bundle of tubes where the heating zone is spread and lower heat flux densities are available. The developed bubble pumps of the DACM pilot plants are basically vertical shell-and-tube heat exchangers where the solution flows inside the tubes of small circular cross-section, forming slug flow at best, and the heating medium flows through baffled tube bundles on the shell side.

The operation of the gas bubble pump is based on internal forced convection boiling, commonly referred to as two-phase flow, and is characterized by rapid changes from liquid to vapour in the flow direction. By external heating of the vertical pipes of the bubble pump, the pipes are surrounded by a heat transfer medium such as a water–glycol mixture or thermo-oil. The ammonia/water solution rests in the bundle of pipes in such a way that when the tube walls are overheated ammonia is expelled from the solution. Thus, bubbles are formed at the inner surface of the tube walls under constant heat supply. In this case, the procedure is called flow boiling. The internal flow boiling has to be distinguished by five two-phase flow regimes (bubbly flow, slug flow, annular flow, transition flow and mist flow) as shown in Figure 5.48.

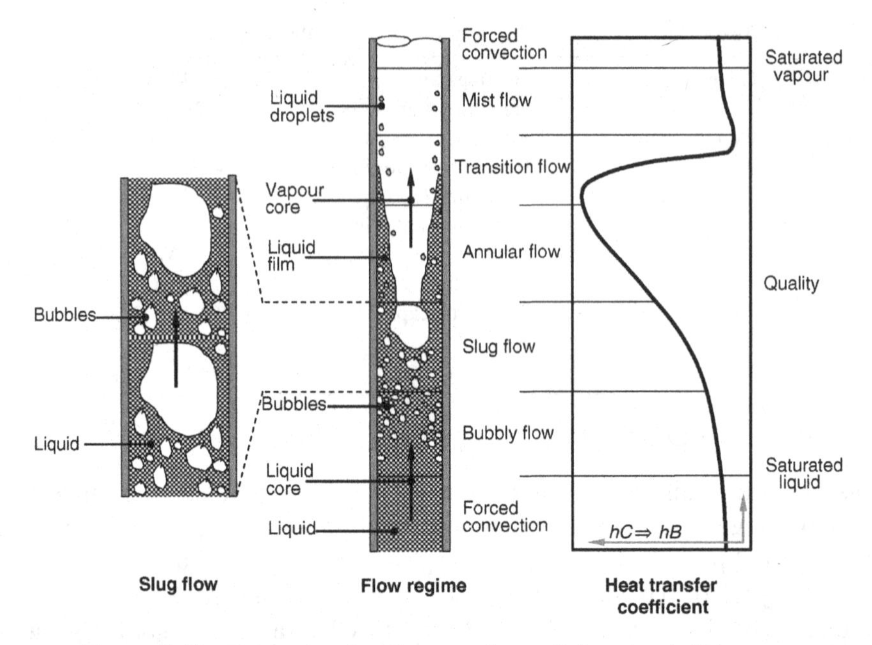

Figure 5.48 Flow regimes and heat transfer coefficients in a bubble pump

For the generators investigated only the slug flow regime is relevant. At low heat flux densities, slug flow is formed by more and more germ cells. The bubbles form plugs which engage the whole circular tube section. This flow regime must be reached in order for the operation of the DACM to be able to guarantee the solution lifting and with that the solution circulation.

For the characterization of the gas bubble pump, five different generators were constructed and compared. The measured total heat transfer capacities were between 0.29 and 0.12 kW K^{-1}, which is too low to obtain small temperature differences across the generator. The main limitations on increasing the heat transfer power are the costs of too many tubes, the high pressure drops on the external side and especially leakage flows around the baffles, which are fixed to the tubes and cannot be closely sealed to the external wall. In the third prototype the number of tubes was therefore increased by 30%. The values of the investigated range of generator heating temperatures and mass flow rates are presented in Table 5.5.

The measured solution concentrations of the first pilot plant were 31% for X_{Sw} and 44% for X_{Sr} with a degassing width of 13%. At the same pressure, but a lower degassing width, the specific solution circulation index in prototype 2 is about 1.5 to 2 times higher than that of the generator of prototype 1.

During the measurements of prototype 1, the volume flow for the generator was determined for different heating inlet temperatures as well as transferred heat capacities. The measuring method for the acquisition of the volume flow of the lifted

Table 5.5 Pressure levels, concentration ranges and solution circulation in the large generator of prototype 1 and the smaller prototype 2 generator

	Prototype 1		Prototype 2	
Total pressure/10^5 Pa	20	18	20	12
NH_3 initial concentration/%	38	30	40	37
Rich solution concentration X_{sr}/%	45–42	30	40	37
Weak solution concentration X_{sw}/%	31–34	24–26	31–33	24–29
Degassing width E/%	14–8	6–4	9–7	13–8
Vapour concentration X_{V1}/%	88–91	82–85	89–91	84–89
Specific solution circulation f/-	4–7	11–15	6.5–8.5	4.5–7.5

weak solution, called the temperature flank method, is described in detail by Biesinger (2002). For the investigation of the generator, the lifting ratio b, which is the ratio of lifted liquid volume $\dot{V}_L$ to simultaneously expelled vapour volume flow $\dot{V}_V$, is an important value used to quantify the performance of the generator (Cattaneo, 1935)

$$b = \frac{\dot{V}_L}{\dot{V}_V} \tag{5.6}$$

Another important characteristic value of the generator is the specific solution circulation index f. This index is defined as the ratio of the mass flow rates of the rich solution $\dot{m}_{Sr}$ to the ammonia vapour $\dot{m}_V$, or alternatively as the ratio of the vapour concentration X_{V1} and weak solution concentration X_{Sw} difference to the degassing width E (which is the difference between the weak and the rich solution concentration X_{Sw} and X_{Sr})

$$f = \frac{\dot{m}_{Sr}}{\dot{m}_V} = \frac{X_{V1} - X_{Sw}}{X_{Sr} - X_{Sw}} \tag{5.7}$$

An evaluation of the measurement results of the volume flows was done with the corresponding measured pressures and vapour concentrations. For the maximum measured volume flow of the lifted weak solution of $53\,\mathrm{l\,h^{-1}}$, the evaluated vapour volume flow is $12.2\,\mathrm{l\,h^{-1}}$ and thus the maximum lifting ratio is 4.3 when the external generator inlet temperature is 165 °C. Figure 5.49 shows the three curves of the measured liquid volume flow, the calculated vapour volume flow and the lifting ratio of prototype 1. The volume flow of the vapour was continually increased because of the changing flow regime. The slug flow collapsed and converted to annular flow, and thus nucleate boiling changed to film boiling. Therefore, at heating temperatures above 165 °C, more vapour is generated but the solution lifting decreases.

Converted to mass flow, the lifted weak solution volume flow shows a maximum of $42.0\,\mathrm{kg\,h^{-1}}$ and a specific solution circulation of 8.0 at an external generator inlet

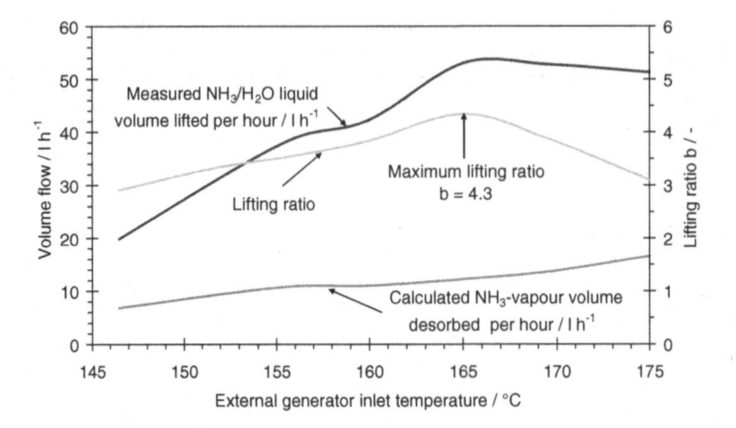

Figure 5.49 Liquid and vapour volume flows and lifting ratios as a function of the external generator inlet temperature

temperature of 165 °C. The design mass flow of the weak solution for a 2.5 kW evaporator cooling capacity is 50 kg h^{-1}, and the ammonia vapour mass flow is 8 kg h^{-1}.

Another way to analyse the performance of the bubble pump without measuring the volume flow is to calculate the solution mass flows. For that, first the ammonia vapour mass flow $\dot{m}_V$ has to be calculated. The vapour mass flow is defined as follows:

$$\dot{m}_V = \frac{\dot{Q}_C}{c_{V,NH_3}\left(T_{VC} - T_{C,s}\right) + h_V - h_L + c_{L,NH_3/H_2O}\left(T_{C,s} - T_{LC}\right)} \tag{5.8}$$

Here, $\dot{Q}_C$ is the condenser capacity, c_{V,NH_3} is the specific heat capacity of ammonia vapour, T_{VC} and T_{LC} are the measured condenser vapour inlet and liquid condensate outlet temperatures, $T_{C,s}$ is the condensation temperature, $c_{L,NH_3/H_2O}$ is the specific heat capacity of liquid ammonia and h_V and h_L are the vapour and liquid enthalpies. The weak $\dot{m}_{Sw}$ and rich $\dot{m}_{Sr}$ solution mass flows are then calculated as follows:

$$\dot{m}_{Sw} = \left(\frac{X_V - X_{Sr}}{X_{Sr} - X_{Sw}}\right)\dot{m}_V \tag{5.9}$$

$$\dot{m}_{Sr} = \left(\frac{X_V - X_{Sw}}{X_{Sr} - X_{Sw}}\right)\dot{m}_V \tag{5.10}$$

X_{Sr} is the rich solution mass concentration, which is given by the initial mass concentration. X_{Sw} is the weak solution mass concentration, which is calculated using a modified Clausius–Clapeyron equation (Bourseau and Bugarel, 1986) relating the generator weak solution concentration, its vapour pressure and the weak solution outlet temperature. X_V is the vapour concentration, which is dependent on the pressure and

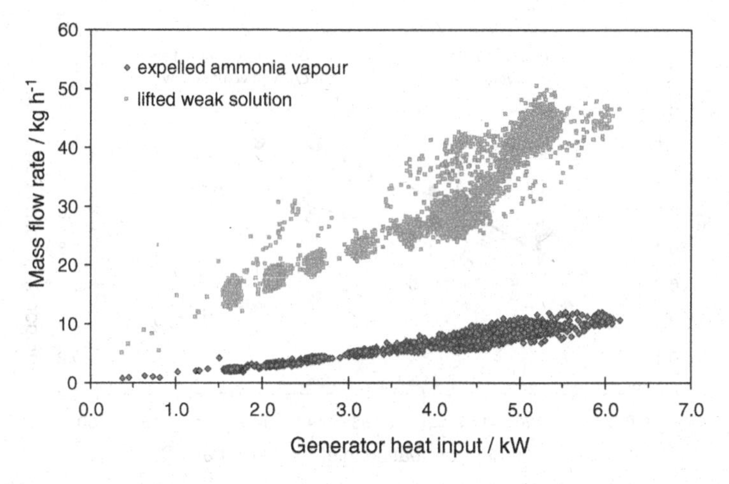

Figure 5.50 Mass flow of ammonia vapour and lifted solution as a function of generator power

the weak solution or the rectified liquid concentration, and has been calculated using empirical equations from Jain and Gable (1971).

For prototype 1, the calculated solution mass flows for external generator inlet temperatures of 147 to 172 °C resulted in values of 20 to 40 kg h^{-1} for the weak solution and 25 kg h^{-1} up to 46 kg h^{-1} for the rich solution. The values were obtained for an external generator volume flow of 21 l min^{-1} and an average total pressure of 20×10^5 Pa. For the operation of prototype 2, lower external generator inlet temperatures of 115 to 150 °C were sufficient. The calculated average mass flow rates were between 14 and 45 kg h^{-1} (see Figure 5.50) for the weak solution and between 17 and 53 kg h^{-1} for the rich solution. The expelled ammonia vapour which was produced by the generator of prototype 1 was determined from all of the measurements between 4 and 8 kg h^{-1}. This should have given sufficient cooling capacity, but in the first pilot plant the auxiliary gas circulation was insufficient and a fast ammonia saturation occurred in the evaporator. Consequently, only limited evaporator cooling capacity could be generated. The ammonia vapour mass flow of the generator of prototype 2 ranged between 1 and 4 kg h^{-1} during the first measurements. After modification of the plate heat solution heat exchanger (SHX) to a coaxial heat exchanger, the ammonia liquid mass flow reached higher values between 2.5 and 10 kg h^{-1} depending on the external generator inlet temperature and the volume flow. Therefore, the design values were also obtained with the last set-up of the second pilot plant.

The developed bubble pumps worked over a wide range of temperatures as well as external heating volume flow rates. Figure 5.51 shows a comparison of COPs and evaporator cooling capacities at a high and a low generator external heating volume flow of prototype 2 with coaxial SHX. The cooling capacities were approximately 30% higher when a high volume flow rate was used, enhancing the heat transfer.

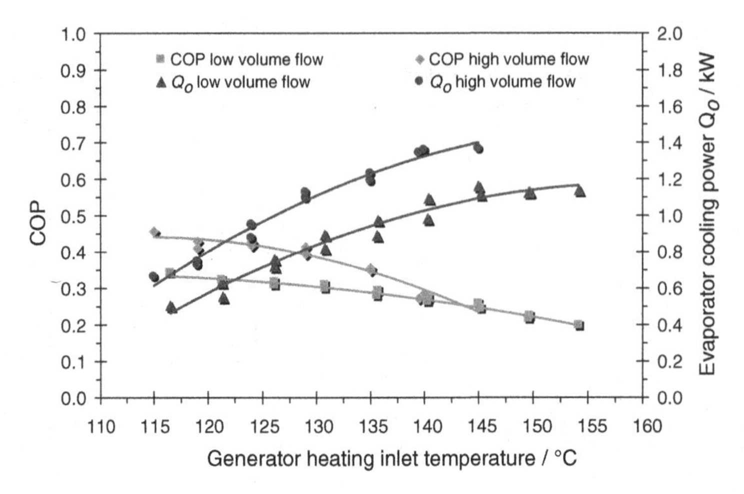

Figure 5.51 Influence of external mass flow in the generator on cooling power and COP

For the indirectly driven generator of the DACM, the limit values for the inner tube diameter are presented in Figure 5.52 for the transition from slug to annular flow and Figure 5.53 for the transition from bubbly to slug flow. The maximum value of the tube diameter for maintaining slug flow increases with increasing surface tension and decreases with liquid density. The minimum tube diameter for preventing bubbly flow increases with the desired vapour volume flow and decreases with increasing surface tension. The equations derived by Chisholm, Taitel and others are summarized in Jakob (2006). Slug flow usually occurs within tubes that have an inner diameter ranging from 5 to 41 mm (Jakob, 2006). All calculations were done for a generator with 19 tubes.

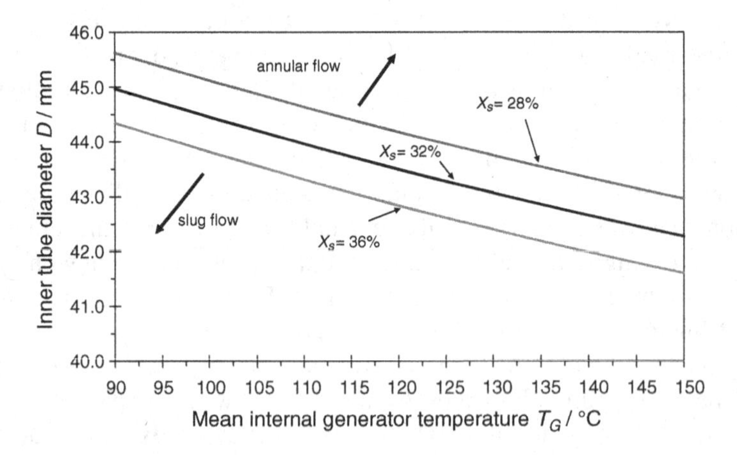

Figure 5.52 Maximum inner tube diameter as a function of generator temperature for the transition between slug and annular flow

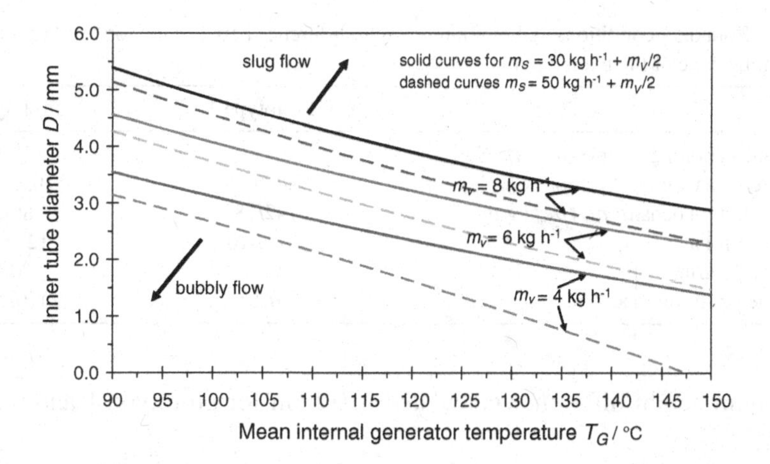

Figure 5.53 Transition from slug to bubbly flow as a function of mean generator temperature for different tube diameters

A simple expression given by Narayankhedkar and Maiya (1985) relates the required minimum height difference Y between the rich solution level in the reservoir and the level of the heating zone at which the bubbles are formed in the bubble pump for the circulation. The derivation of the minimum height difference equation with the coefficients K_1 and K_2 is described in Chen *et al.*, (1996). The minimum height difference increases with the bubble pump lifting height Δh and decreases with the mean generator boiling temperature T_b. The analysis is based on the assumptions that all bubbles are formed at the bottom of the minimum height difference, that there is no relative velocity between the weak solution and the refrigerant vapour bubbles in the tube, and the friction loss and vapour pressure of the absorbent water are negligible:

$$Y = \frac{p\Delta h}{\rho_{Sr}\left(K_1 p + K_2 10^3\right) - p} \tag{5.11}$$

where

$$K_1 = \frac{1 - X_{Sr}}{(1 - X_{Sw})\,\rho_{Sw}}$$

$$K_2 = \frac{(X_{Sr} - X_{Sw})\,RT_b}{(1 - X_{Sw})\,M_{NH_3}}$$

For a relative molecular mass M_{NH_3} of the refrigerant ammonia of 17 kg kmol^{-1} and the given data in Table 5.6 for mean boiling temperatures, rich and weak solution concentrations and densities, total pressures and the two different lifting heights,

Table 5.6 Boundary conditions and minimum height difference between the top level of the reservoir and the heating zone of the bubble pump

	Prototype 1	Prototype 2
Generator mean boiling temperature $T_b/^\circ C$	130	110
NH_3 rich/weak mass concentration X_{Sr}/X_{Sw}	0.42/0.34	0.37/0.27
Rich/weak solution density ρ_{Sr}/ρ_{Sw} kg m^{-3}	742/783	789/837
Total pressure p/Pa	20×10^5	12×10^5
Lifting height Δh/m	1.0	0.5
Minimum height difference Y/m	0.12	0.03

the minimum height difference calculated is 0.12 m for prototype 1 and 0.03 m for prototype 2.

Results from Prototype 1 The first pilot plant with a total height of 3.70 m and approximately 800 kg weight went into operation in November 2000. A series of measurements were taken with an indirect, liquid heating system at generator heating inlet temperatures of 150 to 175 °C and evaporator temperatures between 25 and 0 °C. The measurements were taken with and without the dephlegmator. Maximum COPs measured were 0.2 and the evaporator cooling capacity of the pilot plant was 0.5 to 1.5 kW, but the operation was not continuous. The evaporator capacity decreased with time as shown in Figure 5.54, as the auxiliary gas starts to get saturated with ammonia at an insufficient auxiliary gas circulation. This could not be attributed to excessive pressure drops in the auxiliary gas circuit: at auxiliary gas volume flows below 4 m^3 h^{-1}, the pressure drop calculated and measured after deconstruction of the prototype was only 1 Pa and the driving force calculated from the density difference between ammonia rich auxiliary gas and weak gas after the absorber is about 15 Pa for a 5 °C evaporation temperature and a height difference of 1 m between the evaporator and absorber. The initial peak of cooling power can be explained by evaporation of condensed ammonia within the tubes between the condenser and evaporator and the upper section of the evaporator. For the first hour, the evaporator inlet temperatures are very low at 8 °C from start-up values of 35 °C and then increase again.

A further effect explaining the performance decay with time is due to the absorber: as the solution heat exchanger had an included reservoir with an approximate capacity of 60 litres, the solution inlet to the absorber is initially at room temperature and only heats up slowly. In the second prototype, the reservoir was removed from the solution heat exchanger and reinstalled below the absorber.

Figure 5.55 shows the operating performance and the measured COPs and cooling capacity of prototype 1 for different generator heating water inlet temperatures. The measurements were taken with heating capacities between 10.0 and 13.5 kW at an operating pressure of 20 × 10^5 Pa. The maximum evaporator cooling capacity was 1.5 kW. The evaporator could not evaporate all the available liquid ammonia into

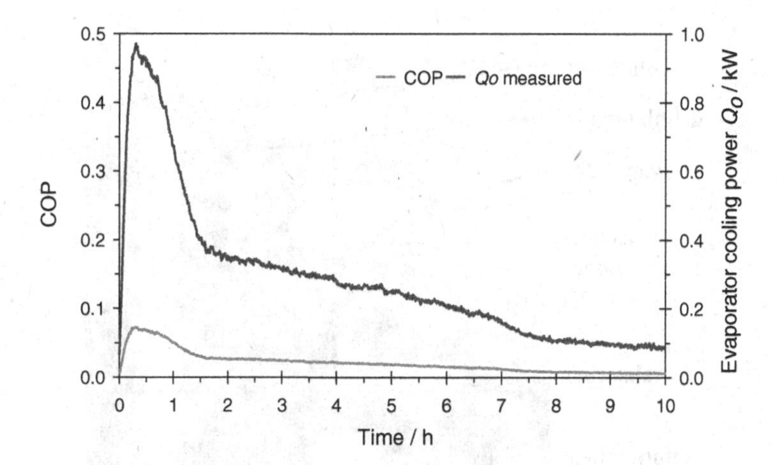

Figure 5.54 Cooling power and COP of the first prototype as a function of time with increasing saturation of the auxiliary gas

the helium gas atmosphere, so the cooling power did not increase with increasing generator power. Due to the low heat recovery factors Φ_{Sr} and Φ_{Sw} of the investigated tubular SHX of 39.6% and 51.1% (the low and high values correspond to the rich and weak solution sides respectively) the measured generator heating capacities were very high and consequently the COPs were very low.

Results from Prototype 2 A second compacted pilot plant was built based on the experiences and results from the first, using partly standard commercial components such as nickel-soldered plate heat exchangers and a coaxial heat exchanger for the condenser and the SHX, respectively. For this prototype, a further indirectly heated

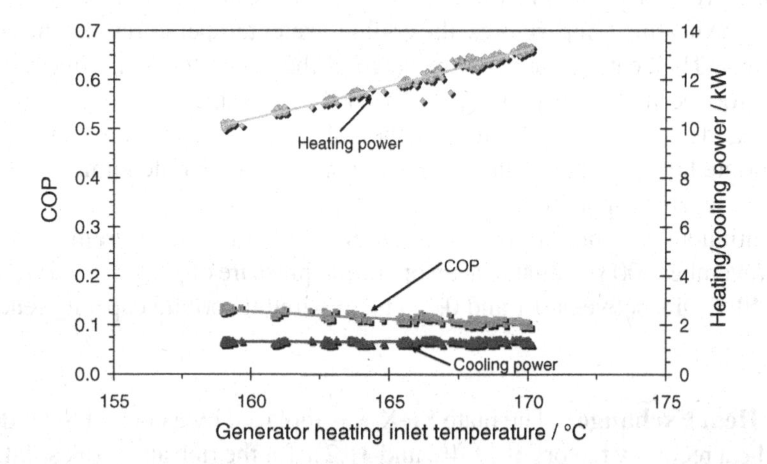

Figure 5.55 COP and heating and cooling power of the first prototype

Figure 5.56 View of the second prototype with its components

bubble pump was developed and the auxiliary gas circuit was constructively reworked to reduce pressure drops. These changes resulted in a weight reduction down to 290 kg and a height reduction down to 2.40 m (see Figure 5.56) which are important factors for a more marketable unit. The second prototype went into operation in July 2003 and ran until July 2005.

Measuring results of stationary temperature, pressure and capacity levels were taken by varying the heating temperatures, the cooling water temperatures and the cold brine temperatures. The heating temperature range of the generator was reduced from 150 to 175 °C for the first pilot plant to 110 to 140 °C for the second pilot plant. This is due to the decreased lifting height of the bubble pump by a half. Combined with an appropriate heat transfer surface, the efficiency of the bubble pump increased and temperature levels dropped.

The continuous evaporator cooling capacities of the first measurements from July 2003 to November 2003 evaluated at an operating pressure of 18×10^5 Pa were around 0.5 kW with COPs between 0.1 and 0.2. The maximum cooling capacity reached was 0.8 kW.

Solution Heat Exchange The plate SHX was replaced by a coaxial SHX due to the very low heat recovery factors of 11.4% and 31.2% for the rich and weak solution sides respectively. The heat recovery factors of the coaxial heat exchanger of prototype 2

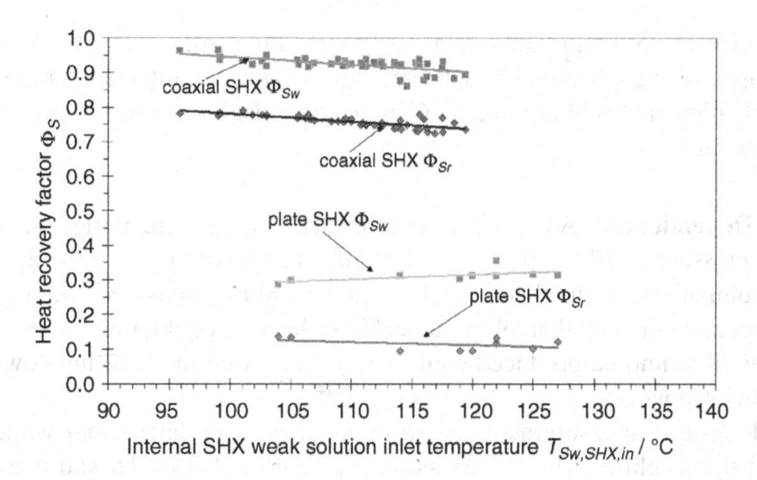

Figure 5.57 Heat recovery factors for plate and coaxial solution heat exchangers as a function of the weak solution inlet temperature

were within an acceptable range of 76% and 92%. Figure 5.57 shows a comparison of the different heat recovery factors measured for different SHX weak solution inlet temperatures.

Due to the low solution mass flow rates, the coaxial SHX had a better heat exchange than the plate heat exchanger and the tubular heat exchanger of prototype 1 which was also too heavy. A series of test runs were carried out following the above changes and the analysis showed an improved COP of 0.2 to 0.3 and continuous evaporator cooling performance of 0.5 to 1.5 kW (see Figure 5.58); this was achieved when it was run at an operating pressure of 20×10^5 Pa, a cooling water inlet to the absorber of 20 °C,

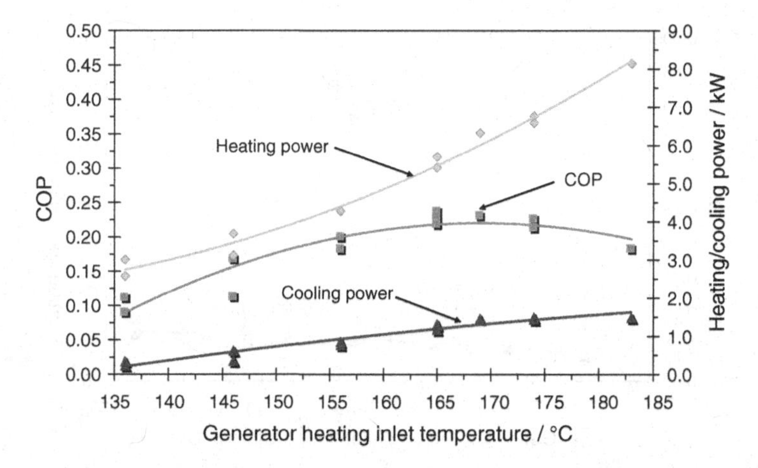

Figure 5.58 COP and heating and cooling power ranges as a function of generator temperature for the second prototype

to the condenser 35 °C and an evaporator temperature inlet of 25 °C. A maximum cooling capacity of 2 kW could be achieved if the evaporator inlet temperature was set to relatively high values of around 25 °C, indicating insufficient heat transfer capacity of the evaporator.

Pressure Dependence A further series of measurements were carried out at a lower operating pressure of 12×10^5 Pa. The COP decreases with rising generator temperature. A comparison of the theoretically possible and the measured cooling capacity of the evaporator showed that, above a generator heating inlet temperature of 125 °C, the amount of ammonia produced could not be converted into cooling power by the falling film evaporator.

Many long-term measurements of up to 11 days were carried out without shutting down the machine. The results showed a COP of 0.3 to 0.5 and a continuous evaporator cooling performance up to 1.6 kW at evaporator outlet temperatures for air-conditioning and cooled ceilings between 22 and 15 °C (see Figure 5.59). The lowest external evaporator outlet temperatures reached were −5 °C at an operating pressure of 8.5×10^5 Pa and a generator heating inlet temperature of 145 °C.

The improvement in performance can be mainly attributed to the lower total pressure of the system, which improves both the bubble formation process and the diffusion rates in the evaporator.

Evaporator Performance From the measured liquid ammonia mass flow, a theoretically possible evaporator cooling capacity was determined between 0.6 and 2.7 kW for heating temperatures between 115 and 155 °C. The measured cooling capacities were only between 0.6 and 1.5 kW. The falling film evaporator could not evaporate all

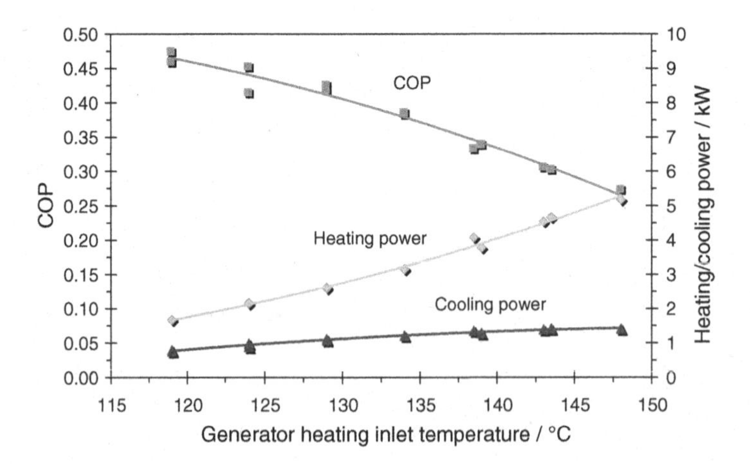

Figure 5.59 COP and heating and cooling power of the second prototype at a lower total pressure of 12×10^5 Pa

of the available liquid ammonia into the helium gas atmosphere, not even with high external evaporator inlet temperatures of around 25 °C. Therefore, the falling film evaporator needed to be rebuilt with either more or longer evaporation tubes so that a larger heat transfer surface would exist, which would lead to a smaller film thickness and, therefore, a longer delay time. Also, the issue of liquid distribution into the evaporator tubes plays an important role in the evaporator performance. At the given very low condensate flow rates it is extremely difficult to obtain an even distribution of liquid into all the tubes of the evaporator. Experiments with metal wicks and annular capillary systems are currently being carried out to improve the liquid distribution.

First Experimental Results from the Third Prototype The experimental investigations of the third prototype with marketable dimensions were focused on the combined evaporator-GHX-absorber unit and the newly designed generator (see Figure 5.60).

The heating temperature range of the generator was reduced from 150 to 175 °C for the first prototype to 110 to 155 °C for the second prototype and then to 100 to 150 °C for the third prototype. Very stable operation was achieved during the first experiments; however, cooling water temperatures and the cooling power obtained were rather low (see Figure 5.61).

The first experimental results gave evaporator cooling capacities from 0.7 up to 3.0 kW and COPs between 0.12 and 0.38. The evaporator temperatures chosen were 12/6 °C and 18/15 °C (see Figure 5.62). The cooling water temperatures at a constant

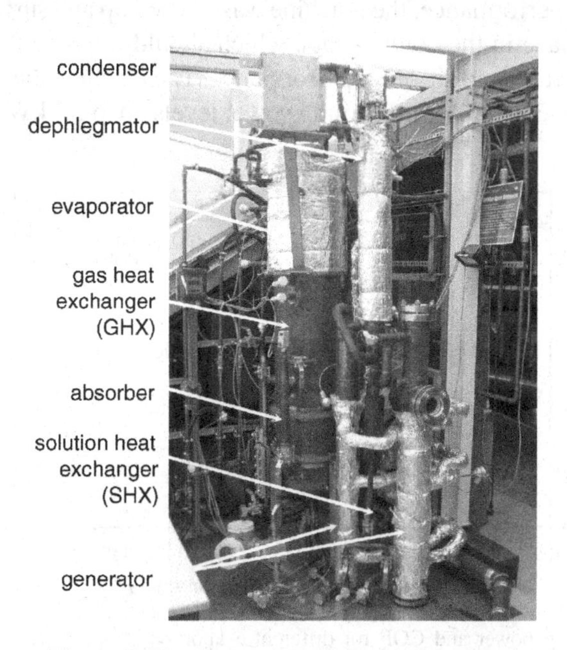

Figure 5.60 Components of the third prototype

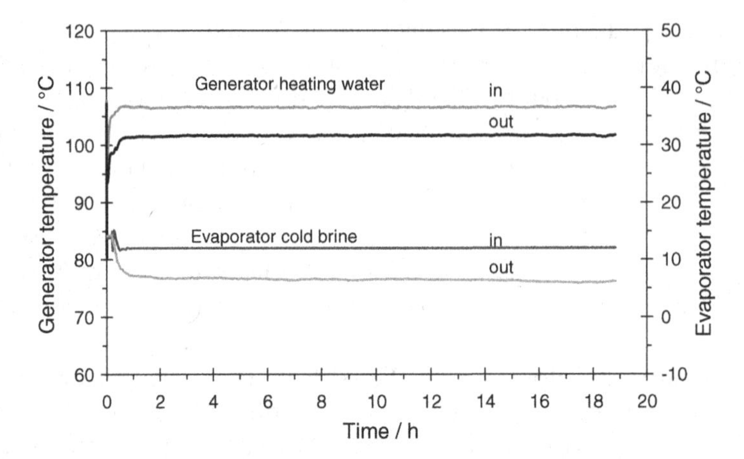

Figure 5.61 Temperature levels at 21.5 °C absorber inlet temperature and 24.4 °C condenser outlet temperature

external mass flow rate to the condenser and absorber slightly increased during the course of the measurements from 23 °C initially at the lowest generator temperature to 26 °C for the condenser and 29 °C for the absorber inlet temperature with temperature differences between the inlet and outlet of 2–3 K. The lowest logged external evaporator outlet temperature was −15 °C at a generator heating inlet temperature of 135 °C.

To improve the performance, the machine was opened up and surface corrosion was allowed to take place in the components, which should improve the surface wetting and thus the evaporator and absorber performance. The machine then operated reliably at COPs between 0.3 and 0.4 with cooling power levels up to 2.7 kW (see Figure 5.63).

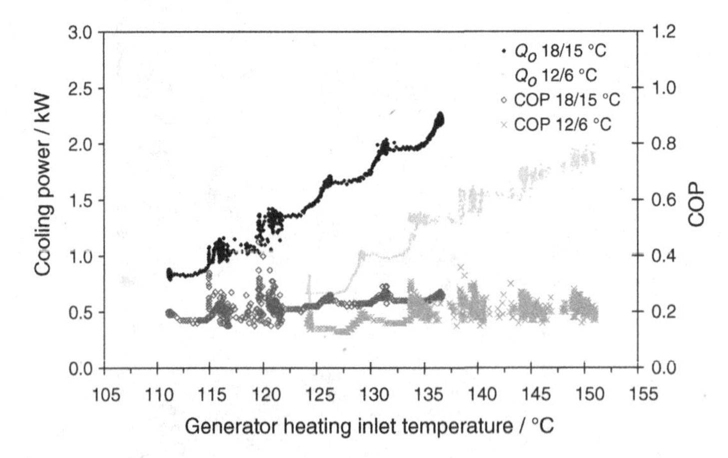

Figure 5.62 Cooling power and COP for different evaporator temperature levels as a function of generator inlet temperature

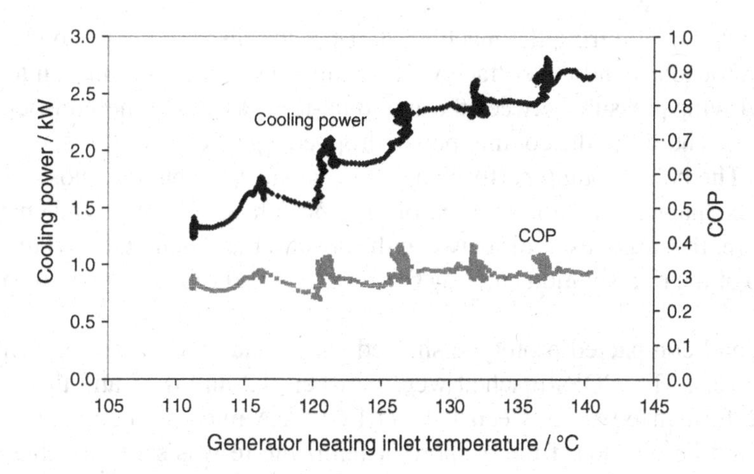

Figure 5.63 Cooling power and COP as a function of generator inlet temperature for the third prototype

During the course of the measurement sequence with generator temperatures increasing from about 110 to 140 °C and rather constant cooling water inlet temperatures of about 23 °C, the evaporator outlet temperatures decreased and the cooling water outlet increased by about 5 K (see Figure 5.64).

Summary of Experimental Performance The first DACM prototype reached COP values between 0.10 and 0.20 with an evaporator cooling capacity up to 1.5 kW. However, the auxiliary gas circulation was not high enough, leading to rapid saturation of evaporated ammonia. While evaporation took place between the condenser

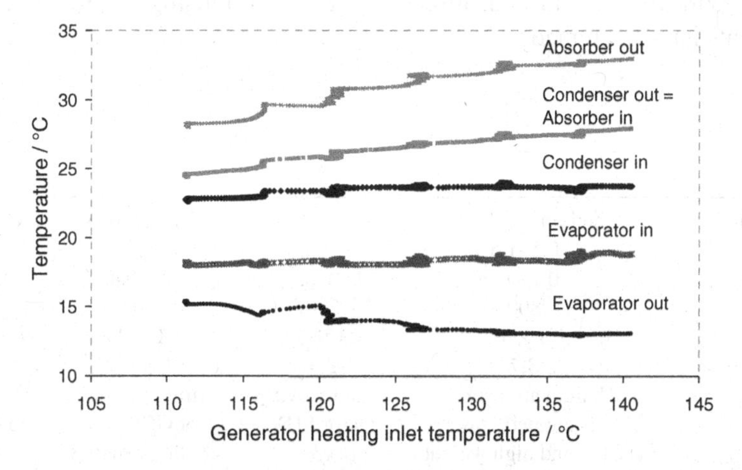

Figure 5.64 Temperature levels of all components during a measurement sequence with increasing generator temperatures

and evaporator inlet during the machine start-up, the distribution of condensate into the evaporator tubes and the surface wetting must have been too uneven to generate sufficient driving pressure between the ammonia-rich gas and the helium-rich gas after the absorber. Therefore the cooling power dropped significantly after about 1 hour of operation. The bubble pump performance was satisfactory, but the chosen shell-and-tube heat exchangers had a maximum of only 50% heat recovery efficiency (SHX). Furthermore, the large reservoir between the absorber and generator led to very slow heating up of the rich solution entering the generator and COPs were correspondingly low.

The second compacted prototype showed stable and continuous temperature and pressure levels. The COPs reached were between 0.2 and 0.45 and the continuous cooling performance was between 1.0 and 1.6 kW. A maximum cooling capacity of 2.0 kW could be reached if the evaporator temperature was set to a value of 25 °C. Again, condensate distribution into the evaporator tubes and surface wetting seemed to be the crucial problems of the second prototype, whereas the bubble pump performance was good. Coaxial solution heat exchangers gave much better performance results than the initially chosen plate heat exchangers.

The third prototype was set up in October 2005; this new prototype now has marketable dimensions. The design cooling power of 2.5 kW was reached at generator temperatures of 120–130 °C at COPs of nearly 0.4 (see Table 5.7).

An expanded, steady-state model of the DACM based on the characteristic equation of sorption chillers is described in the next chapter. It gives a good agreement between experimental and simulated data. Efficient evaporation with high surface wetting factors is essential for high performance.

Further development is required regarding the evaporator cooling capacity and COPs at lower heating inlet temperatures as well as weight and production cost reduction. With the second and third pilot plants, promising steps regarding these requirements have been made.

Table 5.7 Summary of prototype development

Prototype	1	2	2 mod.	3
Heat exchange	Shell-and tube SHX	Plate SHX	Coaxial SHX	Coaxial SHX
Cooling power/kW	0.5–1.5	0.5–0.8	0.5–2.0	0.7–3.0
COP	0.1–0.2	0.10–0.15	0.20–0.45	0.12–0.38
Weight/kg	800	290	240	240
Area/m^2	1.5 × 1.5	0.8 × 0.8	0.8 × 0.8	0.6 × 0.6
Height/m	3.7	2.4	2.4	2.2
Results	Helium atmosphere saturated, low COP and high weight	Heat recovery factor SHX <15%	Improvement of COP and cooling capacity	Design cooling power reached 100–150
Generator temp. / °C	150–175	110–155	110–155	

5.3.2 Liquid Desiccant Systems

As buildings with low energy demand are often equipped with mechanical ventilation systems, it is useful to consider air-based thermal cooling technologies, such as open desiccant cooling systems (DCSs), for low-power applications. Small desiccant rotors, heat exchangers and humidifiers are available on the market for volume flows in the range of typical mechanical ventilation systems (around $300\,m^3\,h^{-1}$ fresh air supply for single family houses). Liquid sorption systems have also been patented for such applications (Bachofen, 1999), but there is no system yet on the market and only limited experimental results are available (Saman and Alizadeh, 2002). Both conventional desiccant cooling units and liquid sorption systems dehumidify the outside air, which is then precooled, humidified and injected into the rooms. The direct humidification of inlet air still causes concerns about hygiene, especially if maintenance is not guaranteed, which is generally the case in residential buildings.

To avoid any hygienic problems for low-power cooling applications, a new system configuration is proposed here, which shifts the whole air treatment to the exhaust air side and uses only sensible cooling for the outside air stream. The room exhaust air is dehumidified by a desiccant wheel or by liquid desiccant systems, precooled in a heat exchanger using an additional humidified air stream and further cooled by direct humidification. Finally the cooled return air is used to cool the supply air in an efficient heat exchanger.

The disadvantage of the proposed system is the requirement of an additional heat exchanger to transfer the generated cooling power and additional air volume flows to provide the precooling after the sorption process. The additional heat exchanger reduces the available cooling power and the additional volume flow increases the pressure drop and fan power. In order to reach a cooling performance comparable with the conventional system, highly efficient dehumidification combined with an effective removal of the condensation/absorption enthalpy is therefore essential.

The driving force for mass transfer in the dehumidification process is the difference in water vapour pressure between process air and liquid desiccant. The best performance is achieved if the desiccant material maintains a low vapour pressure during the whole absorption process. As the water vapour pressure of desiccants, here solid silica gel or concentrated salt solutions, increases with temperature, it is essential to remove the heat generated during the absorption process. Heat removal during the absorption process can be achieved using finned cooling coils as the contact surface in liquid desiccant systems (Lävemann *et al.*, 1993; Öberg and Goswami, 1998a; Park *et al.*, 1994a,1994b,1994c). However, an additional circuit for cooling water combined with a cooling tower is necessary, which increases the costs of the system.

Considering these facts, a cross-flow type of plate heat and mass exchanger used as a direct contactor for liquid desiccants seems to offer the best opportunities (Saman and Alizadeh, 2001). The channels of the proposed heat and mass exchanger are sprayed with liquid desiccants for dehumidification on the return air side and with water for

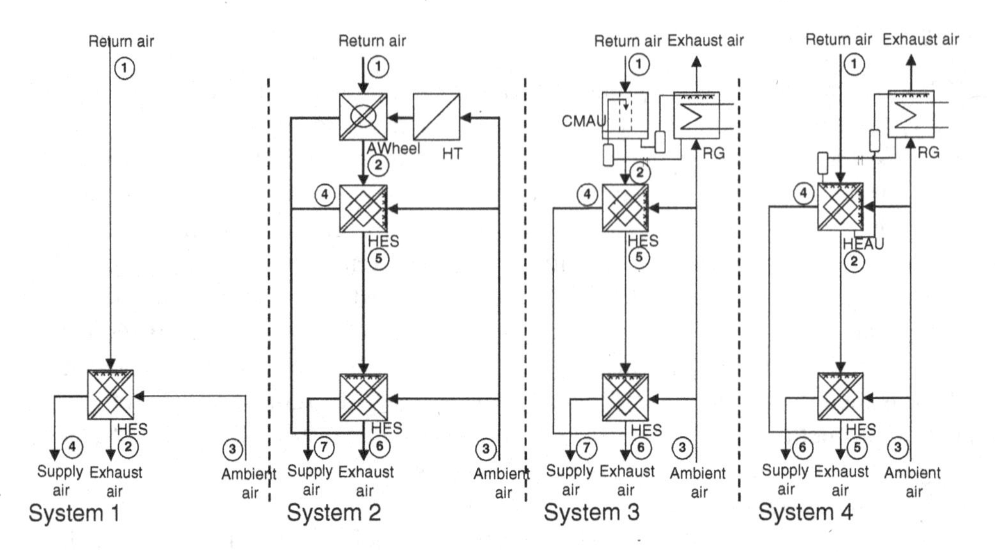

Figure 5.65 Schematics of the indirect supply air cooling system for the exhaust air humidification (system 1), return air desiccant rotor (system 2), contact matrix absorber (system 3) and fully integrated heat exchanger absorber unit (system 4). HES, water-sprayed heat exchanger; HT, heater; HEAU, heat exchanger absorber unit; CMAU, contact matrix absorber unit; A wheel, desiccant rotor; RE, regeneration unit

indirect evaporative cooling on the ambient air side, each liquid flowing parallel to the air stream. A second cross-flow heat exchanger with evaporative cooling combines the heat exchanger and evaporator so that only three instead of six components, as shown in Figure 5.65, remain. The performance of such spray-cooled cross-flow heat exchangers, with and without liquid sorption, is theoretically and experimentally analysed in the following, using different salt solutions.

System Technology Options

Four different technological options for return air cooling were investigated with increasing integration of components and improvement of performance, which are all shown in Figure 5.65.

System 1 In the simplest reference cooling system available on the market, the supply air is cooled by the return air in a cross-flow heat exchanger, which is sprayed with water on the return air side for evaporative cooling. This water-sprayed cross-flow heat exchanger combines the function of a humidifier and a heat exchanger. During the humidification process of the return air in the water-sprayed heat exchanger, heat is transferred from the supply air to the humidified return air. This allows the evaporation of a greater amount of water, resulting in higher cooling performances compared with systems with separated humidification and heat exchanging.

System 2 In the basic desiccant cooling system, a desiccant rotor available on the market is used for return air dehumidification. After dehumidification the return air is precooled by ambient air in a water-sprayed cross-flow heat exchanger to remove the adsorption enthalpy and the heat transferred from the regeneration side of the desiccant rotor. The precooled return air is then injected into a second water-sprayed cross-flow heat exchanger, where the return air is humidified by the water and heated by the supply air, which is thereby cooled in the channels of the cross-flow heat exchanger.

System 3 Here the desiccant rotor is replaced by a liquid desiccant sprayed onto a contact matrix absorber unit. A regeneration unit for the liquid desiccant replaces the air heater of system 2; all other components stay the same as described for system 2. The concentrated liquid desiccant is sprayed onto the contact matrix absorber unit, where it is brought into contact with the return air for dehumidification. During the absorption process the concentration of the liquid desiccant decreases, while the temperature increases. After passing the contact matrix absorber unit the weak liquid desiccant is stored in a storage tank before it is sprayed onto a regeneration unit, where the solution is heated and the absorbed water is removed. After regeneration the humidified ambient air (exhaust air) after the first water-sprayed heat exchanger can be used to cool the hot and strong liquid desiccant in a liquid to air heat exchanger before it is stored in a second storage tank.

System 4 In the most integrated system, the liquid desiccant is sprayed directly into a cross-flow heat exchanger, which also integrates a spray humidifier on the secondary side. The temperature reduction of the liquid sorption material through secondary water evaporation during the absorption process improves the absorption performance and lowers the supply air temperature even further. Together with the second water-sprayed heat exchanger already used in systems 2 and 3, the number of components is now reduced to three. The circuit of the salt solution is similar to system 3.

Experimental Set-up for Performance Analysis of Integrated Components

An experimental set-up was built to condition air (temperature, humidity) for both the outside and return air side. Experiments were carried out for return air from the conditioned space at 26 °C dry bulb air temperature and 50 to 70% relative humidity and for ambient air at summer conditions at 32 °C dry bulb air temperature and 40% relative humidity.

To compare the performance of the four systems described in the previous sections, the return air dehumidification of the desiccant wheel, the contact matrix absorber unit and the heat exchanger absorber unit were measured for return air at 26 °C and 55% relative humidity and constant air and liquid flow rates (200 $m^3 h^{-1}$ return and ambient air, 100 h^{-1} liquid desiccant and water).

For the validation of the developed theoretical models additional operating conditions were investigated for the contact matrix and heat exchanger absorber unit. During the first set of experiments the air flow and liquid flow rates were kept constant ($200\,m^3\,h^{-1}$ return and ambient air, $100\,l\,h^{-1}$ liquid desiccant and water), while the relative humidity of the return air was varied from 50 to 70%. During the second set, the air flow rates of the return and ambient air were varied from 100 to $300\,m^3\,h^{-1}$, while all other parameters were kept constant. To examine the performance of two different liquid desiccants, all experiments were carried out using both lithium chloride and calcium chloride solutions, each at a concentration of 43% mass fraction. Before starting the experiments the solutions and water were heated to a temperature level of about $27 \pm 2\,°C$ to demonstrate the most realistic summer conditions.

Desiccant Rotor (AWheel) A silica gel desiccant rotor with a diameter of 350 mm was used as a reference desiccant system (system 2), which is already commercially available. The rotation velocity used for the experimental analysis was 20 rotations per hour, with a volume flow rate of $200\,m^3\,h^{-1}$ on the return air side and $185\,m^3\,h^{-1}$ on the regeneration air sides. For regeneration, ambient air was heated to 70 °C, resulting in a relative humidity of 6%. Four thermocouples ($\pm$ 0.1 K accuracy) were used to monitor the inlet and outlet temperatures of the air on the return air and regeneration air sides as well as the inlet and outlet temperatures of the water and the solution. Additionally, on each air side the relative humidity was measured using four capacity-based humidity sensors ($\pm$ 1% accuracy).

Contact Matrix Absorber Unit (CMAU) A cellulose matrix ($300 \times 200 \times 150$ mm) offering a specific contact area of $600\,m^2\,m^{-3}$ was used for the direct contact matrix absorber unit (system 3). For dehumidification, the return air is brought into contact with the liquid desiccant running down the cellulose matrix of the absorber unit in cross-flow to the air stream as shown in Figure 5.66. Eight water spray nozzles with a spray angle of 60° are placed above the cellulose matrix at a distance of 6 cm, to distribute the liquid desiccant uniformly over the matrix, allowing a minimum solution flow rate of $100\,l\,h^{-1}$. The contact matrix absorber unit is designed for a maximum air

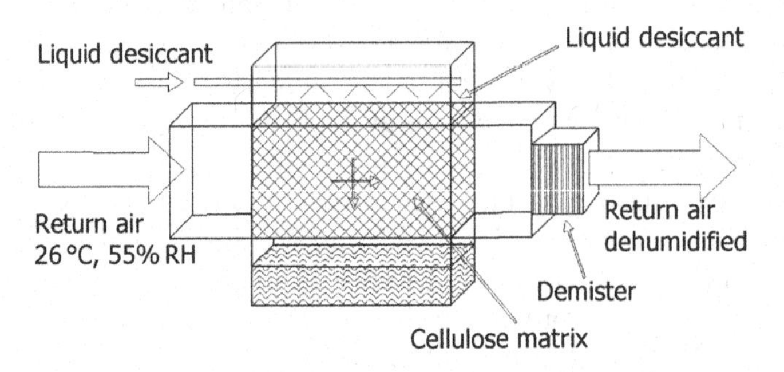

Figure 5.66 Contact matrix absorber unit for drying the room exhaust air

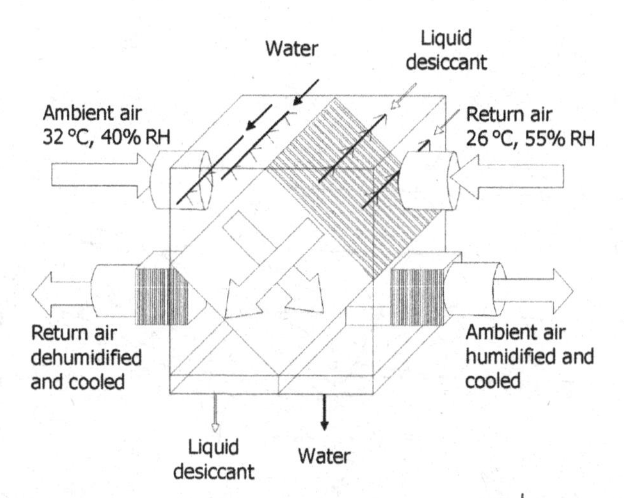

Figure 5.67 Liquid desiccant drying unit integrated in a water-cooled heat exchanger for return air drying

flow rate of 300 $m^3 h^{-1}$. Four thermocouples were used to monitor the inlet and outlet temperatures of the return air and the salt solution. Additionally, two capacity-based humidity sensors were used to measure the relative humidity of the return air at the inlet and outlet of the contact matrix absorber unit.

Heat Exchanger Absorber Unit (HEAU) A cross-flow plate heat exchanger (300 × 300 × 250 mm) with 54 channels in each flow direction and a heat transfer efficiency of 70% was used to build the heat exchanger absorber unit for the most integrated unit of system 4. The spacing between each plate (0.2 mm thickness) of the heat exchanger is 2 mm. The heat exchanger is designed for a maximum air flow rate of 300 $m^3 h^{-1}$. As shown in Figure 5.67, the liquid desiccant or water is sprayed onto the heat exchanger channels by spray nozzles. After passing through the heat exchanger, the liquids are collected in separate basins at the bottom of the absorber unit case. Eight spray nozzles on the return air side and on the ambient air side allow a solution/water flow rate of 100 $l h^{-1}$. The first prototype was made of aluminium, which will not be a long-term solution due to corrosion problems. Eight thermocouples were used to monitor the inlet and outlet temperatures of the air on the return and ambient side as well as the inlet and outlet temperatures of the water and the solution. Additionally, on each air side the relative humidity was measured using four capacity-based humidity sensors.

Cooling System Analysis and Supply Air Temperatures

The different dehumidification processes were compared for a constant return air condition of 26 °C and 55% relative humidity and ambient inlet air conditions of

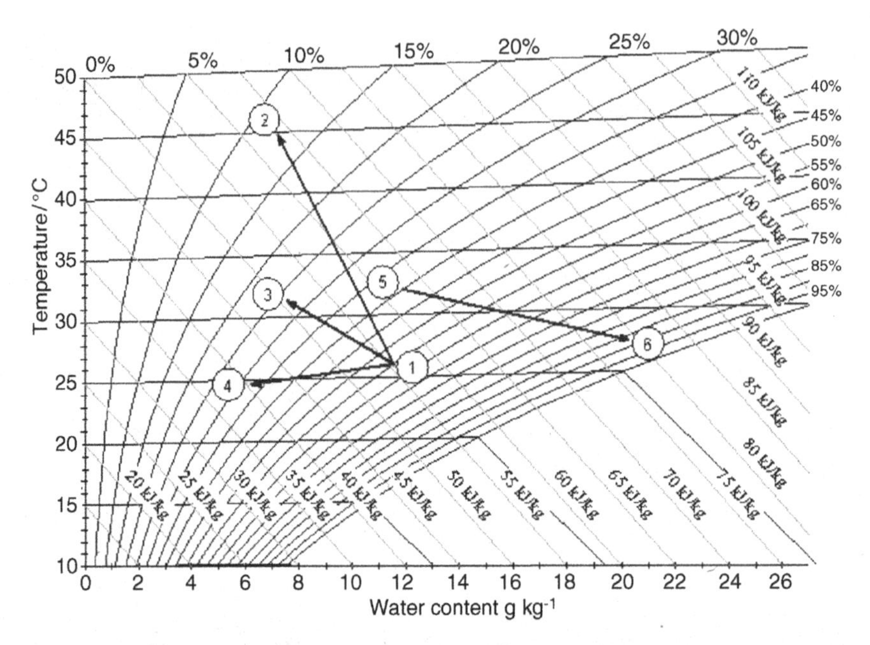

Figure 5.68 Paths of return air dehumidification for different process options

32 °C and 40% relative humidity. The results of the measurements taken at an air volume flow of 200 $m^3 h^{-1}$ are shown in Figure 5.68.

Path 1–2 describes the measured dehumidification process in the investigated desiccant rotor (AWheel) using regeneration air with a 12 $g_{water}\ kg_{air}^{-1}$ humidity ratio, at 70 °C and volume flows of 187 $m^3 h^{-1}$. The humidity of the return air is reduced in the desiccant rotor by about 4.3 $g_{water}\ kg_{air}^{-1}$, while the temperature of the return air is increased by about 19 °C. This significant rise in temperature results from the adsorption enthalpy and the heat transferred from the regeneration side.

Path 1–3, which indicates the dehumidification of the return air in the contact matrix absorber unit (CMAU), shows a dehumidification of about 4.2 $g_{water}\ kg_{air}^{-1}$ combined with an increase in the return air temperature of about 5 °C. A part of the absorption enthalpy is removed by the liquid desiccant so that the process also reduces the enthalpy.

Path 1–4 shows the measured dehumidification of the return air in the heat exchanger absorber unit (HEAU). The humidity is reduced by about 5.7 $g_{water}\ kg_{air}^{-1}$, while the temperature of the return air is decreased by about 1 °C, that is, the process is just slightly better than isothermal.

Path 5–6 shows the humidification and evaporative cooling process of the ambient air in the cooling channels of the heat exchanger absorber unit. The humidification process is also not isenthalpic. This is a consequence of the simultaneous removal of the absorption enthalpy from the return air channels in the heat exchanger absorber unit.

The main objective for all systems is to reach high cooling performances, which can be calculated from the reachable supply air temperatures. The supply air temperature was obtained using the experimental results from the different sorption units and a theoretical model for the water-sprayed cross-flow heat exchangers (HES).

The paths describing the processes in the different systems are shown in the following Mollier diagrams. The lowest supply air temperature of 18.8 °C can be reached with the heat exchanger absorption unit (HEAU) in system 4 (figure 5.69). The liquid flow rates were $100\,l\,h^{-1}$ LiCl solution and $100\,l\,h^{-1}$ water on the spray cooling side. Path 1–2 describes the dehumidification of the return air in the HEAU, path 3–4 the humidification of ambient air, path 2–5 the humidification of return air in the second spray-cooled heat exchanger (HES) and path 3–6 the cooling of the supply air in the HES.

The second lowest supply air temperature is reached by the contact matrix absorption unit (CMAU) at 19.4 °C in system 3 (Figure 5.70). Again, $100\,l\,h^{-1}$ of LiCl solution was used. Path 1–2 describes the dehumidification of return air in the CMAU, path 3–4 the humidification of ambient air in the HES, path 2–5 the cooling of return air in the HES, path 5–6 the humidification of return air in the HES and path 3–7 the sensible cooling of supply air in the second HES.

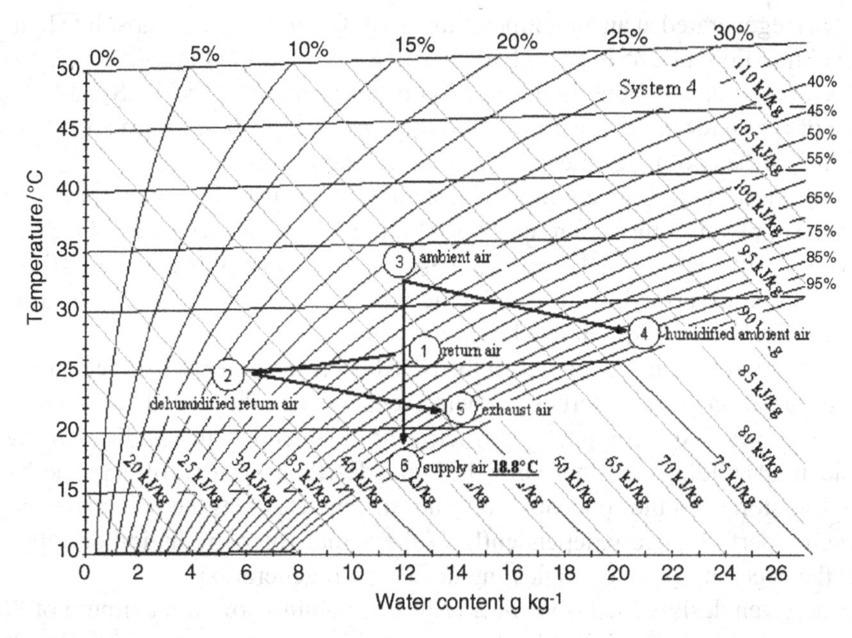

Figure 5.69 Supply air cooling of system 4 with a spray-cooled liquid desiccant heat exchanger (HEAU)

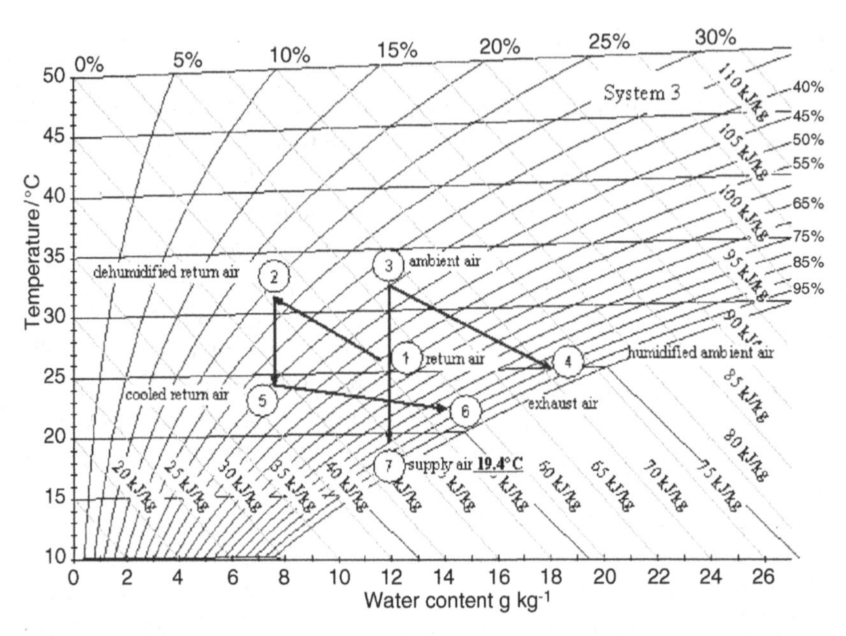

Figure 5.70 Supply air cooling of system 3 with the contact matrix absorber unit followed by two spray humidifier heat exchangers

The adsorption wheel reached 19.6 °C in system 2 (Figure 5.71). The adsorption wheel was regenerated at an air temperature of 70 °C with 12 g kg^{-1} absolute humidity and a volume flow of 185 m^3 h^{-1}. Path 1–2 describes the dehumidification of return air in AWheel, path 3–4 the humidification of ambient air in the HES, path 2–5 the cooling of return air in the HES, path 5–6 the humidification of return air in the HES and path 3–7 the cooling of the supply air in the second HES.

The simplest supply air cooling in system 1, with just one water-sprayed HES, reaches a supply air temperature of 22 °C (Figure 5.72). Path 1–2 shows the humidification of return air in the HES and path 3–4 the cooling supply air in the HES. This process can be described as an isenthalpic humidification followed by heating and humidification at constant relative humidity.

However, the essential advantage of the HEAU over the other two desiccant systems is the combination of the absorber, the heat exchanger and the humidifier in one single unit. This allows a very compact construction of the intended air-conditioning system for residential buildings. Furthermore, the liquid desiccant systems offer the possibility of loss-free storage of the concentrated solution, which enables the systems to produce the cooling performance independently of the actual solar irradiation and therefore avoids the necessity of an extra heating device for regeneration.

For the given design conditions of 200 m^3 h^{-1} volume flow, a maximum of 886 W cooling power can be achieved using the most integrated system 4. The results for the summer design conditions of 32 °C, 40% relative humidity outside air, and 26 °C, 55% relative humidity room air, are summarized in Table 5.8.

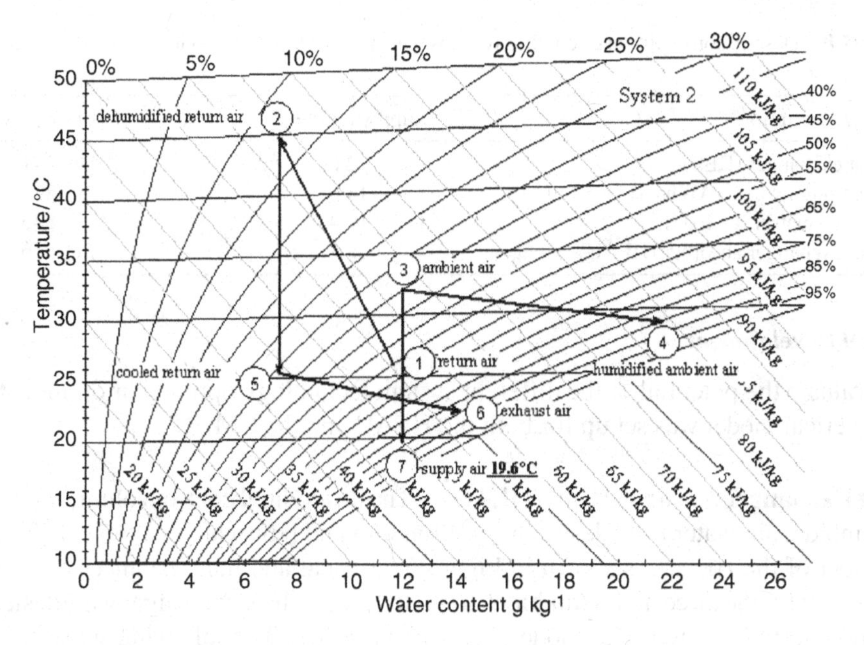

Figure 5.71 Supply air cooling of system 2 with adsorption wheel

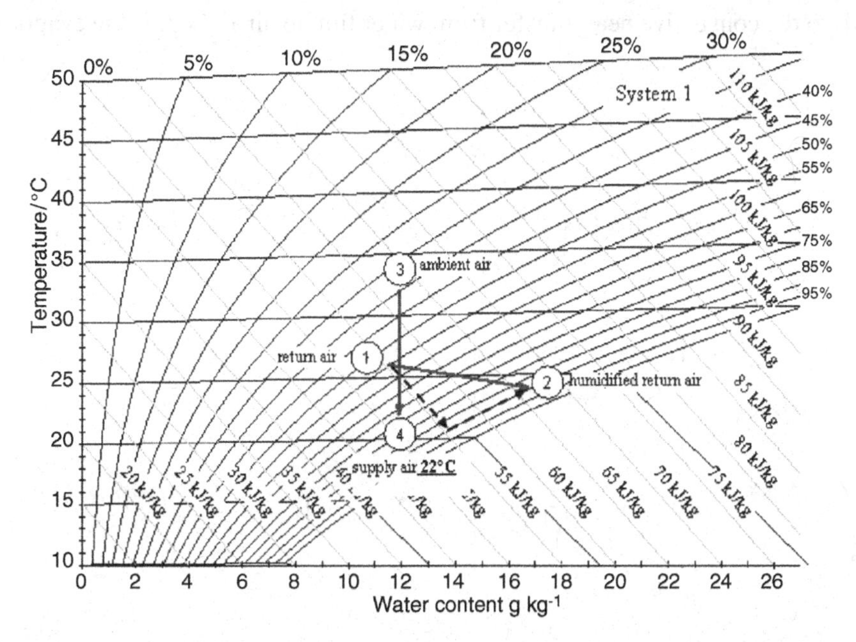

Figure 5.72 Supply air cooling of system 1 with an exhaust air spray humidifier heat exchanger

Table 5.8 Description of air-based cooling systems with achievable supply air temperatures and cooling power

System description	Supply temp./°C	Cooling power/W
1: Heat exchanger (HES)	22.0	671
2: Adsorption wheel (AWheel)	19.6	832
3: Contact matrix (CMAU)	19.4	846
4: Heat exchanger absorber (HEAU)	18.8	886

Model Development

To evaluate the potential of the different system technology options for optimization, a numerical model was set up for both the HEAU and the CMAU.

Heat Exchanger Absorber Unit (HEAU) The differential control volume including the ambient air, water film, desiccant solution film and return air, in a typical absorber chamber of the HEAU, is shown in Figure 5.73. Heat and mass balances developed separately for the three nodes (ambient air node 1, water/heat exchanger wall/desiccant solution node 2 and return air node 3) are shown below. The following subscripts are used: A, ambient air; R, return air; W, liquid water in film; D, water vapour; S, salt solution; a, air.

Ambient Air Node 1 Energy balance: the differential enthalpy of the air Δh_A corresponds to the convective heat transfer from water film to air $Q_{c,A}$ plus the evaporation

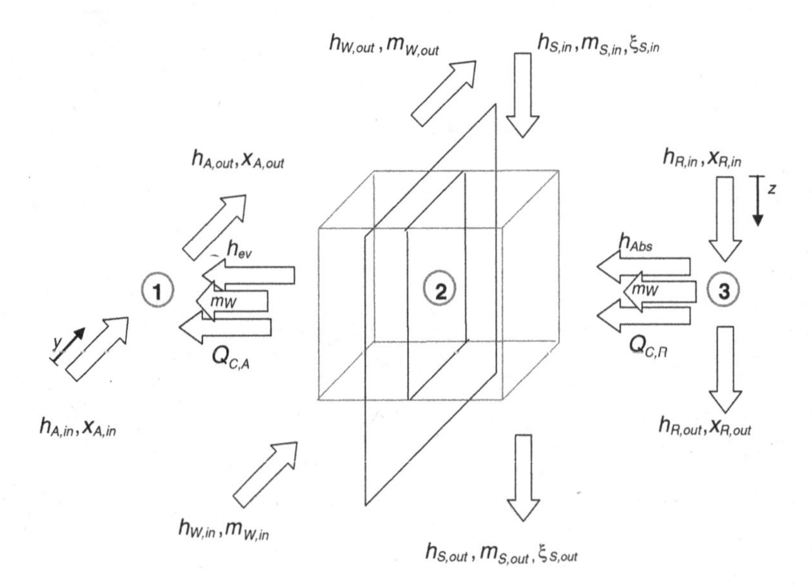

Figure 5.73 Temperature nodes, enthalpy and heat transfer coefficients of the HEAU

enthalpy h_{ev} :

$$\Delta h_A = Q_{c,A} + h_{ev} \tag{5.12}$$

$$c_{P,a}\left(T_{A,out} - T_{A,in}\right) + c_{PD}\left(T_{A,out}x_{A,out} - T_{A,in}x_{A,in}\right) + \Delta x_A\, h_V = \frac{h_{c,A}A}{\dot{m}_A}\left(\overline{T_W} - \overline{T_A}\right) + \left(c_{P,W}\Delta x_A T_{W,in} + \Delta x_A h_V\right)$$

Mass balance: the increase in water content of the ambient air Δx_A corresponds to the evaporated liquid water in the film $\Delta \dot{m}_W$ referenced to the ambient air mass flow rate $\dot{m}_A$:

$$\Delta x_A = \frac{\Delta \dot{m}_W}{\dot{m}_A} \tag{5.13}$$

Liquid/Wall/Liquid Node 2 Energy balance: the differential enthalpy of the liquid water Δh_W and the salt solution Δh_S corresponds to the sum of the heat given off to the air and the heat supplied convectively by the return air $Q_{c,R}$ plus the absorption enthalpy h_{ab}:

$$\Delta h_W + \Delta h_S = -Q_{c,A} - h_{ev} + Q_{c,R} + h_{ab} \tag{5.14}$$

$$c_{P,S}\left[\left(1 + \frac{\Delta \dot{m}_S}{\dot{m}_S}\right)T_{S,out} - T_{S,in}\right] + c_{P,W}\left[\left(1 + \frac{\Delta \dot{m}_W}{\dot{m}_W}\right)T_{W,out} - T_{W,in}\right]$$
$$= -\frac{h_{c,A}A}{\dot{m}_A}\left(\overline{T_W} - \overline{T_A}\right) - \left(c_{P,W}\Delta x_A T_{W,in} + \Delta x_A h_V\right)$$
$$+ \frac{h_{c,R}A}{\dot{m}_R}\left(\overline{T_R} - \overline{T_S}\right) + \left[c_{P,W}\Delta x_R T_{R,in} + \Delta x_R(h_V + h_d)\right]$$

Mass balance:

$$\Delta \dot{m}_W + \Delta \dot{m}_S = \dot{m}_A \Delta x_A + \dot{m}_R \Delta x_R \tag{5.15}$$

Return Air Node 3 Energy balance: the differential enthalpy of the return air Δh_R corresponds to the convective heat transfer to the liquid $Q_{c,R}$ plus the absorption enthalpy h_{ab}:

$$\Delta h_R = Q_{c,R} + h_{ab} \tag{5.16}$$

$$c_{P,a}\left(T_{R,in}-T_{R,out}\right)+c_{PD}\left(T_{R,in}x_{R,in}-T_{R,out}x_{R,out}\right)+\Delta x_R h_V$$
$$=\frac{h_{c,R}A}{\dot{m}_R}\left(\overline{T_R}-\overline{T_S}\right)+\left[c_{P,a}\Delta x_R T_{R,in}+\Delta x_R(h_V+\Delta h_d)\right]$$

Mass balance:

$$\Delta x_R=\frac{\Delta\dot{m}_S}{\dot{m}_R} \tag{5.17}$$

Boundary Conditions The boundary conditions for temperatures T, humidity ratios x ($kg_{water}\ kg_{air}^{-1}$), mass flow rates $\dot{m}$ and concentration of the salt solution ξ_S (e.g. kg_{LiCl} $kg_{H_2O}^{-1}$) are:

$$\text{at } y=0,\quad T_{A,in}=T_{A,0};x_{A,in}=x_{A,0};\dot{m}_{W,in}=\dot{m}_{W,0}$$
$$\text{at } z=0,\quad T_{R,in}=T_{R,0};x_{R,in}=x_{R,0};\dot{m}_{S,in}=\dot{m}_{S,0};\quad \xi_{S,in}=\xi_{S,0}$$

To reduce the complexity of the given problem, the following assumptions and simplifications have been made in the developed model:

- There is no heat transfer to the surroundings from the absorber unit.
- The temperature gradient between the water and the desiccant film across the thin separation plate is negligibly small ($T_{W,out}=T_{S,out}$).
- The transferred specific enthalpy, during the absorption process h_{ab} or during the evaporation process h_{ev} is calculated using the element inlet temperature of the return air $T_{R,in}$ or of the water $T_{W,in}$.
- The heat and mass transfer coefficients on the return and ambient air sides are calculated using the element inlet temperatures of the return air $T_{R,in}$, liquid desiccant $T_{S,in}$, ambient air $T_{A,in}$ and water $T_{W,in}$.

Good wetting of the heat exchanger plates with liquid desiccant on the return air side and with water on the exterior air side is critical for achieving a good performance of the HEAU. Since the wetting of the surfaces depends on surface properties and heat and mass transfer conditions, a correct model prediction of the wetting factors is nearly impossible. For the absorber unit a wetting factor ϵ for each surface has been defined and iteratively calculated during the model validation using the experimental results.

The study of the flow pattern of the desiccant solution and water film within the heat exchanger unit requires the film thickness and velocity of the liquid film to be known. Under the assumption of laminar continuous film flow, the film thickness can be calculated from the Nusselt falling film theory by neglecting the contribution of the free surface. The equation for computing the solution film thickness δ_S (m) is a function of the solution mass flow, the viscosity μ_S ($kg\,m^{-1}\,s^{-1}$), the density ρ_S

(kg m^{-3}) and the angle φ of the heat exchanging walls to the horizontal (in this case, 90°):

$$\delta_S = \left(\frac{3\dot{m}_S \mu_S}{\rho_S g \sin \varphi} \right)^{\frac{1}{3}} \tag{5.18}$$

The mass transfer of water vapour from the liquid water film to the adjacent air flow depends on the mass transfer coefficient and on the wetting of the surface (Kast, 1988). The driving force in the transport process is the difference in vapour pressure between the air and the water film interface. The mass transfer (kg s^{-1}) in flow direction y can be calculated from the following equation:

$$\Delta \dot{m}_W = A\epsilon \frac{\beta_A M}{R_m T} \left[p_{H_2O} \left(T_{A,in} \right) - p_{S,H_2O} \left(T_{W,in} \right) \right] \tag{5.19}$$

The mass transfer coefficient β_A (m s^{-1}) depends on the flow pattern in the heat exchanger channels and can be calculated from the heat transfer coefficient $h_{c,A}$, using the thermal diffusivity a_a (m^2 s^{-1}), the diffusion conduct coefficient δ_a (m^2 s^{-1}) and the parameter $n = 1/3$ for laminar air flow:

$$\beta_A = \frac{h_{c,A}}{c_{P,a} \rho_a} \left(\frac{a_a}{\delta_a} \right)^{-(1-n)} \tag{5.20}$$

The mass transfer of water vapour from the return air to the surface of the liquid desiccant $d\dot{m}_S/dy$ can be calculated from Equation 5.19, whereby the water vapour pressure p_{H_2O} is calculated at the inlet temperature of the return air and the water vapour saturation pressure is calculated at the inlet temperature of the salt solution. The mass transfer coefficient β_R is obtained from Equation 5.20, using the heat transfer coefficient $h_{c,R}$ between the return air and liquid desiccant film.

Since the water vapour pressure $p_S(T_{S,in}, \xi_{S,in})$ above the salt solution film depends on the concentration of the salt solution at the solution/return air interface, the concentration gradient across the desiccant film thickness in the absorber unit has to be known. The resistance to water vapour transport from the surface of the solution film into the solution is described by a diffusion coefficient $D(\xi_S)$ (m^2 s^{-1}) for water transport in salt solutions, which can be calculated as a function of the salt concentration and temperature (Conde, 2003):

$$D\left(\xi_S\right) = D_{H_2O} \left\{ 1 - \left[1 + \left(\frac{\sqrt{\xi_s}}{\tau_1} \right)^{\tau_2} \right]^{\tau_3} \right\} \tag{5.21}$$

Table 5.9 Diffusion coefficient parameters for water transport in salt solutions

	LiCl-H_2O	$CaCl_2$-H_2O
τ_1	0.52	0.55
τ_2	−4.92	−5.52
τ_3	−0.56	−0.56

For the salt solutions used, the parameters τ_1, τ_2 and τ_3 are as given in Table 5.9. The distribution of the molar water concentration $C(X)$ across the liquid desiccant film thickness can be approximated by a linear function (see Figure 5.74).

With the boundary conditions, $C(X = 0) = C_2$ and $C(X = \delta_S) = C_1$, the concentration $C(X)$ (mol/l) is

$$C(X) = \frac{C_1 - C_2}{\delta_S} X + C_2 \tag{5.22}$$

The concentration C_2 at $X = 0$ is assumed to be equal to the average concentration in the previous differential volume element of the HEAU. Furthermore, it is assumed that the concentration at $X = \delta_S$ is constant within each differential volume element.

The diffusion coefficient for water in salt solutions ($m^2\,s^{-1}$) can be described as a function of the molar water concentration (Klopfer *et al.*, 1997):

$$D(C) = D(C_2) \exp\left(\frac{C(X) - C_2}{C_1 - C_2} \ln \frac{D(C_1)}{D(C_2)}\right) \tag{5.23}$$

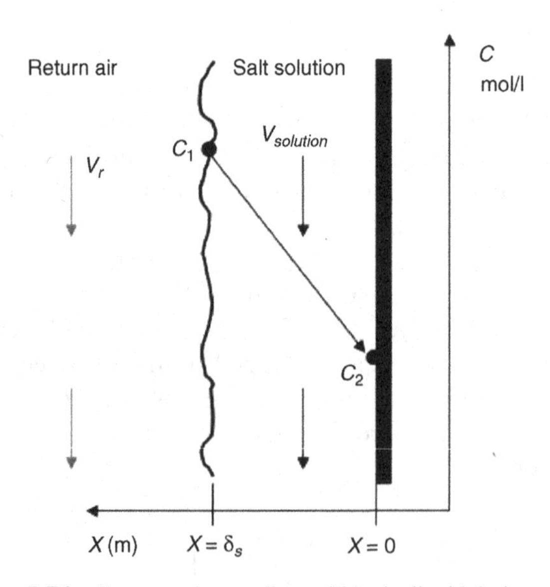

Figure 5.74 Concentration gradient within the liquid desiccant film

With Equation 5.22 the diffusion coefficient for water in salt solutions ($m^2\,s^{-1}$) can be described as a function of the film thickness X:

$$D(X) = D(C_2)\exp\left(\frac{X}{\delta_S}\ln\frac{D(C_1)}{D(C_2)}\right) \tag{5.24}$$

The mass transfer of water vapour from the surface of the liquid desiccant into the solution film ($kg\,s^{-1}$) in direction X perpendicular to the desiccant flow direction can be described as a function of the concentration drop within the solution film and depends on the heat/mass transfer surface area A, the surface wetting factor ϵ and the concentration gradient:

$$\Delta\dot{m}_S = D(C_2)\exp\left(\frac{X}{\delta_S}\ln\frac{D(C_1)}{D(C_2)}\right) A\epsilon M_{H_2O}\frac{dC}{dX} \tag{5.25}$$

Following the mass balance, the amount of water vapour transferred from the return air to the surface of the salt solution film calculated from Equation 5.19 must be equal to the amount of water vapour transferred from the surface into the solution film calculated from Equation 5.25: $\Delta x_R \dot{m}_A = \Delta\dot{m}_S$.

Both sides of the equation still depend on the unknown water concentration C_1 on the surface of the liquid desiccant film. As a consequence of the numerous dependencies, the water concentration C_1 can only be calculated iteratively. Lithium chloride and calcium chloride were employed in the absorber unit as desiccant solutions. The thermodynamic flow properties of both solutions were calculated from equations developed and compiled by Conde (2004) and Chaudhari and Patil (2002). The moist air properties were reported by Glück (1991) and Hering *et al.* (1997).

Contact Matrix Absorber Unit (CMAU) The differential control volume including the return air and the desiccant solution film in a typical absorber chamber of the CMAU is shown in Figure 5.75. Heat and mass balances are developed separately for the two nodes (ambient air node 1, cellulose matrix wall/desiccant solution node 2). All other calculations concerning the absorption process are similar to those already mentioned above for the HEAU and thus the equation set is not explicitly noted. To solve the developed equation systems for heat and mass transfer within the two absorber units, each absorber was divided into a finite number of control volumes in two orthogonal directions (finite element method), namely the return and ambient air flow directions, in the case of the HEAU, and the return air and liquid desiccant flow direction, in the case of the CMAU. The equations for the finite control volumes, each including the heat exchanger wall and half of the two adjacent flow chambers, were solved by the Gaussian elimination method, using the simplifications described above.

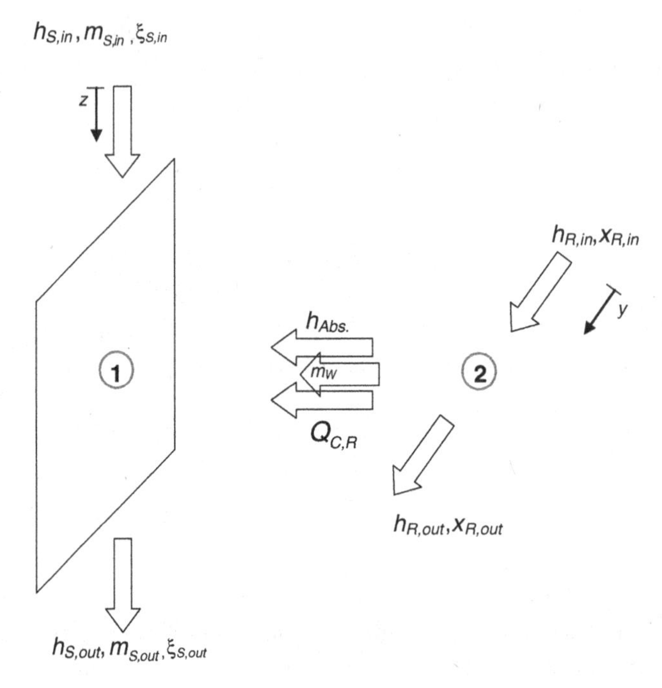

Figure 5.75 Nodes for energy and mass balances in the CMAU

Model Validation and Performance Analysis

Absorber Tests with Variation of the Return Air Relative Humidity The dehumidification performance of the two developed absorber units has been tested by varying the relative humidity of the return air (26 °C) from 50% to 70%. Volume flows were 200 $m^3 h^{-1}$ and the ambient air conditions for the spray humidifier side in the HEAU were 32 °C and 40% relative humidity. The results depicted in Figure 5.76 show the obtained return air dehumidification and return air outlet temperature of both the examined absorber units versus the inlet relative humidity of the return air. The HEAU reaches a considerable higher dehumidification, combined with lower outlet temperatures of the return air. This positive effect can be explained by the very effective indirect evaporative cooling function of the HEAU.

With increasing inlet relative humidity of the return air from 50 to 70%, the dehumidification increases from about 5 to 7 $g_{water}\,kg_{air}^{-1}$ if LiCl solution is used as the liquid desiccant. The simple CMAU reaches about 1.5 $g_{water}\,kg_{air}^{-1}$ lower dehumidification rates. In both absorber units $CaCl_2$ solution offers a significantly lower dehumidification potential.

A comparison of the return air outlet temperatures in Figure 5.76 illustrates the good performance of the integrated evaporative cooling function of the HEAU. Whereas the return air outlet temperature of the HEAU is lower than the inlet temperature,

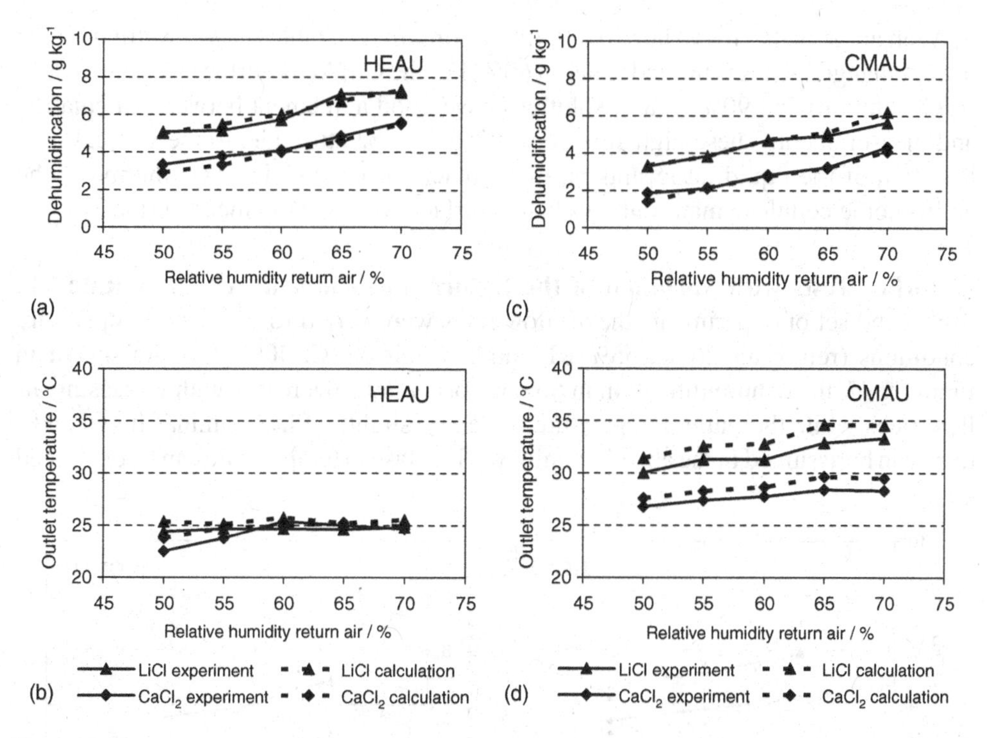

Figure 5.76 Comparison of calculations with experiments for return air dehumidification vs. return air inlet relative humidity; a) + b) return air dehumidification and outlet temperature, HEAU; c) + d) return air dehumidification and outlet temperature, CMAU

the return air temperature increases considerably during the absorption process in the CMAU. The small fluctuations within the experimental results are due to imprecise control of the two humidifiers used for the conditioning of return and ambient air, combined with measurement inaccuracies (thermocouples, humidity sensors, air flow measuring device, etc.).

The experimental results were then compared with the developed theoretical models. The calculated return air dehumidification (dashed lines) corresponds very well to the results of the experiments (solid lines) for both the HEAU and CMAU. The calculated outlet temperatures are slightly higher than the experimental results. This can be attributed to the neglect of heat losses in the model. Since the prototypes of the absorber units are not insulated, this simplification is visible in the obtained results.

The surface wetting of the HEAU and CMAU was calculated iteratively. For the HEAU, good agreement between calculated and measured dehumidification is reached for surface wettings of 35% in the case of the LiCl solution and 45% in the case of the $CaCl_2$ solution. The worse surface wetting of the LiCl solution can be explained as a result of the higher surface tension of concentrated LiCl solution.

At an angle of 45° to the horizontal, the maximum reachable surface wetting on the heat exchanger plates is limited to about 60%. For the CMAU, a surface wetting of 80% (LiCl solution) and 90% ($CaCl_2$ solution) gave good agreement between calculation and measurement. These high surface wettings can be attributed to the vertical flow direction of the liquid desiccants on the contact absorber surface. Furthermore, the hygroscopic cellulose matrix allows better surface wetting than metal surfaces.

Absorber Tests with Variation of the Return and Ambient Air Flow Rate In the second set of experiments the air flow rates were varied for constant temperature conditions (return air 26 °C, 55% RH; ambient air 32 °C, 40% RH). As shown in Figure 5.77, the dehumidification in both absorber units decreases with increasing air flow rates while the outlet temperature increases slightly. Since laminar flow conditions can be assumed in the absorber units, the increase in temperature can be explained

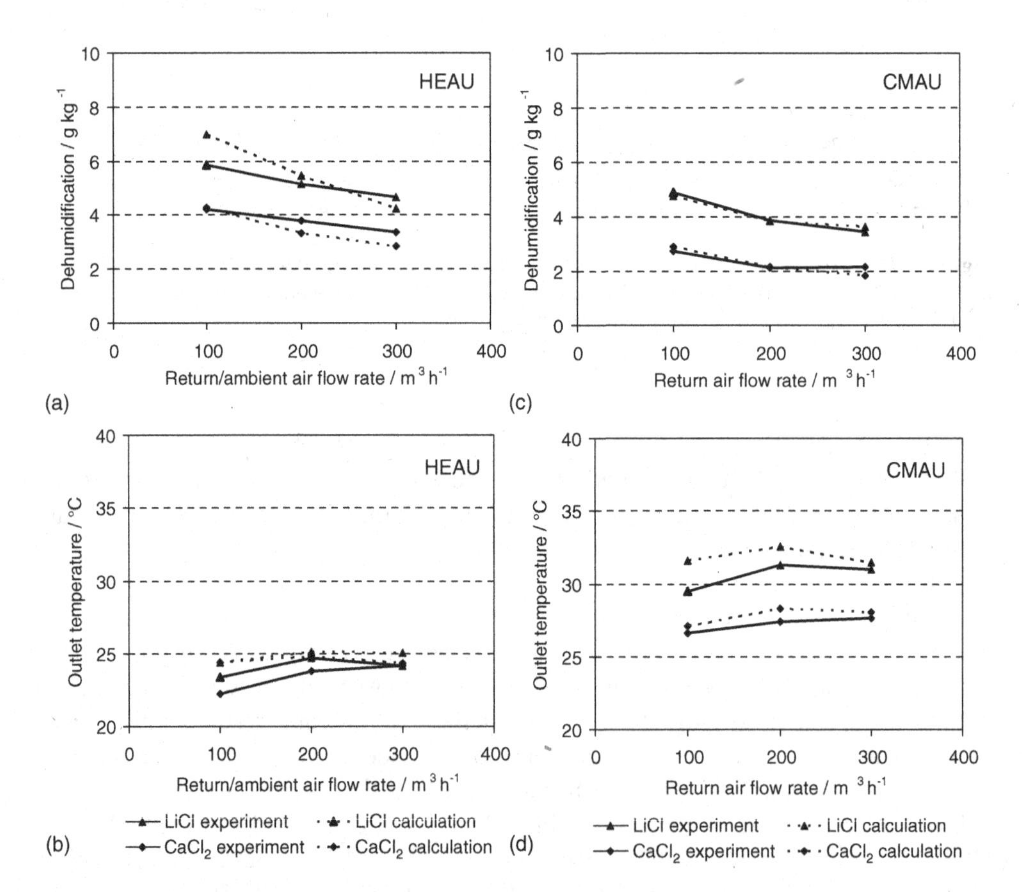

Figure 5.77 Comparison of calculations with experiments for return air dehumidification vs. return and ambient air volume flow rate: a) + b) return air dehumidification and outlet temperature, HEAU; c) + d) return air dehumidification and outlet temperature, CMAU

by the increase of the heat transfer coefficients. This should usually result in rising mass transfer coefficients within the absorber units. However, the mass transfer of water vapour from the air to liquid desiccant is mainly influenced by the mass transfer conditions within the solution film. Since the resistance to water vapour transport is significantly higher within the solution film than in air, the mass transfer coefficient depends mainly on these slower transport mechanisms. Furthermore, the decrease of dehumidification with increasing air flow rates can be attributed to the time reduction for dehumidification within the absorber. As can be seen from Figure 5.77, the calculated dehumidification and outlet return air temperatures fit the experimental results well. This indicates that the heat and mass transfer conditions are well described within the two developed models. However, for the HEAU the calculated decrease of dehumidification is higher than the measured one, especially for LiCl. This can be attributed to inaccuracies in the volume flow regulation (portable measuring instrument) combined with imprecise control of the two humidifiers, used for the conditioning of return and ambient air, and further measurement inaccuracies (thermocouples, humidity sensors, etc.).

Optimization Potential of the Heat Exchanger Absorber Unit (HEAU)

The experimental results demonstrate the good performance of the developed HEAU. The high dehumidification together with the low outlet temperature of the return air can be attributed to the integrated indirect evaporative cooling function. However, the spray nozzles used for the distribution of the liquid desiccant over the return air channels allow only high solution flow rates of at least $100\,\mathrm{l\,h^{-1}}$. In combination with the bad surface wetting of 35–45%, just a small amount of the dehumidification potential of the utilized salt solutions is used in the HEAU. If for example, the LiCl solution flows once through the HEAU, under the examined conditions, the concentration is reduced only from 43% to about 41.5% LiCl mass fraction. The flow rate of the solution could be considerably reduced without decreasing the dehumidification rate, if the same or higher surface wettings are reached. To illustrate this optimization potential, further calculations have been carried out, where the surface wetting and the desiccant flow rate were varied. The results are displayed in Figure 5.78, which illustrates the reachable dehumidification of the return air versus the desiccant flow rate and the reached surface wetting. The results were obtained for volume flows of $200\,\mathrm{m^3\,h^{-1}}$, return air conditions of 26 °C, 55% RH and ambient air conditions of 32 °C, 40% RH.

The reachable dehumidification increases with increasing solution flow rate and increasing surface wetting. However, depending on the surface wetting percentage, a dehumidification threshold value is reached with increasing solution flow rate. Considering the results shown in Figure 5.78, the main objective for further developments of the HEAU should be the realization of an optimum solution flow rate in the range of 20 to $40\,\mathrm{l\,h^{-1}}$ combined with a surface wetting of $\geq 60\%$. Under these conditions the reachable dehumidification can be improved from $5.7\ g_{water}\ kg_{air}^{-1}$ to a maximum

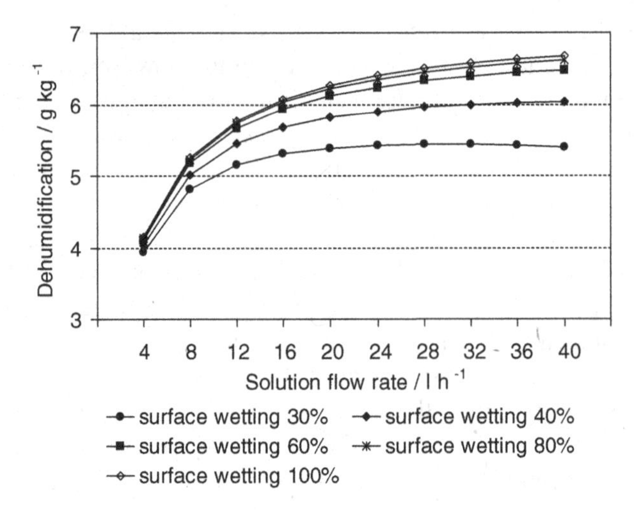

Figure 5.78 Calculated dehumidification of the return air in the HEAU by varying the desiccant flow rate and surface wetting

of 6.5 g_{water} kg_{air}^{-1}. Also, for the construction of further absorber units the utilization of non-corrosive materials, like polyethylene or polypropylene, is imperative. Since these materials are extremely hydrophobic, the main objective of a high surface wetting is harder to reach than for metal surfaces. Special surface structures or mineral surface coatings may offer a practicable solution under these circumstances. Another challenge, for high surface wetting and the distribution of the liquid desiccants, is the intended low desiccant flow rate. This problem may be solved using special spray nozzles or eventually thin membrane tubes on top of the heat exchanger. Considering the results of Saman and Alizadeh (2001), further improvements can be expected if a counter-flow arrangement between the liquid desiccant and the return air is realized, instead of the current parallel flow arrangement. On the other hand, an acceptable distribution of liquid desiccants with spray nozzles is then only practicable for very low air velocities. This requires low air flow rates, leading to large absorber units.

Conclusions on Liquid Desiccant Cooling

A new desiccant cooling system is proposed, which shifts the air drying and humidification process completely to the return air side. The supply air is cooled using an efficient heat exchanger, so that hygienic problems are avoided. The system solution requires a highly efficient desiccant process to compensate the transfer losses through the additional heat exchanger. Different technological solutions were experimentally and theoretically analysed for the return air drying. A solid material rotating desiccant wheel can dry best in an isenthalpic process with a corresponding temperature increase, leading to dehumidification rates around 4 g_{water} kg_{air}^{-1}. Liquid desiccants sprayed on a contact matrix give similar performance results, as the absorption heat cannot be efficiently removed during the process. The best performance was achieved for a cross-flow heat exchanger for simultaneous dehumidification and indirect evaporative cooling (system 4) with nearly 6 $g_{water}\,kg_{air}^{-1}$ dehumidification at no temperature increase. Concentrated lithium chloride and calcium chloride solutions have been used as liquid desiccants in the two absorber units. The lithium chloride solution gave 40–50% higher dehumidification rates over a wide relative humidity range. The measured dehumidification drops by about 1 $g_{water}\,kg_{air}^{-1}$ if the volume flow rates are tripled from 100 to 300 $m^3\,h^{-1}$.

For the summer design conditions of 32 °C, 40% relative humidity, the integrated system reaches supply air temperatures of 18.8 °C, which corresponds to a cooling power of 886 W at the measured volume flow of 200 $m^3\,h^{-1}$. This is 32% more cooling power than reached by a water-sprayed cross-flow heat exchanger with directly humidified return air (system 1).

The developed numerical models have been validated using the experimental data and good agreement was found between experiments and simulations. The model of the heat exchanger absorber unit was then used to study the optimization potential of the system. If the surface wetting can be improved from currently only 35% to 60%, about 15% higher dehumidification rates can be achieved at significantly lower solution volume flows. A further improvement to system 4 can be reached if the efficiency of the water-sprayed cross-flow heat exchanger is increased from 70% to about 76%, by the implementation of heat transfer ribs between the heat exchanger walls. With these improvements to system 4, the supply air temperatures could be reduced to 18 °C.

The new system offers a high potential for low-cost solar thermal cooling applications in residential buildings using liquid desiccants in an air-based ventilation system.

6

Sustainable Building Operation Using Simulation

Dynamic system simulation tools are widely used in the planning stage for complex buildings and their energy supply and distribution system. As shown in the overview of solar cooling projects, detailed simulation models are not always available for new technologies such as solar thermally driven cooling plants. The interaction of the solar thermal cooling system with the building will therefore be analysed in this chapter.

Dynamic simulation models often contain algorithms to control heating, ventilation or cooling plants (fan and pump volume flows, valves and flaps, etc.), renewable energy systems or building components (sun shading systems, lights, etc.). These simulation models, however, have no great significance for the real operation of buildings, during which the energy is consumed. A 60-building case study by Lawrence Berkeley National Laboratories of commercial buildings showed that 50% of the buildings had control problems. Analysing the savings through operation and maintenance improvements in 132 further buildings demonstrated that 77% of the savings were obtained by correcting control problems (Claridge *et al.*, 1994). Energy savings in such buildings are usually in the range of 10–25%, sometimes as high as 44% (Hicks and Neida, 2000). This demonstrates the need for the improved design, implementation and online supervision of control systems in buildings. Simulation systems can provide a tool to improve building control during operation.

The integration of simulation models and the building automation system also offers an opportunity to test the hardware controller within the simulation system. So-called hardware-in-the-loop testing is widely used in the automotive industry, but practically unknown in the building energy community. The simulation model provides the environment to alternate full system simulations and real hardware testing of the automation stations.

Finally, during the operation of the building, simulation models can be used for online fault detection and diagnosis and even model-based control of energy plants and building components. Simple control systems for radiators in test room configurations have been analysed online using the building simulation tool ESP-r (Clarke *et al.*, 2002) and experiments have been carried out for model-based control of heat exchangers and humidifiers in air-conditioning systems in car-painting processes (Uchihara *et al.*, 2002). However, the simulation models have not yet been fully integrated into commercial building management environments.

A strong argument for model-based controllers is the fact that conventional feedback controllers do not have much information on the system they are controlling and nonlinearities in the plant characteristics can easily lead to non-ideal controller actions. If the performance of a plant slowly deteriorates, the controller does not necessarily recognize the system faults, as long as setpoints can still be reached. Salsbury and Diamond (2001) investigated the advantages of using model-based feedforward controllers simply to control the flaps and ventilators of an air handling unit. Not only could the model be used to recommission the air handling unit (i.e. detect implementation faults), but also the controller improved accuracy by 10–43% (i.e. deviation between setpoint and operating point). Models are provided for heating and cooling coils and a mixing box.

Physical models are also very useful for controlling nonlinear systems, where the different control loops are strongly coupled and single input, single output relations do not give satisfactory results. Shah *et al.* (2004) developed physical models for compression chillers to simulate pressure levels and vapour superheating as a function of the expansion valve and evaporator cooling rates. The model-based control resulted in better efficiency and capacity control of the cooling unit.

Both roles of simulation tools in the planning and operation phase of a sustainable building will be shown in the following chapter. New developments in the modelling of solar cooling systems will be presented first, which are necessary for the correct dimensioning of a solar thermal plant. The analysis is followed by implementations of online simulation systems for buildings and energy plants.

6.1 Simulation of Solar Cooling Systems

Solar or waste-heat-driven absorption cooling plants can provide summer comfort conditions in buildings at low primary energy consumption. For the often used single effect machines, the ratio of cold production to input heat (COP) is only in the range of 0.5–0.8, while electrically driven compression chillers today work at COPs around 3.0 or higher. Solar fractions therefore need to be higher than about 50% to start saving primary energy (Mendes *et al.*, 1998). The exact value of the minimum solar fraction required for energy saving depends not only on the performance of the thermal chiller, but also on other components such as the cooling tower: a thermal cooling system with an energy-efficient cooling tower performs better than a compression chiller at a solar fraction of 40%; a low-efficiency cooling tower increases the required solar fraction to 63%. These values were calculated for a thermal chiller COP of 0.7, a compressor COP of 2.5 and an electricity consumption of the cooling tower between 0.02 and 0.08 kWh_{el} per kWh of cold (Henning, 2004b).

Double effect absorption cycles have considerably higher COPs around 1.1–1.4, but require significantly higher driving temperatures between 120 and 170 °C (Wardono and Nelson 1996), so that the energetic and economic performance of the solar thermal cooling system is not necessarily better (Grossmann, 2002).

Several authors have published detailed analyses of the absorption cycle performance for different boundary conditions, which showed the very strong influence of cooling water temperature, but also of chilled water and generator driving temperature levels on the COP(Engler *et al.*, 1997; Kim and Machielser, 2002). Most models resolve the internal steady-state energy and mass balances in the machines and derive the internal temperature levels from the applied external temperatures and heat transfer coefficients of the evaporator, absorber, generator and condenser heat exchangers. To simplify the performance calculations, characteristic equations have been developed (Ziegler, 1998),which are an exact solution of the internal energy balances for one given design point and which are then used as a simple linear equation for different boundary conditions. For larger deviations from design conditions or for absorption chillers with thermally driven bubble pumps, one single equation does not reproduce accurately the chiller performance (Albers and Ziegler, 2003). The disadvantages of the characteristic equation can be easily overcome if dynamic simulation tools are used for the performance analysis and internal enthalpies are calculated at each time step – an approach which is chosen in this work using the simulation environment INSEL (Schumacher, 1991). Dynamic models taking into account the thermal mass of the chillers are also available and can be used for the detailed optimization of control strategies such as machine start-up (Kim *et al.*, 2003; Willers *et al.*, 1999). However, for an energetic analysis of annual system performance, steady-state models are sufficiently accurate.

The available steady-state models of the absorption chillers provide a good basis for planners to dimension the cooling system with its periphery such as fans and pumps, but they do not give any hints for the dimensioning of the solar collectors or any indication of the solar thermal contribution to total energy requirements. This annual system performance depends on the details of the collector, storage and absorption chiller dimensions and efficiencies for the varying control strategies and building load conditions. An analysis of the primary energy savings for a given configuration with renewable energy heat input requires a complete model of the cooling system coupled to the building load with a time resolution of at least 1 hour. If stratified storage tanks are modelled, the time resolution has to be even higher (10 minutes or less).

Very few results of complete system simulations have been published. In the IEA task 25 design methods have been evaluated. In the simplest approach a building load file provides hourly values of cooling loads and the solar fraction is calculated from the hourly produced collector energy at the given irradiance conditions. Excess energy from the collector can be stored in available buffer volumes without considering specific temperature levels. Different building types were compared for a range of climatic conditions in Europe with cooling energy demands between 10 and 100 $kW\ h\,m^{-2}\ a^{-1}$. Collector surfaces between 0.2 and 0.3 m^2 per square metre of conditioned building space combined with 1–2 kWh of storage energy gave solar fractions above 70% (Henning, 2004a).

System simulations for an 11 kW absorption chiller using the dynamic simulation tool TRNSYS gave an optimum collector surface of only 15 m^2 for a building with a 196 m^2 surface and 90 $kWh\,m^{-2}$ annual cooling load, or less than 0.1 m^2 per square metre of building surface. A storage volume of 0.6 m^3 was found to be optimum (40 litres per square metre of collector), which at 20 K useful storage temperature difference only corresponds to 14 kWh or 0.07 kWh per square metre of building surface (Florides *et al.*, 2002). Another system simulation study (Atmaca and Yigit, 2003) considered a constant cooling load of 10.5 kW and a collector field of 50 m^2; 75 litres of storage volume per square metre of collector surface was found to be optimum. Larger storage volumes were detrimental to performance. The main limitation of these models is that the storage models are very simple and only balance energy flows, but do not consider stratification of temperatures. This explains the performance decrease with increasing buffer volume, if the whole buffer volume has to be heated up to reach the required generator temperatures. Also, attempts have been made to relate the installed collector surface to the installed nominal cooling power of the chillers in real project installations. The surface areas varied between 0.5 and 5 m^2 per kilowatt of cooling power with an average of 2.5 $m^2\,kW^{-1}$. In the present chapter it will be shown that this rule of thumb is unsuitable for solar cooling system design and that the required collector surface correlates much better with the annual cooling energy than with the nominal cooling power.

To summarize the available solar cooling simulation literature, there are detailed thermal chiller models available, which are mainly used for chiller optimization and design and not for yearly system simulation. Solar thermal systems, on the other hand, have been dynamically modelled mainly for heating applications.

In the following, three different chiller power ranges were analysed, which cover the current absorption chiller market up to 100 kW. The lowest power chiller based on NH_3/H_2O technology produces 2 kW in cooling power at a COP of 0.5 and is currently in a prototype state of development at the University of Applied Sciences in Stuttgart (Jakob *et al.*, 2003). A medium-size $LiBr/H_2O$ machine with 15 kW nominal cooling power is now available on the German market (EAW, 2003) and there are several manufacturers which produce machines around 100 kW cooling power. For all three machines, theoretical models were developed based on hourly steady-state energy balance equations and validated against experimental data in the case of the 2 kW machine or against manufacturers' data for the 15 and 100 kW machines.

The heating input power to the thermal chillers is provided by solar thermal collectors, which are connected via an external heat exchanger to a buffer storage volume. The required temperature level for the chillers is taken from the storage tanks and series-connected auxiliary heaters, which depends on the control strategy chosen (fixed or variable temperature levels). The cold produced is obtained from the thermal chiller model and used to cover the building's cooling load with the option of including a cold storage tank.

The building construction (insulation standards, orientation, glazing fraction, size, etc.) was chosen so that a given chiller power is sufficient to maintain room temperature levels at a given setpoint of 24 °C for at least 90% of all occupation hours. To evaluate the influence of the time-dependent building cooling loads on the solar fraction, building load files were calculated with a predominance of internal loads through people or equipment or with dominating external loads through glazed façades.

For the economic analysis, a market survey of thermal chillers up to 200 kW cooling power and for solar thermal collector systems was carried out. The annuity was calculated for different system combinations and cooling energy costs were obtained.

6.1.1 Component and System Models

Absorption Chiller Models

Steady-state absorption chiller models are based on the internal mass and energy balances in all components, which depend on the solution pump flow rate and on the heat transfer between external and internal temperature levels. Several problems are associated with a single characteristic equation which calculates all internal enthalpies only for the design conditions: if bubble pumps are used, as in diffusion–absorption machines or in some single effect absorbers, the solution flow rate strongly depends on the generator temperature. Also, if the external temperature levels differ significantly from design conditions, the internal temperature levels change and consequently the enthalpies. In the current work, a comparison was carried out between the constant characteristic equation and a quasi-dynamic characteristic equation based on changing internal enthalpies and – if necessary – changing solution flow rates.

The characteristic equation is based on a double temperature difference $\Delta\Delta t$ between the mean external generator and absorber temperatures t_G and t_A on the one hand and the external condenser and evaporator temperatures t_C and t_E on the other hand:

$$\Delta\Delta t = (t_G - t_A) - (t_C - t_E) \cdot B \tag{6.1}$$

The constant B is the Dühring factor determined by Equation 6.2, which is the ratio of the slope of the isosteres of the pure refrigerant to the solution. It is determined by

Dühring's rule for the solution field, where the internal temperatures of the generator T_G, condenser T_C, absorber T_A and evaporator T_E are combined:

$$B = \frac{(T_G - T_A)}{(T_C - T_E)} \tag{6.2}$$

For single effect water/lithium bromide absorption chillers the Dühring factor ranges between 1.1 and 1.2 for normal operating conditions. For a single effect ACM or DACM with the ammonia/water, working pair the Dühring factor is between 1.6 and 2.4.

The cooling power $\dot{Q}_E$ delivered by the evaporator of the machine is then a simple function of the double temperature difference $\Delta\Delta t$

$$\dot{Q}_E = s\,(\Delta\Delta t - \Delta\Delta t_{min}) \tag{6.3}$$

The slope s in Equation 6.4 contains the transferred power UA between external and internal circuits and the enthalpy differences in each component related to the specific evaporation enthalpy. For example, for the condenser C_E is obtained from the enthalpy difference between the incoming refrigerant vapour and the exiting liquid refrigerant: $C_E = (h_{entry,cond} - h_{exit,cond})/q_{evaporation}$. Likewise in the generator, G_E is calculated using the enthalpy difference between the expelled refrigerant and the solution outlet and A_E with the enthalpy difference between the incoming refrigerant vapour and the incoming weak solution for the absorber. The intersection $\Delta\Delta t_{min}$ is given by the performance of the solution heat exchanger:

$$s = \left[\frac{G_E}{UA_G} + \frac{A_E}{UA_A} + B\left(\frac{C_E}{UA_C} + \frac{1}{UA_E}\right)\right]^{-1} \tag{6.4}$$

$$\Delta\Delta t_{min} = \left(\frac{\dot{Q}_{gx}}{UA_G} + \frac{\dot{Q}_{ax}}{UA_A}\right) \tag{6.5}$$

The solution heat loss $\dot{Q}_{gx}$ describes the enthalpy difference between generator outlet and solution inlet, that is after the heat exchanger:

$$\dot{Q}_{gx} = \dot{m}_{sr}\,\left(h_{solution,outlet} - h_{solution,inlet}\right) \tag{6.6}$$

Likewise $\dot{Q}_{ax}$ contains the enthalpy difference between the solution entering the absorber, that is returning from the heat exchanger, and the solution exiting the absorber, multiplied by the rich solution mass flow. The generator power is also a function of s and the double temperature difference $\Delta\Delta t$. In addition the solution heat loss $\dot{Q}_{gx}$

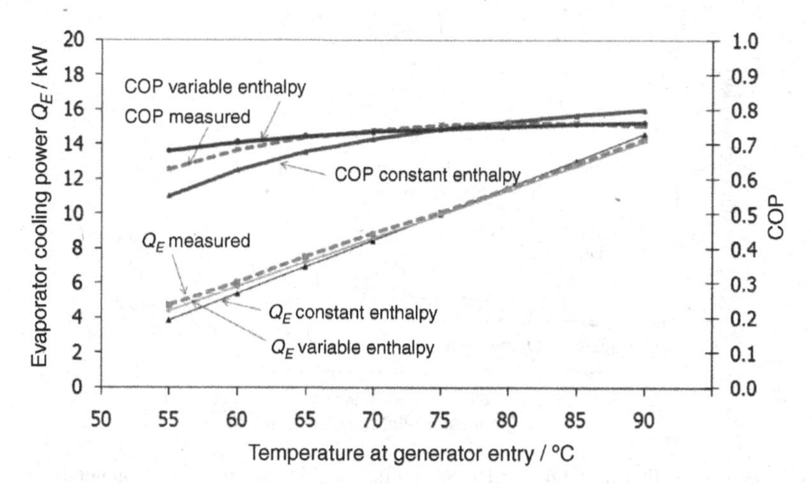

Figure 6.1 Measured and simulated cooling power as a function of generator entry temperature for a 10 kW absorption chiller

and the generator enthalpy balance G_E is needed:

$$\dot{Q}_G = G_{ES}(\Delta\Delta t - \Delta\Delta t_{min}) + \dot{Q}_{gx} \tag{6.7}$$

Instead of using the external mean temperatures, the energy balance equations can also be resolved as a function of the external entrance temperatures, so that changing mass flow rates can be considered in the model.

For a given cold water temperature, the cooling power and the COP mainly depend on the generator and the cooling water temperatures for the absorption and condensation process. If the temperatures deviate significantly from the design point, the variable enthalpy model reproduces the measured results much better than the constant enthalpy equation. This is shown in Figure 6.1 for a new 10 kW LiBr/H_2O machine produced by the German company Phoenix, where the slopes for both evaporator cooling power and COP are better met at low and high generator temperatures. The design generator entry temperature is 75 °C at an absorber cooling water entry temperature of 27 °C and an evaporator temperature of 18 °C.

At lower evaporator temperatures, the COP is lower and the generator temperature has to be increased. This is shown for a 15 kW LiBr/H_2O machine produced by the German company EAW with an absorber cooling water entry temperature of 27 °C, an evaporator entry temperature of 12 °C and exit of 6 °C (see Figure 6.2). The external mass flow rates were 2 $m^3 h^{-1}$ for the generator and evaporator and 5 $m^3 h^{-1}$ for the absorber/condenser cooling water circuit. For a 6 °C evaporator exit temperature and generator temperatures between 70 and 95 °C the cooling power was then calculated as a function of cooling water temperature (see Figure 6.3). Both the power and COP drop with increasing cooling water temperature level. To compare performances

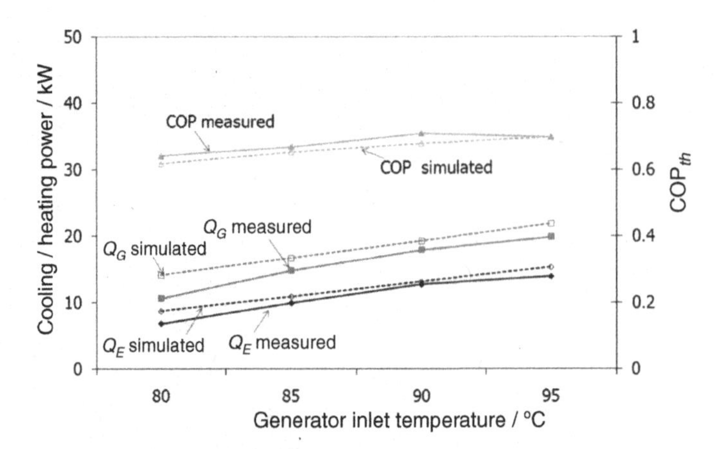

Figure 6.2 Power and thermal COP of a 15 kW cooling machine as a function of generator temperature

of different absorption chillers, it is therefore essential to compare the same three temperature levels of evaporator, absorber/condenser and generator.

To model the DACM an expanded characteristic equation is set up for the NH_3/H_2O working pair. Additional components such as the dephlegmator, the gas heat exchanger (GHX) and the bubble pump with the variable mass flow rates, are included in the model. Equations for ammonia/helium are used to calculate the enthalpies in the auxiliary gas circuit.

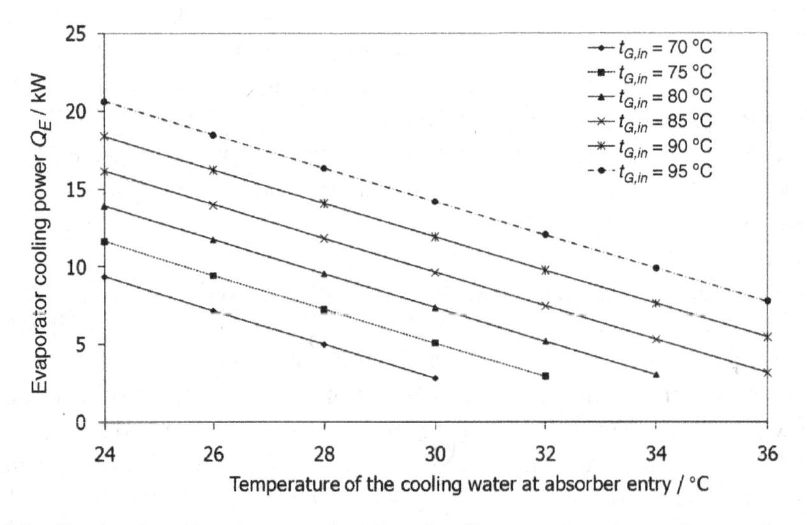

Figure 6.3 Simulated cooling power as a function of cooling water temperature for a market-available 15 kW absorption chiller

The characteristic equation for the evaporator cooling capacity $\dot{Q}_E$ of the DACM now includes the cooling losses in the auxiliary gas circuit Q_{AUX}.

$$\dot{Q}_E = s_E(\Delta\Delta t - \Delta\Delta t_{min}) - \dot{Q}_{AUX} \tag{6.8}$$

The intersection $\Delta\Delta t_{min}$ is determined by Equation 6.5 using the efficiency of the solution heat exchanger and thus the dissipated energy which results from the solution circulation between the absorber $\dot{Q}_{ax}$ and the generator $\dot{Q}_{gx}$. The slope s_E again contains the enthalpy differences of each component and the heat transfer coefficient UA between the external and internal circuits and includes as additional components the dephlegmator and the auxiliary gas circuit (generator G, condenser C, absorber A, evaporator E):

$$s_E = \left[\left(\frac{G_E + G_{E,deph}}{UA_G} + \frac{A_E + A_{E,aux}}{UA_A}\right) + B\left(\frac{C_E}{UA_C} + \frac{1}{UA_E}\right)\right]^{-1} \tag{6.9}$$

Also, the pumping performance of the bubble pump as a function of generator temperature has to be taken into account. The rich solution mass flow $\dot{m}_{Sr}$ (kg h^{-1}) of the generator/bubble pump performance has been empirically determined from various series of measurements depending on the external generator inlet temperature and external mass flow heating circuit as follows:

$$\dot{m}_{Sr} = \left(2.901\,426 \times 10^{-2}\, t_{g,in}^2 - 5.413\,764\,26\, t_{g,in} + 301.499\,335\,1\right)\dot{m}_g \tag{6.10}$$

The load behaviour of the DACM can be determined when changing the characteristic temperature difference. For given cold water temperature, the cooling capacity and the COP mainly depend on the generator heating and the cooling water temperatures for the absorption and condensation processes. In contrast to the LiBr/H_2O absorption systems, the NH_3/H_2O diffusion absorption machine has decreasing COPs at increasing generator temperatures. This is due to the rising heat losses in the rectifier, if more water is driven out of the solution at high generator temperatures. Figure 6.4 shows a comparison of measured and simulated evaporator cooling power and the COP for the current prototype. The variable enthalpy model fits the experimental data points better than the constant enthalpy model.

The main optimization potential of the prototype is the falling film evaporator, which has an insufficient distribution capacity for the liquid ammonia. The evaporation efficiency is related to the surface wetting factor, which in this case is only 0.5 (Jakob, 2006). Also, the heat transfer power of both evaporator and absorber is too low.

A parameter study was carried out to determine how to improve the performance of the DACM at different evaporator inlet and cooling water temperatures, together with evaporator surface wetting factors and GHX recovery factors. The COP and the

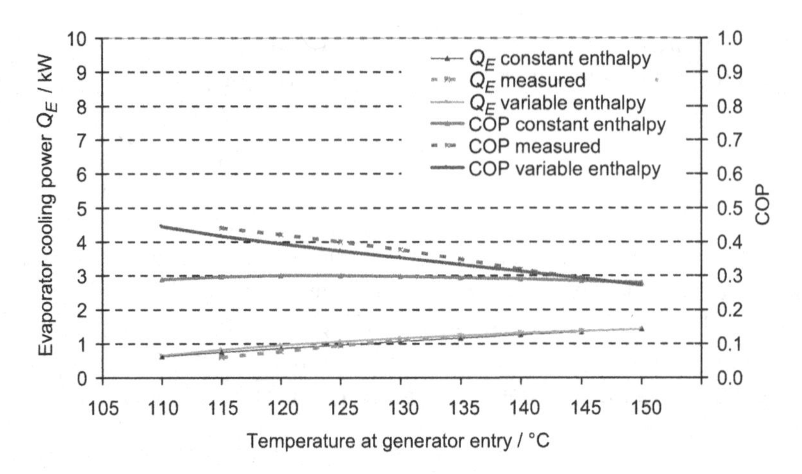

Figure 6.4 Measured and simulated cooling power and COP for the second Stuttgart prototype

evaporator cooling capacity decrease at lower evaporator temperatures. For a DACM with coaxial SHX, a surface wetting factor ϵ_W of 1.0 and a GHX efficiency Φ_{GHX} of 0.3, the design cooling capacity of 2.5 kW could be reached for evaporator inlet temperatures of 12 and 24 °C, and at generator heating inlet temperatures of 162 and 123 °C. The corresponding COPs were 0.38 and 0.85. Furthermore, the lower the absorber cooling inlet temperature, the higher the resulting evaporator cooling capacity. Figure 6.5 presents the performance at different GHX heat recovery factors for an evaporator inlet temperature of 12 °C and a constant generator temperature

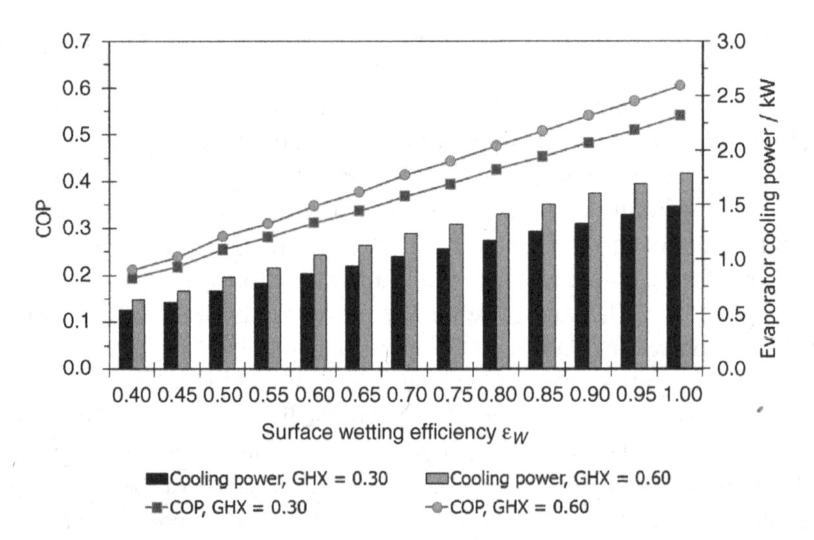

Figure 6.5 Simulated COP and evaporator cooling power as a function of the surface wetting factor and the heat recovery efficiency of the auxiliary GHE

of 130 °C. If the heat recovery factor increases from the measured value of 0.30 to 0.60, both the cooling capacity and COP increase by a factor of 1.1 to 1.2. To achieve this improvement it is necessary to optimize the heat transfer inside the GHX using constructive steps to reach higher heat recovery factors.

Other important influences on the performance of the DACM are the heat transfer coefficients UA and the heat losses $\dot{Q}_{ax}$ and $\dot{Q}_{gx}$, which depend on the rich solution mass flow $\dot{m}_{Sr}$; that is, on the bubble pump performance.

A cooling capacity of 2.0 kW can be reached in the next prototypes to be built, if the solution mass flow is increased by a factor of 1.8 and the evaporator wetting factor reaches 1.0; that is, full evaporation takes place. Together with an increase of heat transfer surface area by a factor of 1.5 for the absorber and 1.8 for the evaporator, the total heat transfer power UA is three times higher for both evaporator and absorber. At a generator entry temperature of 120 °C and a cooling water entry temperature of 27 °C, the evaporator outlet temperature is 6 °C with a COP of nearly 0.5.

As the absorption chiller model is part of a more complex dynamic model including varying meteorological conditions, there is a need to use dynamic system simulation tools anyway. The simplification of the easy-to-use characteristic equation with constant slopes is therefore not advantageous and the more exact quasi-stationary model should be used.

6.1.2 Building Cooling Load Characteristics

To evaluate the energetic and economic performance of solar cooling systems under varying conditions, different building cooling load files were produced with the simulation tool TRNSYS. The methodology for choosing the building shell parameters is as follows. For a given chiller power of 15 kW an adequate building size was selected, for example a south-orientated office building with 425 m^2 surface area and rectangular geometry. The orientation of the building was varied to study the influence of daily fluctuations of external loads. The dimensions and window openings of the buildings were adjusted, so that the given chiller power could keep the temperature levels below a setpoint of 24 °C for more than 90% of all operating hours.

The air exchange rates were held constant at 0.3 h^{-1} for the office throughout the year. This limited air exchange rate leads to cooling load files, which in some cases contain cooling power demand during winter and transition periods for south European locations. Only if the air exchange rate can be significantly increased either by natural ventilation or by using a mechanical ventilation system can such a cooling power demand be reduced by free cooling. In the buildings analysed in this work, heat was always removed by a water-based distribution system, which was fed by cold water from the cooling machines. At low ambient temperatures, the cooling tower alone provides the required temperature levels for the cold distribution system.

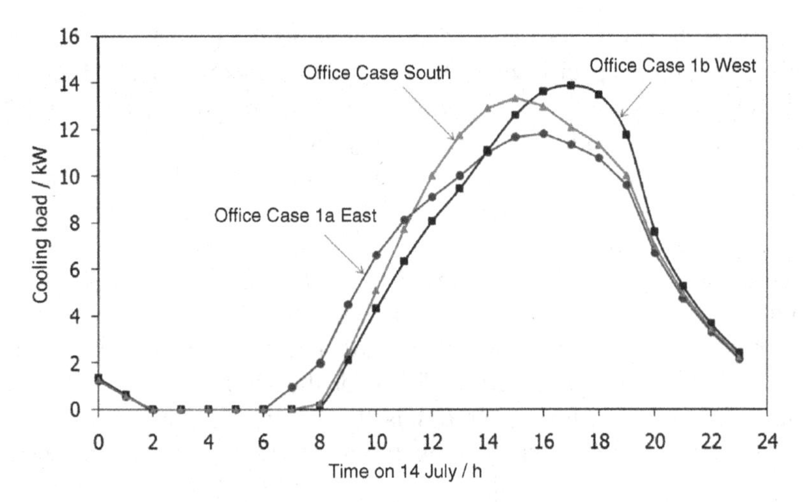

Figure 6.6 Daily cooling loads of a small office building with the main window front facing east, south or west

To evaluate the influence of the specific time series of the building cooling load, two cases were simulated:

- Case 1: Cooling load dominated by external loads through solar irradiance for unshaded windows and low internal loads of $4\,W\,m^{-2}$ (see Figure 6.7).
- Case 2: Cooling load dominated by internal loads of $20\,W\,m^{-2}$ with good sun protection of all windows (see Figure 6.9 and Table 6.2).

In addition, Case 1a is an office building with the same geometry, but the main window front is to the east instead of south. Case 1b has nearly 40% of the windows facing west. The peak values of the daily cooling loads are highest for the office with a western window front and lowest for the eastern offices. The phase shift between the curves is about one hour (see Figure 6.6).

The cooling load file with high internal loads shows less daily fluctuations and is dominated by the external temperature conditions. For the office building, the load is between 12 and 14 MWh for low internal loads up to 30 MWh for the same building, but higher internal loads. The same building at a different geographical location (Stuttgart in Germany instead of Madrid in Spain) has a cooling energy demand of only 4.7 MWh (Figures 6.7 and 6.8). A wide range of specific cooling energies is covered, ranging from about $10\,kWh\,m^{-2}$ for an office with low internal loads in a moderate climate up to $70\,kWh\,m^{-2}$for the same building in Madrid (Figure 6.9) and high internal loads (see Figure 6.10).

The specifications for the buildings with the different cases are summarized in Tables 6.1 and 6.2.

Table 6.1 Building geometry used for cooling load calculations

Building type	Surface/ m^2	Volume/ m^3
Office 1	425	1275
Office 1a	550	1650
Office 1b	450	1350
Office 2	450	1350
Office 3 Stuttgart	500	1500

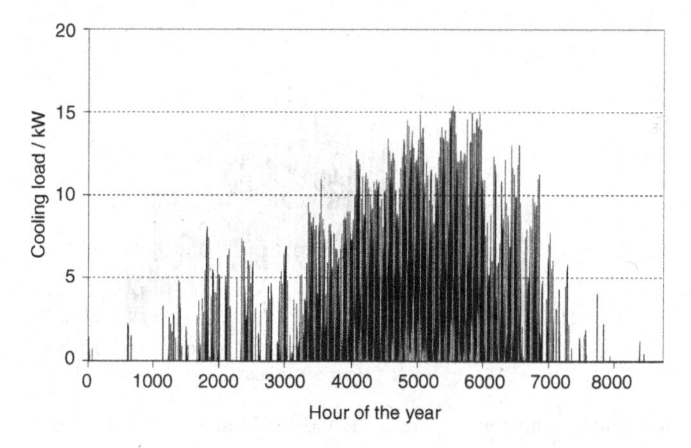

Figure 6.7 Office building dominated by external loads located in Madrid, Spain (office 1)

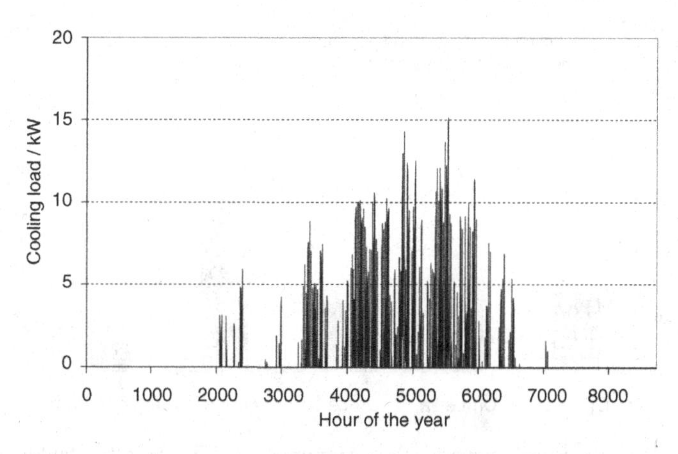

Figure 6.8 Office building dominated by external loads located in Stuttgart, Germany (office 3)

Table 6.2 Window size and shading fractions for cooling load calculations

Building Case	Window surface fraction / %				Shading fraction / %			
	North	South	East	West	North	South	East	West
1 Madrid/ 3 Stuttgart	39	39	11	11	0	0	0	0
1a	39	11	39	11	0	0	0	0
1b	39	11	11	39	0	0	0	0
2	39	39	11	11	90	90	90	90

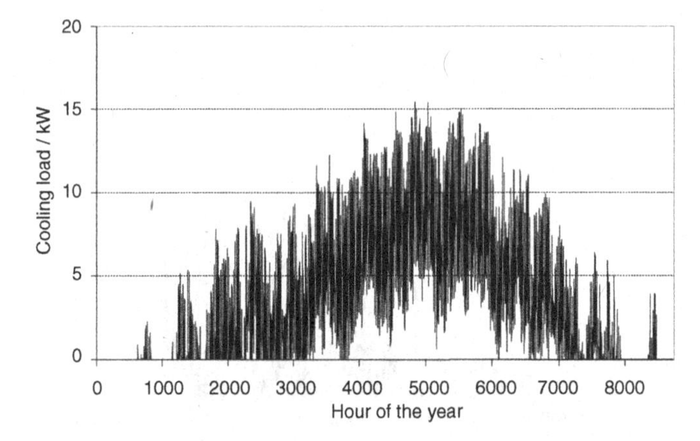

Figure 6.9 Office building with high internal loads and good sun protection (office 2) located in Madrid

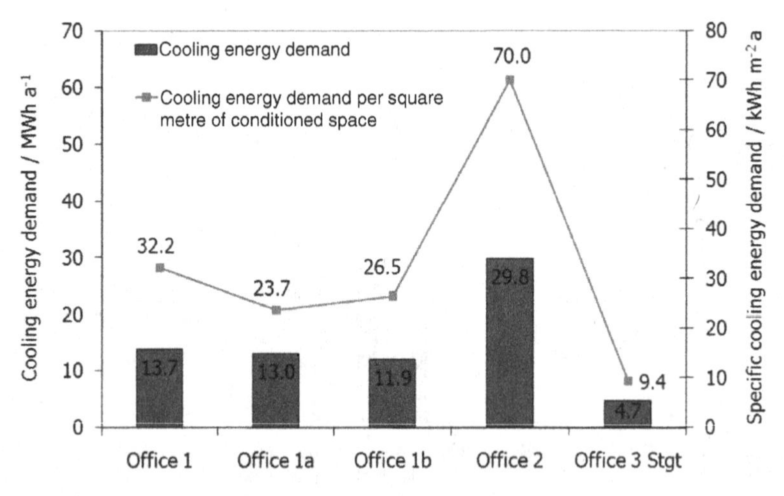

Figure 6.10 Specific and total annual cooling energy demand for the office building with different orientations and locations and dominated either by external (Case 1) or internal (Case 2) loads

Table 6.3 Internal loads and full load hours for the building in different locations

Building type	Internal load/ W m^{-2}	Full load hours/ h
Office 1 Madrid	4	913
Office 1a Madrid	4	866
Office 1b Madrid	4	793
Office 2 Madrid	20	1986
Office 3 Stuttgart	4	313

A crucial factor for the economics of the solar cooling installations is the operating hours of the machines, whereby buildings with higher full load operation have lower cooling costs for the investment part. The load hours for the buildings considered are between 680 and 2370 for the location in the Madrid and only 313 for the building in Stuttgart (Table 6.3).

Solar Thermal System Models

The solar thermal system model includes a collector, a stratified storage tank, a controller and a back-up heater. The thermal collector is simulated using the steady-state collector equation with an optical efficiency $\eta_0 F'$ and the two linear and quadratic heat loss coefficients U_{L1} and U_{L2}. A commercial vacuum tube collector was used for all simulation runs, which has an optical efficiency of 0.775, with U_{L1} at 1.476 W m^{-2} K^{-1} and U_{L2} at 0.0075 W m^{-2}K^{-1}. The storage tank has 10 temperature nodes to simulate stratification. The collector injects heat into the storage tank via a plate heat exchanger, if the collector temperature is 5 K above the lowest storage tank temperature. The return to the collector is always taken from the bottom of the tank; the load supply is taken from the top of the tank; the collector outlet and the load return are put into the storage tank at the corresponding stratification temperature level. To validate the collector model using experimental data, a dynamic model was also developed, which includes heat capacity effects. However, for the yearly simulation, the steady-state model was considered as sufficiently accurate.

6.1.3 System Simulation Results

Before evaluating the influence of the building cooling load, the thermal cooling system performance was studied in detail for one chosen load file, namely the office building dominated by external loads (office 1). The main influences on the cooling system performance are the external temperature levels in the generator, evaporator, absorber and condenser. In the case of the evaporator, the temperature level depends on the type of cold distribution system, for example fan coils with 6 °C/12 °C or chilled or activated ceilings with higher temperature levels of 15 °C/21 °C. The higher the temperature of the cold distribution, the better the system performance. In the case of

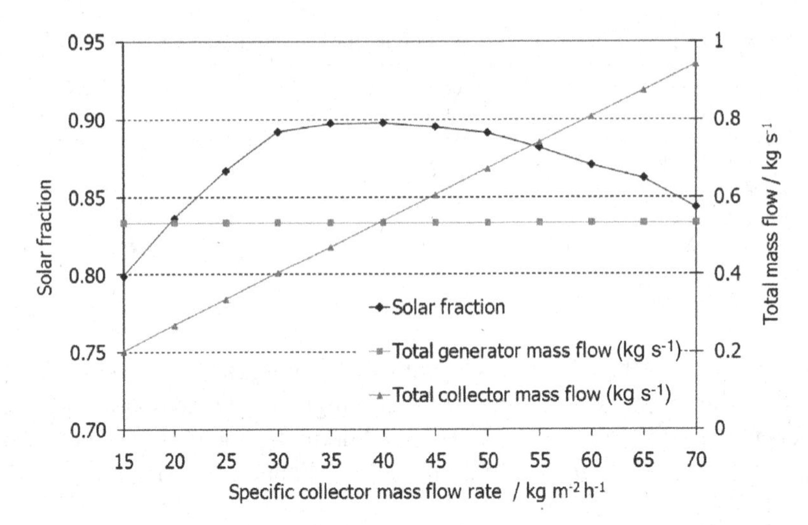

Figure 6.11 Influence of collector mass flow on solar fraction for constant generator inlet temperature

the absorber and condenser temperature levels, the type and control of the recooling system are decisive for performance. In this work, wet and dry cooling towers have been modelled. The generator temperature level mainly depends on the chosen control strategy. In the simplest case, the inlet temperature to the generator is fixed, which means high collector temperatures and poorer performance. An improved control strategy allows a temperature reduction in partial load conditions.

For an annual cooling energy demand of nearly 14 MWh and an average COP of 0.7, the system requires about 20 MWh of heating energy. To achieve a solar fraction of 80% for the given cooling load profile, a collector surface area of 48.5 m^2 and a storage tank volume of 2 m^3 is required, if the generator is always operated at an inlet temperature of 85 °C. For the constant generator inlet temperature level of 85 °C, the specific collector energy yield is only 393 kWh m^{-2} a^{-1} for an annual irradiance of 1746 kWh m^{-2} a^{-1}, that is the solar thermal system efficiency is 22% (Case 0). If the collector field were also used for warm water heating and heating support the annual yield could be significantly increased (to about 1000 kWh m^{-2} a^{-1}).

For such constant high generator inlet temperatures and a low flow collector system of 15 kg m^{-2} h^{-1} mass flow, the temperature levels in the collector are often above 100 °C if the cooling demand is low and the solar thermal energy production high. An increase in mass flow reduces the problem: by doubling the collector mass flow and thus lowering average solar collector temperatures, the solar fraction for the same collector surface area rises to 90% (see Figure 6.11). Using the higher mass flow of 30 kg m^{-2} h^{-1} and constant generator temperatures, the collector surface area can be reduced to 33.1 m^2 and the collector energy yield rises to 584 kWh m^{-2} a^{-1} (Case 1).

If the controller allows a reduction of generator temperature for partial load conditions, the performance improves due to lower average temperature levels and low-flow

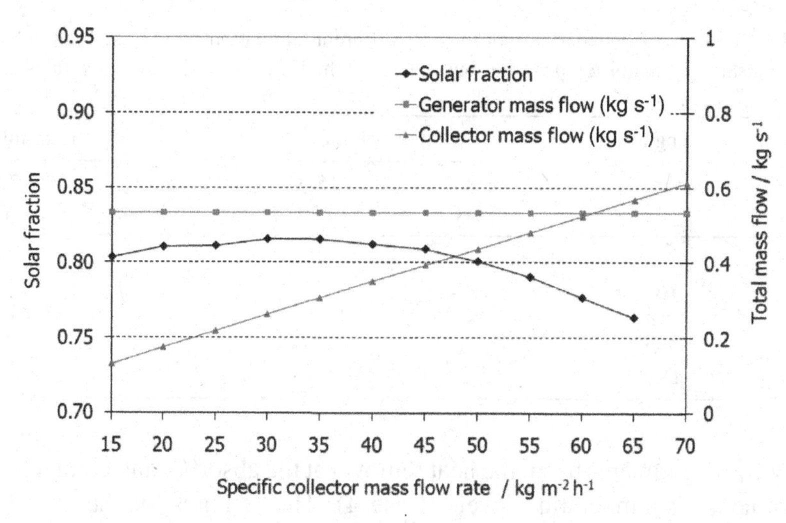

Figure 6.12 Influence of collector mass flow on solar fraction for variable generator inlet temperature

systems can again be used (see Figure 6.12). The collector surface area required to cover 80% of the demand is now reduced to 31 m^2, that is only 2 $m^2\,kW^{-1}$. The cold water temperatures were still at 12 °C/6 °C and a wet cooling tower was used (Case 2).

If the cold is distributed using chilled ceilings or thermally activated concrete slabs, the temperature levels can be raised and performance improves (Case 3). For cold water temperatures of 21 °C/15 °C the required collector surface area is only 27 m^2, that is 1.8 $m^2\,kW^{-1}$ (see Figure 6.13).

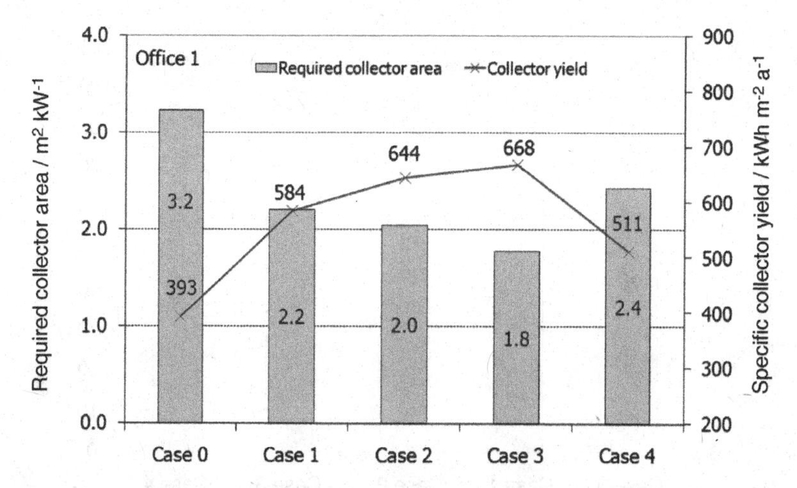

Figure 6.13 Required collector area per kilowatt of cooling power to achieve 80% solar fraction for different generator and evaporator inlet temperatures applied to the office, Case 1

Table 6.4 Summary of boundary conditions for different simulation runs. Case 0 and Case 1 both operate at constant generator temperature, but in Case 0 the solar thermal system operates at low-low conditions

Cases	Cooling tower type	Cold distribution 6 °C/ 12 °C	15 °C/ 21°C	Generator inlet 85 °C/ Const.	70–95 °C Variable
0	Wet	×		×	
1	Wet	×		×	
2	Wet	×			×
3	Wet		×		×
4	Dry		×		×

If a dry cooling tower is used, the heat removal at the absorber and condenser occurs above ambient air temperature levels (Case 4). The setpoint for the absorber inlet temperature is 27 °C, which cannot always be reached for the dry cooling tower. This leads to an increase of required collector surface area to 36 m^2, that is 2.4 m^2 kW^{-1}. The cases are summarized in Table 6.4.

For the improved control strategy, the collector energy delivered to the storage tank reaches between 511 and 670 kWh m^{-2} a^{-1} depending on the cold and recooling water temperature levels.

The thermal COP of the absorption chiller is highest (0.76) if the cold water temperature level is high (21 °C/15 °C) and stays high, even if a dry recooler is used (0.73). Low cold water temperatures give COPs of 0.67–0.7 (see Figure 6.14). The solar thermal efficiency is calculated from the energy produced and delivered to the hot storage tank divided by the solar irradiance.

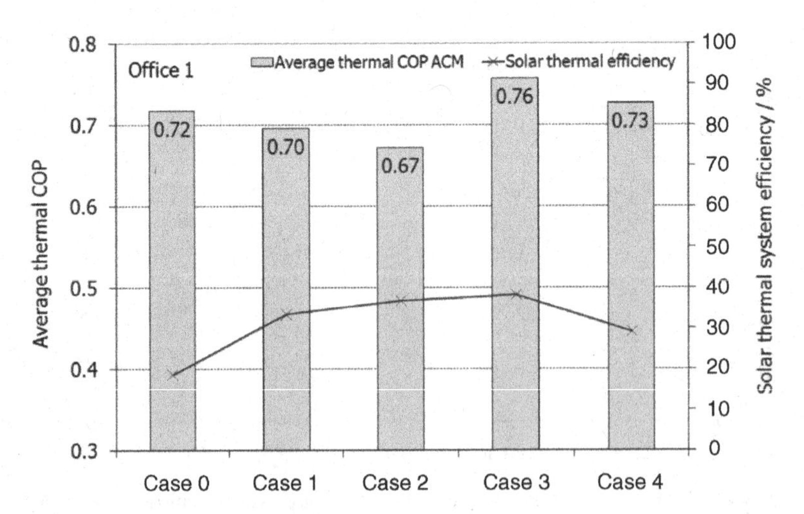

Figure 6.14 Average annual COP and solar thermal collector efficiency for different operating modes

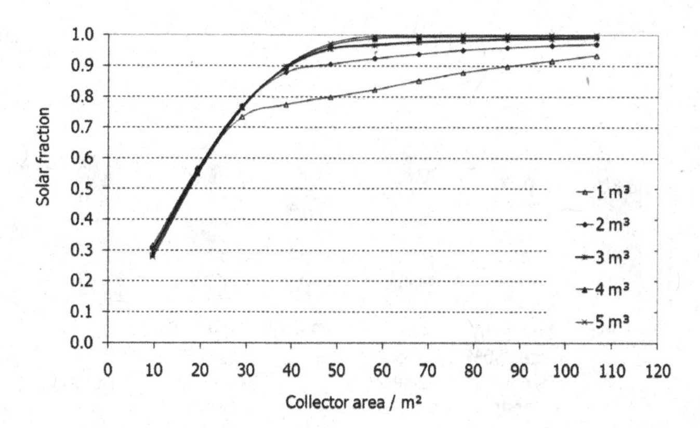

Figure 6.15 Solar fraction as a function of collector surface area for a 15 kW solar cooling system with constant generator temperatures (Case 1)

The storage tank volume only becomes important for high solar fractions of 80% and higher. At constant generator temperatures and a collector mass flow of 30 kg m^{-2} h^{-1} the solar fraction drops by a maximum of 10 percentage points, if the specific storage volume is reduced from 50 to 25 l m^{-2}. For specific storage volumes above 0.06 m^3 m^{-2}, the solar fraction hardly changes (see Figure 6.15).

As usual, the specific collector yield is highest for low solar fractions, that is for small collector surface areas. For the location in Madrid, it varies between 230 and 880 kWh m^{-2} a^{-1} (see Figure 6.16).

In the following, the influence of storage tank volume, insulation thickness and heat exchanger size is evaluated for Case 2 conditions, that is improved control with varying generator temperatures. The solar fraction to the total heat demand is reduced by

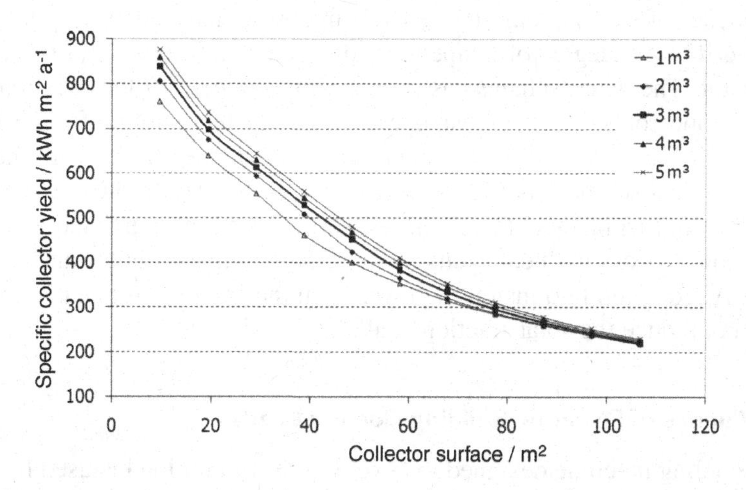

Figure 6.16 Collector yield as a function of collector surface area for constant generator temperature (Case 1). The storage volume is varied from 1 to 5 m^3

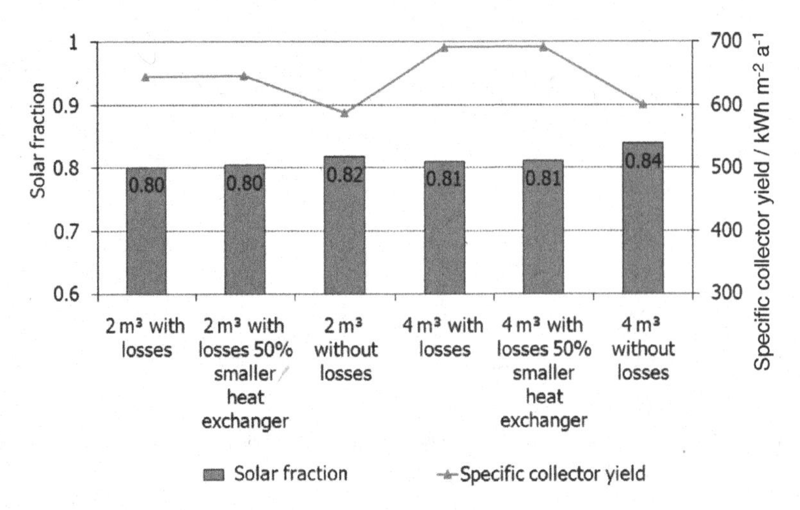

Figure 6.17 Influence of storage tank insulation and heat exchanger size on solar fraction and specific collector yield

2 percentage points for a typical insulation thickness of 10 cm compared with an ideal loss-free storage tank (see Figure 6.17). This corresponds to an additional auxiliary heating demand of 340 kWh or 9% more. Doubling the storage volume increases the solar fraction by 1 percentage point, which corresponds to a reduction of auxiliary heating energy of 200 kWh or 5% less. The larger the storage tank, the more important the good insulation quality. The specific collector energy delivered to the storage tank even drops if the insulation quality improves, as the storage tank is generally hotter. However, the energy delivered from the storage tank to the absorption chiller increases, so that in total the solar fraction improves.

The influence of the solar circuit heat exchanger was analysed by varying the transferred power *UA* per degree of temperature difference between the primary and secondary circuit. The heat exchanger is usually dimensioned for the maximum power of the solar collector field. At a mean operating design temperature for the collectors of 85 °C and an ambient air design temperature of 32 °C the efficiency of the vacuum tube collectors chosen here is 67.5%, that is, the collectors produce a maximum of 675 W m^{-2} at full irradiance. For the given surface area of 31 m^2 and a set temperature difference across the heat exchanger of 3 K, this results in a transfer power of 7 kW K^{-1}. A reduction in transferred power from the solar circuit heat exchanger by 50% does not reduce the solar fraction at all.

6.1.4 Influence of Dynamic Building Cooling Loads

If a given cooling machine designed to cover the maximum load is used for different cooling load profiles, the influence of the specific load distribution and annual cooling energy demand can be clearly seen. The boundary conditions for the control were set

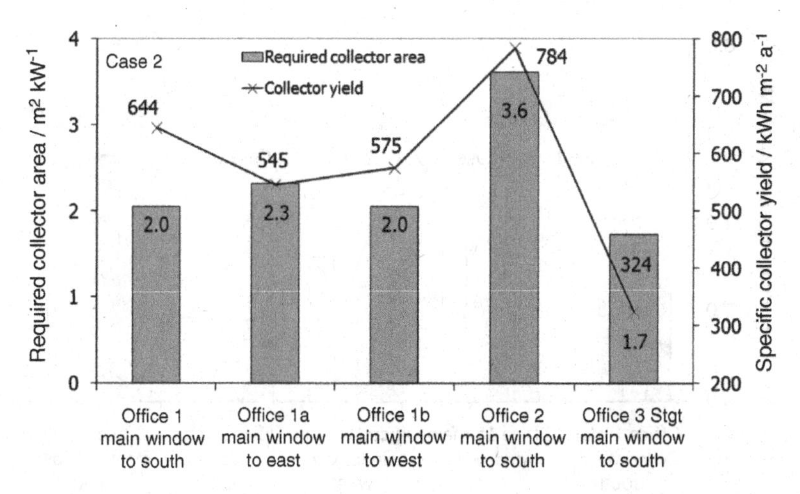

Figure 6.18 Correlation between required collector area and maximum building cooling load

to Case 2 conditions, namely a wet cooling tower, a low cold distribution temperature network of 6 °C/12 °C and variable generator inlet temperatures. The solar fraction was always at 80%. The office building with low internal loads and the main windows facing south requires 2 $m^2 kW^{-1}$ of solar thermal collector surface area (office 1). The same building with a different orientation to the east can be about 20% bigger in size to fit the 15 kW maximum cooling power. It has a 15% higher collector surface area and 15% less collector yield. If the building is orientated with the main window front to the west, the building surface area is only slightly higher than the south-orientated building with a lower total cooling energy demand. The collector surface area is the same as for a south-orientated building, which at lower operating hours means higher total costs.

The same building now dominated by internal loads (office 2) needs a collector surface area of 3.6 $m^2 kW^{-1}$, which is 80% higher than for office 1, although the required maximum power is still only 15 kW. Due to the longer operating hours of the solar thermal cooling system, the specific collector yield is 22% higher at 784 $kWh\, m^{-2} a^{-1}$ at the location in Madrid, so that the solar thermal efficiency is 45% for solar cooling operation alone. If the office building with low internal loads is located in Stuttgart with its more moderate climate, the collector yield drops to 324 $kWh\, m^{-2} a^{-1}$ and the required surface area is 1.7 $m^2 kW^{-1}$ (see Figure 6.18).

For the location in Madrid, the collector surface area required to cover 1 MWh of cooling energy demand varies between 1.6 and 3.5 $m^2 MWh^{-1}$, the depending on the building orientation and control strategy chosen. The lower the cooling energy demand, the higher the required surface area per megawatt hour. This is very clear for the building in Stuttgart with its low total energy demand of 4.7 MWh, where between 4.6 and 6.2 m^2 of solar thermal collector surface area per megawatt hour is necessary to cover the energy demand (see Figure 6.19). The ratios between collector surface and cooling energy demand vary by about 25% for the same location and control strategy.

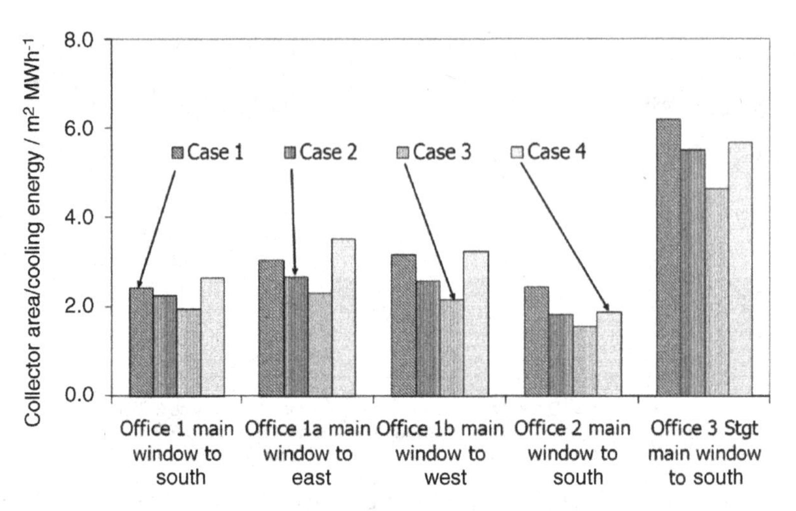

Figure 6.19 Required collector surface area as a function of the cooling energy requirement of office buildings. All buildings are located in Madrid, Spain, apart from office 3

In locations with lower annual irradiance such as Stuttgart, the required collector surface per megawatt hour of cooling energy demand is higher.

The storage volumes are comparable with typical solar thermal systems for warm water production and heating support and vary between 40 and 110 litre per square metre of collector surface area, depending on the control strategy and cooling load file. They increase with cooling energy demand for a given location. In moderate climates with only occasional cooling energy demand, the storage volumes are generally higher (see Figure 6.20).

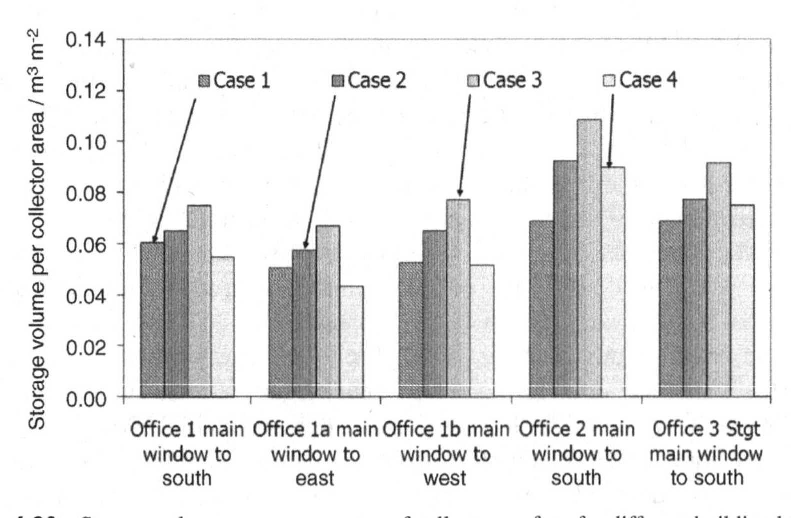

Figure 6.20 Storage volume per square metre of collector surface for different building load files

6.1.5 Economic Analysis

To plan and project energy systems such as solar cooling systems, economic considerations form the basis for decision making. The costs in energy economics can be divided into three categories: capital costs, which contain the initial investment including installation; operating costs for maintenance and system operation; and the costs for energy and other material inputs into the system.

The analysis presented here is based on the annuity method, where all cash flows connected with the solar cooling installation are converted into a series of annual payments of equal amounts. The annuity a is obtained by first calculating the net present value of all costs occurring at different times during the project, that is by discounting all costs to the time $t = 0$, when the investment takes place. The initial investment costs $P\,(t = 0)$ as well as further investments for component exchange in further years $P\,(t)$ result in a capital value CV of the investment, which is calculated using the inflation rate f and the discount or basic interest rate d. The discount rate chosen here was 4% and the inflation rate was set at 1.9%:

$$CV = \sum P\,(t)\,\frac{(1+f)^t}{(1+d)^t} \tag{6.11}$$

Annual expenses for maintenance and plant operation EX, which occur regularly during the lifetime N of the plant, are discounted to the present value by multiplication of the expenses by the present value factor PVF. Thermal chillers today can expect a lifetime of 20 years:

$$PVF\ (N, f, d) = \frac{1+f}{d-f}\left[1 - \left(\frac{1+f}{d-f}\right)^N\right] \tag{6.12}$$

In the case of solar cooling plants, no annual income is generated, so that the net present value NPV is simply obtained from the sum of discounted investment costs CV and the discounted annual expenses. It is defined here with a plus sign to obtain positive annuity values:

$$NPV = CV + EX \times PVF\,(N, f, d) \tag{6.13}$$

Annual expenses include the maintenance costs and the operating energy and water costs. For maintenance costs, some standards like VDI 2067 use 2% of the investment costs. Some chiller manufacturers calculate maintenance contracts at 1% of the investment costs. For large thermal chillers, some companies offer constant cost maintenance and repair contracts: the costs vary between 0.5% for large machines (up to 700 kW) and up to 3% for smaller ones. Repair contracts are even more expensive at 2% for larger machines and up to 12% for a 100 kW machine. In the calculations shown here, 2% maintenance costs are used.

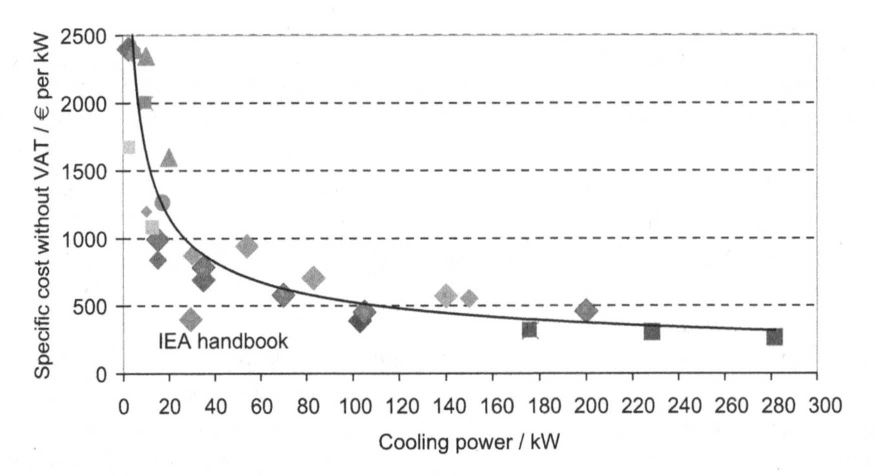

Figure 6.21 Investment costs of low-power thermal absorption chillers

To obtain the annuity a as the annually occurring costs, the NPV is multiplied by a recovery factor r_f, which is calculated from a given discount rate d and the lifetime of the plant N. The cost per kilowatt hour of cold is the ratio of the annuity divided by the annual cooling energy produced:

$$a = NPV \times r_f(N, d) = NPV \times \frac{d(1+d)^N}{(1+d)^N - 1} \tag{6.14}$$

The investment costs for the cooling machines were obtained from our own market study (see Figure 6.21). The costs are from manufacturers based in Germany and from a survey of the International Energy Agency (Henning, 2004a). A regression through the data points was used to obtain the costs for the given power used in the calculations. With a discount rate of 4% and 1.9% inflation costs over a service life of 20 years, the annuity for the cooling machine alone was €1518 per year.

In addition to the chiller investment costs, the annuity of the solar thermal system was calculated from the surface-area-dependent collector investment costs, the volume-dependent storage costs and a fixed percentage of 12% for system technology and 5% mounting costs. Cost information for the solar thermal collectors and storage volumes was obtained from a German database for small collector systems, from the German funding programme Solarthermie 2000 for flat plate collector surface areas above 100 m^2 and for vacuum tube collectors from different German distributors (see Figures 6.22 and 6.23).

Maintenance costs and the operating costs for electric pumps were set at 2%. Major unknowns are the system integration and installation costs, which depend a lot on the building situation, the connection to the auxiliary heating or cooling system, the type of cooling distribution system and so on. Due to the small number of installations,

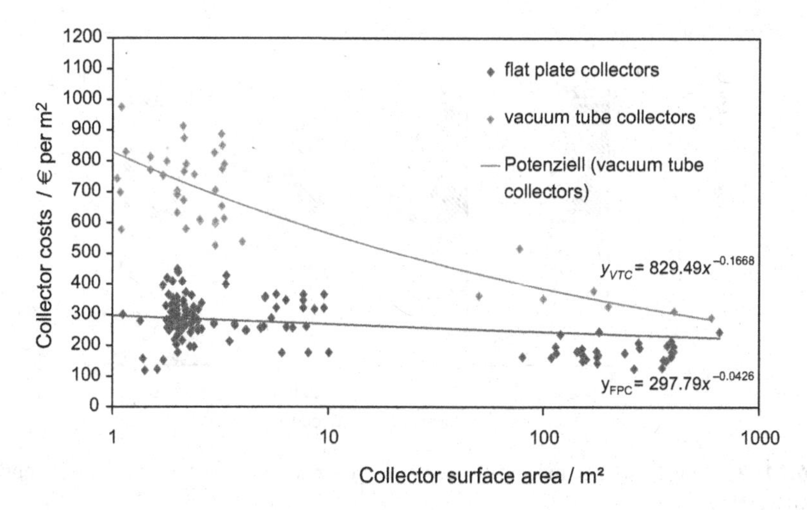

Figure 6.22 Specific collector costs without VAT as a function of size of the installation

it is difficult to obtain reliable information about installation and system integration costs. Therefore, two simulation runs were done with different cost assumptions for installation and integration. The first simulations were done with very low installation costs at 5% of total investment plus 12% for system integration. The results are shown in Figures 6.24, 6.25 and 6.26. A second round of simulations is based on 25% installation and 20% system integration costs (for results see Figure 6.27).

The total costs per megawatt hour of cold produced C_{total} are obtained by summing the chiller cost $C_{chiller}$, the solar costs C_{solar}, the auxiliary heating costs C_{aux} and the costs for cooling water production $C_{cooling}$. The costs for heating have to be divided

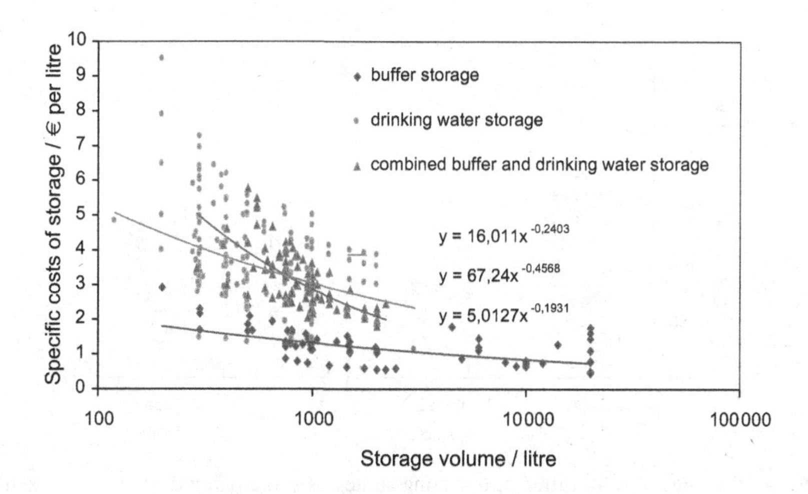

Figure 6.23 Costs for different storage tank systems

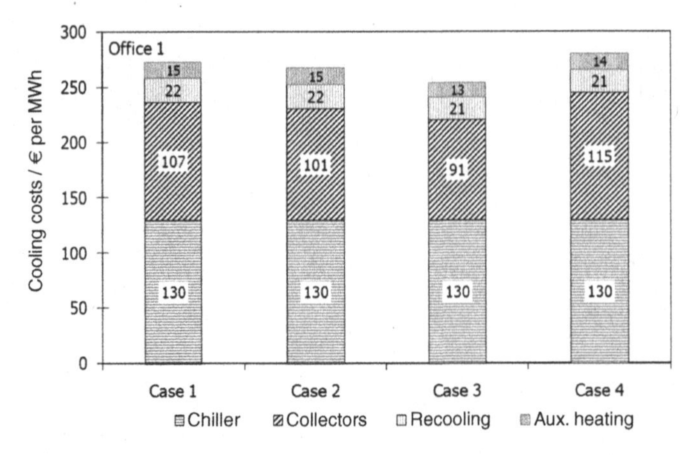

Figure 6.24 Cooling costs per megawatt hour of cold for different system technology options and control strategies

by the average COP of the system to refer the cost per megawatt hour of heat to the cold production and multiplied by the solar fraction s_f for the respective contributions of solar and auxiliary heating. For the cooling water, the costs per megawatt hour of cooling water were taken from the literature (Gassel, 2004) and referred to the megawatt hour of cold by multiplication by $1 + (1/COP)$ for removing the evaporator heat (factor 1) and for a factor of $(1/COP)$ the generator heat:

$$C_{total} = C_{chiller} + \frac{s_f C_{solar}}{COP} + \frac{(1 - s_f)\ C_{aux}}{COP} + C_{cooling}\left(1 + \frac{1}{COP}\right) \qquad (6.15)$$

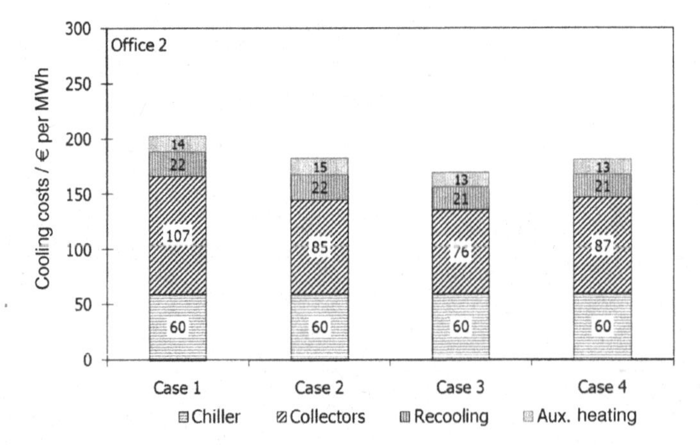

Figure 6.25 Cooling costs for different operating strategies and cooling distribution systems for the office with high internal loads

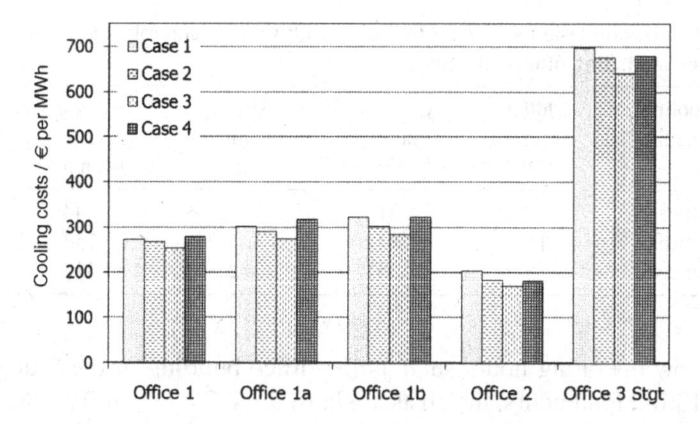

Figure 6.26 Cooling costs for different load situations and system technologies for the locations in Madrid and Stuttgart

A value for $C_{cooling}$ of €9 per MWh of cooling water was used and the auxiliary heating costs $C_{heating}$ were set to €50 per MWh of heat.

The chosen system technology (dry or wet chiller, low or high temperature distribution system, control strategy) influences the costs only slightly (7% difference between the options) if the operating hours are low (such as in the office 1 example with low internal loads, see Figure 6.24). If the operating hours increase, the advantage of improving the control strategy (Case 2) or increasing the temperature levels of the cooling distribution system (Case 3) become more pronounced (16% difference between the different cases, see Figure 6.25).

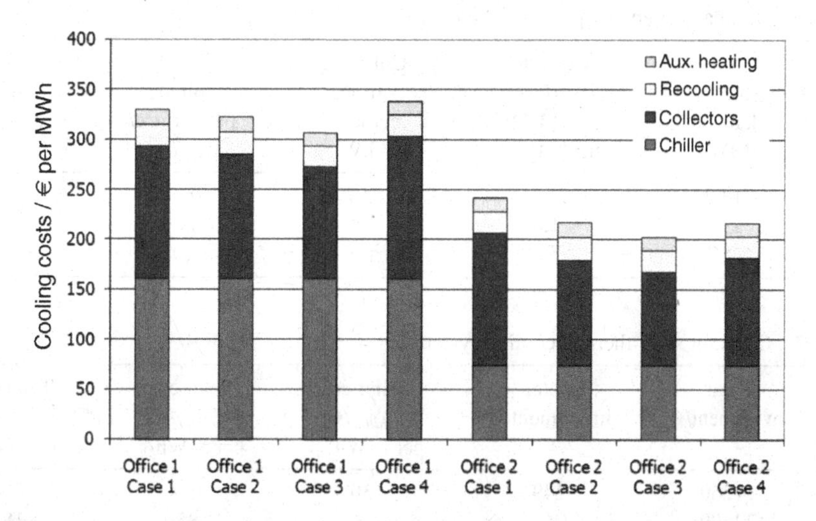

Figure 6.27 Cost distribution for a solar thermal absorption chiller system with mounting and integration costs of 45% of total investment costs

Table 6.5 Summary of design values for the office with variable control strategy (Case 2). Offices 1 and 2 are located in Madrid, office 3 in Stuttgart

Case	Cooling load file	Chiller power/ kW	Collector area/ m^2	Storage volume/ m^3	Solar yield/ kWh m^{-2}	Average COP –
1	Office 1	15	31	2	644	0.67
2	Office 2	15	54	5	784	0.67
3	Office 3	15	25	2	324	0.64

For very low operating hours such as the office building in the Stuttgart climate with only 313 full load hours, the costs are between €640 and 700 per MWh, 60% of which are due to the chiller investment costs alone (see Figure 6.26).

The calculated solar thermal system costs were between €85 and 258 per MWh for solar cooling applications, depending on the operating hours and the location. They go down as far as €76 per MWh for the office in Madrid with high internal loads and a high-temperature cooling distribution system. These costs are getting close to economic operation compared with fossil fuel heating supply. The main dimensioning results for the buildings with a good control strategy and a low-temperature fan coil distribution system (Case 2) are summarized in Tables 6.5 and 6.6.

The total costs per megawatt hour of cold produced by the thermal chillers are given in Table 6.7.

If the system integration and mounting costs are assumed to be 45% of total investment costs instead of 17%, the costs per megawatt hour of cold are in the range

Table 6.6 Summary of energy performance data

Case	Cooling energy demand/ MWh	Collector surface per MWh/ m^2 MWh^{-1}	Collector surface per kW/ m^2 kW^{-1}	Storage volume per surface/ m^3 m^{-2}	Solar thermal efficiency/ %
1	13.7	2.2	2.0	0.07	37
2	29.8	1.8	3.6	0.09	45
3	4.7	5.5	1.7	0.08	26

Table 6.7 Costs for solar thermal cooling systems

Case	Solar investment/€	Total investment/€	Chiller cost $C_{chiller}$/€ per MWh_{cold}	Solar cost C_{solar}/€ per MWh_{heat}	Total cost C_{total}/€ per MWh_{cold}
1	14260	32490	130	101	268
2	22400	40630	60	85	183
3	12140	30370	378	258	676

of €300–390 per MWh for the office in Madrid with low internal loads and €200 per MWh for the best case of the office with longer operating hours (see Figure 6.27).

In comparison, Schölkopf and Kuckelkorn (2004) calculated the cost of conventional cooling systems for an energy-efficient office building in Germany with €180 per MWh: 17% of the costs were for the electricity consumption of the chiller. The total annual cooling energy demand for the 1094 m^2 building was 31 $kWh\,m^{-2}a^{-1}$. Our own comparative calculations for a 100 kW thermal cooling project showed that the compression chiller system costs without cold distribution in the building were between €110 and 140 per MWh.

Henning (2004a) also investigated the costs of solar cooling systems compared with conventional technology. The additional costs for the solar cooling system per Megawatt hour of saved primary energy were between €44 per MWh in Madrid and €77 per MWh in Freiburg for large hotels. It is clear that solar cooling systems can only become economically viable if both the solar thermal and the absorption chiller costs decrease. This can be partly achieved by increasing the operating hours of the solar thermal system and thus the solar thermal efficiency by using the collectors also for warm water production or heating support.

6.1.6 Summary of Solar Cooling Simulation Results

In this work the design, performance and economics of solar thermal absorption chiller systems were analysed. Different absorption chillers were modelled under partial load conditions by solving the steady-state energy and mass balance equations for each time step. The calculated cooling power and coefficient of performance fit the experimental data better than a constant characteristic equation. The chiller models were integrated into a complete simulation model of a solar thermal plant with storage, chiller and auxiliary heating system. Different cooling load files with a predominance of either external or internal loads for different building orientations and locations were created to evaluate the influence of the special load time series for a given cooling power. The investigation showed that to achieve a given solar fraction of the total heat demand requires largely different collector surfaces and storage volumes, depending on the characteristics of the building load file and the chosen system technology and control strategy. To achieve a solar fraction of 80% at the location in Madrid, the required collector surface area is above 3 $m^2\,kW^{-1}$ if the generator is operated at constant high temperature of 85 °C and the solar thermal system operates under low-flow conditions. In this case, it is highly recommended to increase the collector field mass flow, so that temperature levels cannot rise too much. Doubling the mass flow decreases the required collector surface area and thus solar thermal system costs by 30%.

For buildings with the same maximum cooling load but different load time series, the required surface area varies by a factor of 2 to obtain the same solar fraction. The influence of building orientation with the same internal load structure is about

15%. More important are the different internal loads, which can increase the required collector surface area by nearly a factor of 2. The best solar thermal efficiency was 45% for a high full load of nearly 2000 hours. For the location in Madrid, 80% solar fractions are possible for surface areas between 2 and 4 m^2 per kilowatt of cooling power, the high values occurring for larger full load hours. For each megawatt hour of cooling energy demand, between 1.6 and 3.5 m^2 of collector surface are required for the Spanish site and between 4.6 and 6.2 m^2 for the German installation. The total system costs for commercially available solar cooling systems are between €180 and 270 per MWh, again depending on the cooling load file and the chosen control strategy. The total costs are dominated by the costs for the solar thermal system and the chiller itself. For a more moderate climate with low cooling energy demand, the costs can rise as high as €680 per MWh. The work shows that dynamic system simulations are necessary to determine the correct solar thermal system size and to reach a given solar fraction of the total energy requirement.

6.2 Online Simulation of Buildings

After considering the use of simulation tools for the planning of solar cooling plants, the issue of online simulation and control of buildings and plants will be discussed next. This is especially important in large, complex buildings, which often are equipped with building management systems. Today, building management systems (BMS) are commonly designed to control the technical building equipment in order to reach comfortable climatic conditions. This setpoint-orientated control strategy does not normally contain any active supervisory instruments to control the energy consumption of the building. As a consequence, no error messages will appear as long as the setpoints are reached, sometimes even if in the worst case the cooling and heating systems are working against each other. Furthermore, standard BMS control algorithms are only able to detect abrupt changes in the conditions of, for example, HVAC systems, but they do not offer detailed fault detection and diagnostic information and are unable to detect gradual degradations in system performance (Buswell *et al.*, 2003). On the other hand, dynamic simulation tools have so far only been used during the planning phase for the building design and the dimensioning of technical equipment like the heating and cooling system. Control algorithms can then be well developed when the simulation results and the planned equipment are considered, but the implementation is later normally done by the system engineers of the chosen BMS company. Due to information losses, interpretation and implementation errors are nearly unavoidable, which leads to suboptimal system control. To reduce such problems new strategies are required to implement directly well-tested control algorithms from simulation programs into the building control system. During building operation the simulation tools can then be used for instance in the energy management system for online simulation and control, and to check the control actions and measurement data against the simulation results.

By means of information technologies, control and simulation actions can also take place at different locations, so that there are no additional requirements on site for the building automation system. Only the software interfaces for communication need to be provided.

6.2.1 Functions and Innovations in Building Management Systems

Energy Consumption Control

The MS are commonly used to control the building and its energy supply systems and only in some cases also for monitoring the cooling and heating energy consumption of the building. Since the energy consumption depends strongly on ambient conditions like temperature and solar irradiation, a simple monitoring provides no detailed information on whether the system and the building control work properly and are energy efficient. If a standard energy management system is installed, at least the yearly or even monthly energy consumptions of the building can be compared with historical energy consumption data using degree day normalization methods. However, since historical data only for long time periods is analysed, this is more a passive than a real active energy management tool.

Model-Based Control

For the implementation of an active energy management system the monitoring data of the building and plant performance should be compared with predicted values at daily, hourly or even shorter intervals using real operating conditions like weather data and time schedules for room and building occupancies. Instead of comparing measured and predicted energy consumption, the models can also be used for active control of the building and the heating, cooling and ventilation systems. Such applications are for example:

- Optimization of heating up/cooling down periods after night-time or weekend energy-saving time or for partly used rooms when the internal thermal mass of the building is being considered.
- Energy-optimized strategies for the control of the sun shading system in order to increase solar gains in winter and to prevent overheating in summer when the building envelop properties and internal mass are being considered.
- Control of night ventilation cooling by regulating the mechanical ventilation system as a function of predicted weather or by automated window openings for natural ventilation.
- Control of solar-driven absorption/adsorption cooling systems using weather forecast data. For example, the room temperature can be set at 22 °C instead of 24 °C if the weather conditions are optimal for a solar cooling system.

For the implementation of such a model-based control system, online simulation models of the building and the plants have to be developed and validated against measurement data. The main question which has to be considered and clarified is the necessary level of modelling detail to meet the required accuracy for the planned control implementation. Are dynamic models required? How simple can a dynamic model be? Which time steps are required for the planned control action etc.? These are the questions to be answered. In the case of the building, detailed static calculation methods according to DIN EN 832 or DIN V 18599 can be used for calculating the energy demand of the building in monthly or, with small adjustments in the calculation methods, even daily mean values. For shorter periods in time steps of 1 hour the building dynamic caused by the internal mass has to be considered. This requires more or less detailed dynamic building simulations. In a first approach the following simple model, based on an extension of the steady-state calculation methods, has been developed and tested.

6.2.2 Communication Infrastructure for the Implementation of Model-Based Control Systems

For the implementation of a model-based control system a software interface to the installed BMS is required. Such software can run on a PC separate from or parallel to the BMS and can be located on site or centrally in the office of an energy management agency. In any case, the measured data and setpoints collected by the BMS need to be transferred to the model-based control system. The possible solutions for the data transfer strongly depend on the installed BMS and the available interfaces. In the easiest case the BMS provide an interface which transfers the measured data directly to a socket server (e.g. DataSocket or OPC server). If the BMS provides only an interface which loads the measured data into a database on the PC, then additional database reader and writer software is necessary to read the necessary data from the database and write it to an online server. For the transfer of measured data to the online simulation models, a socket client (reader) and a corresponding server have to be installed on the PC which is connected to the BMS and to an Internet/modem connection. Socket clients (reader and writer) have to be installed on a PC at the energy management agency as well. The client reads the necessary measurement data via the Internet or modem connection from the socket server running on the PC which is directly connected to the BMS on site and passes the data to the online simulation tools running on the PC at the energy management agency. The simulation results are then written back to the server running on the PC on site. Additionally, the calculated output performance is compared with the measured performance of the building. If the deviation between measurement and simulation exceeds a certain threshold, additional warning messages and control actions can be written back to the socket server running on the PC on site. DataSocket and OPC are provided by National Instruments amongst others.

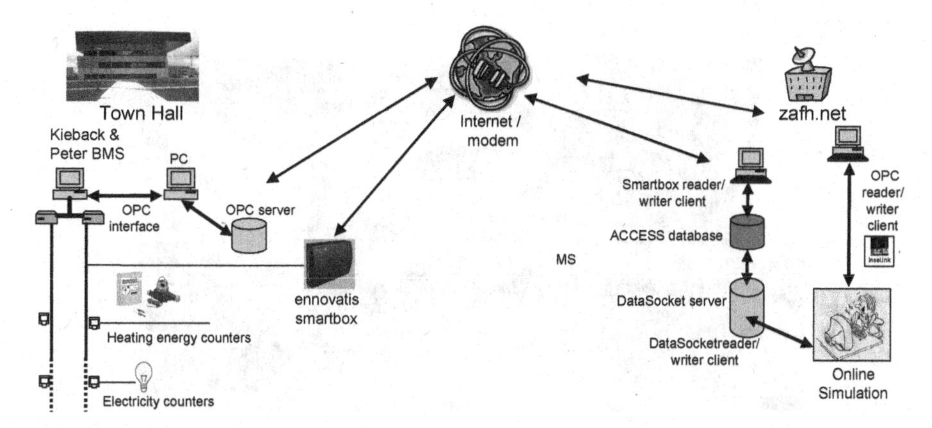

Figure 6.28 Communication infrastructure for online simulation and energy management

An example project for online simulation was carried out in the town of Ostfildern, Germany. Most public buildings are equipped with BMS from the Kieback & Peter company. Since the installed Kieback & Peter BMS use proprietary protocols and system software, software interfaces, OPC clients and an OPC server were developed by Kieback & Peter for the online data transfer. These software tools are installed on a separate PC which is directly connected to the PC of the BMS via a network card. The energy management tool runs on a different PC located at the author's research centre zafh.net in Stuttgart. For the online data transfer between both locations, OPC clients (reader and writer) are installed on the energy management PC which passes the measured data to the online simulation tool. A modem connection is used for the data transfer.

In order to test the BMS at different levels of integration, a simple microcontroller unit called smartbox produced by the ennovatis company with an integrated modem and Internet Interface was additionally installed in the Town Hall and connected to the energy meters for heating and electricity. The software interface and clients for the data transfer from the smartbox to the energy management PC were developed by zafh.net. The implemented communication structures of both systems are shown in Figure 6.28. The model-based energy management tool is tested using different types of building models as described below.

6.2.3 Building Online Simulation in the POLYCITY Project

The POLYCITY project is a European-funded research and demonstration project, which focuses on large-scale, low-energy urban development in the time frame until 2010. POLYCITY comprises three projects in Torino, Stuttgart and Barcelona and is coordinated by the author. The urban conversion project in Ostfildern on the outskirts of Stuttgart, named Scharnhauser Park, covers 178 000 m^2 of newly constructed surface area for 10 000 people and provides high building standards combined with

Figure 6.29 Town Hall, Ostfildern, as an example project for online simulation

biomass cogeneration. In the project a communal energy management system is implemented. For the energy management, simulation tools for the supply side (cogeneration unit, absorption cooling systems) and the demand side (buildings) and the necessary communication structures and interfaces to different types of BMS have been developed.

As the first implementation project the Town Hall of Ostfildern, located within Scharnhauser Park, was chosen (see Figure 6.29). This low-energy standard building was completed in 2003 and is connected to the district heating system fed by the biomass cogeneration unit and is mostly naturally ventilated without active cooling devices.

The building has a total net volume of 14 437 m^3, of which 3724 m^3 is heated to a temperature setpoint of 20 °C (at a used net area of 1027 m^2); 7713 m^3 net volume and a surface area of 2250 m^2 are kept at a lower temperature of 15 °C. There is also an unheated cellar space of 3000 m^3 volume and 818 m^2 surface area. The U-values of the walls are between 0.23 and 0.26 $W\,m^{-2}\,K^{-1}$ and the windows have U-values of 1.1 $W\,m^{-2}\,K^{-1}$ and a g-value of 0.6 with a total window fraction of 20%. The surface-related heat capacity is 90 $Wh\,m^{-2}\,K^{-1}$.

During the first operating year of 2004 the heating energy consumption of the building was read manually from the installed energy meters and documented every month. A model according to DIN EN 832 was used to calculate the energy demand of the building for the same time period under consideration using real weather data in monthly mean values. The results show that the calculated energy demand exceeds the measured data significantly in the winter months (December, January and February) if a constant air exchange rate of 0.7 h^{-1} is used in the model. However, if the air exchange rate is reduced to a value of 0.4 h^{-1} the calculated energy demand fits quite well for the winter months but falls below the measured values in autumn and spring (see Figure 6.30). The calculations with reduced air exchange were done with the new German prestandard DIN V 18599, which essentially uses the same algorithms as the monthly

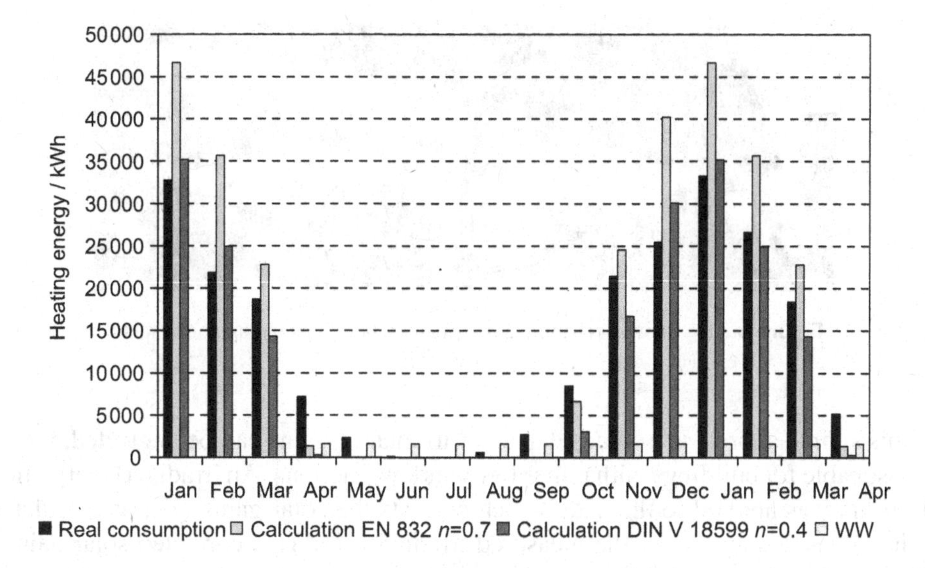

Figure 6.30 Monthly heating energy consumption degree day corrected and compared with calculated consumption for air changes of 0.7 and 0.4 h^{-1}. In addition the measured warm water consumption (WW) is shown

energy balance from EN 832, but which allows more detailed zoning of the building. The results indicate that the real air exchange rate of the naturally ventilated building is related to the outside temperature, which should be considered in further models.

The same effect was analysed for rehabilitated residential buildings in the Scharnhauser Park area (see Figure 6.31), where winter energy consumption could be well modelled with a reduced air exchange of 0.35 h^{-1}, whereas in spring months a higher air exchange of 0.7 h^{-1} gave good agreement with measurement data. Before the buildings were renovated, the higher air exchange rates gave good results; however, the higher airtightness of the new windows led to low air exchange rates especially in the cold winter months. During spring and autumn, on the other hand, the consumption is underestimated if the air exchange rate is kept at the lower value of 0.35 h^{-1} (see Figure 6.32).

To compare measured consumption results $E_{c,measured}$ from several years, it is standard practice to normalize the measured heating energy consumption using the actual degree day $G15$ compared with the long-time average degree day $G15m$ (see e.g. the German standard VDI 3807). If such a degree day correction is carried out, based on temperature differences relative to 15 °C, the maximum difference in annual heating energy consumption reduces from 21% for the uncorrected values to 15%:

$$E_{c,normalized} = E_{c,measured}\frac{G_{15m}}{G_{15}} \tag{6.16}$$

Figure 6.31 View of rehabilitated residential buildings facing north–south

This simple approach assumes that solar irradiance gains can be neglected, which is reasonable for buildings with rather low window fractions. An irradiance correction algorithm was applied to the data, which corrects the solar gains Q_{SG} as calculated in the EN 832 standard by the measured irradiance G. This corrected solar gain is added to the measured heating energy consumption and then degree day corrected. The normalized consumption value is then obtained by subtracting the standard solar gains from the heating energy calculations:

$$E_{c,normalized} = \left(E_{c,measured} + Q_{SG} \frac{G}{G_m} \right) \frac{G_{15m}}{G_{15}} - Q_{SG} \tag{6.17}$$

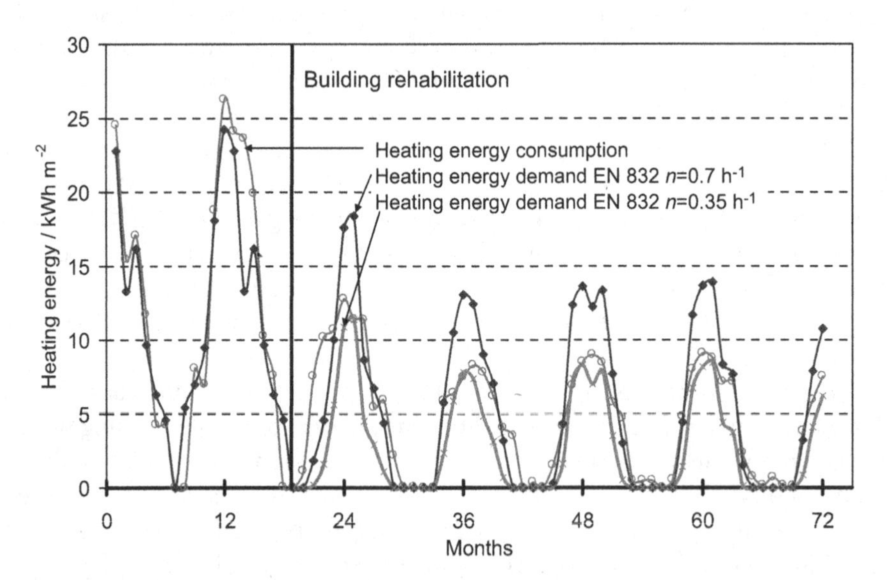

Figure 6.32 Heating energy consumption and demand calculated using the European standard EN 832 with air exchange rates of 0.35 and 0.7 h^{-1}

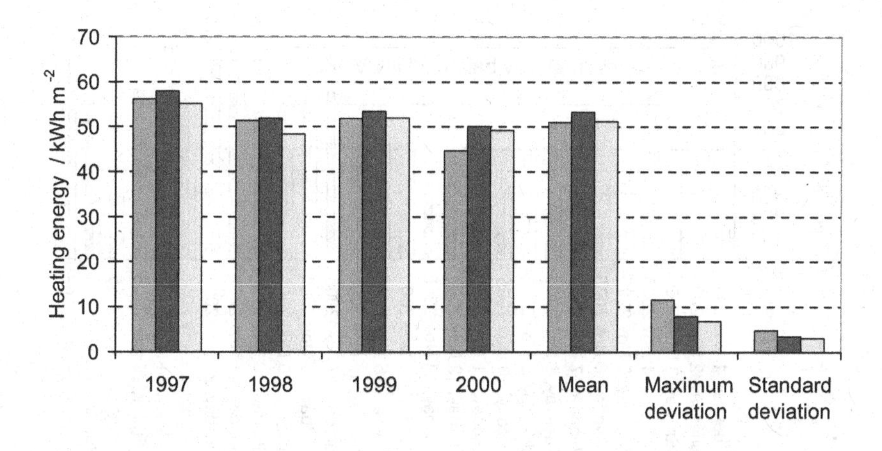

Figure 6.33 Heating energy consumption measured and corrected by degree days and solar gains

A year with very low solar gains therefore has a low contribution to the degree-day-corrected sum of consumption plus gains. After subtraction of the standard gains the normalized consumption value is lower. Conversely, a year with very high solar gains results in a higher corrected heating energy consumption. The improvement in the normalized value using the different corrections is shown in Figure 6.33. The standard deviation is slightly reduced, but the effect is not very significant.

For a higher time resolution in the simulation, for example a daily basis, the heating and cooling energy algorithms of DIN V 18599 have been implemented in the simulation environment INSEL. Daily heating energy demand can be calculated from the building's energy balance for transmission and ventilation losses and heat gains from solar irradiance and internal loads. The building's storage mass influences the efficiency η, which determines the useful heat gains for that day from the ratio of gains to losses. However, stored energy cannot be transferred from one day to the next, as the ratio of gains to losses is calculated anew each day.

Although the error between the measured and simulated monthly energy consumption is only 5%, the daily errors are significant (see Figure 6.34). This is partly due to constant air change rates and internal loads for every day, so that weekend days are often overestimated in the calculation, but also partly due to the missing dynamic storage.

From the measured data, it can be seen that for a given ambient temperature large fluctuations of daily heating energy consumption occur (see Figure 6.35). This can be mainly attributed to the occupancy and user behaviour: for example, consumption drops strongly at weekends independent of ambient temperature (see Figure 6.36).

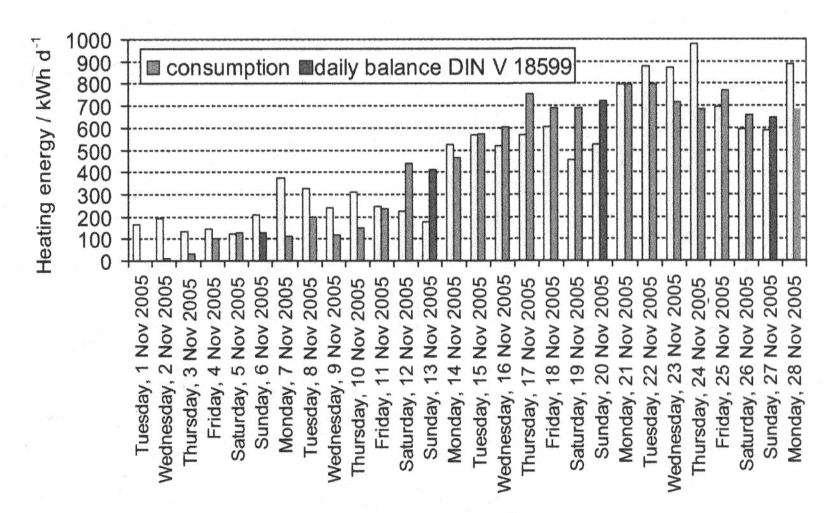

Figure 6.34 Daily measured and calculated heating energy balance of the Town Hall in Scharnhauser Park at a constant air change rate of $0.4\,h^{-1}$. The calculated values for Sunday are highlighted

The dependence of heating energy on daily irradiance is small, which is shown for the months of January and February in Figure 6.37.

To achieve a better agreement between measurement and simulation, a simple dynamic model was then used (Keller, 1997). The building can have several zones at different temperature setpoints Θ and effective heat capacities C. As a consequence, heat fluxes Φ_{12} between two different zones, the zone and the environment Φ_e and into

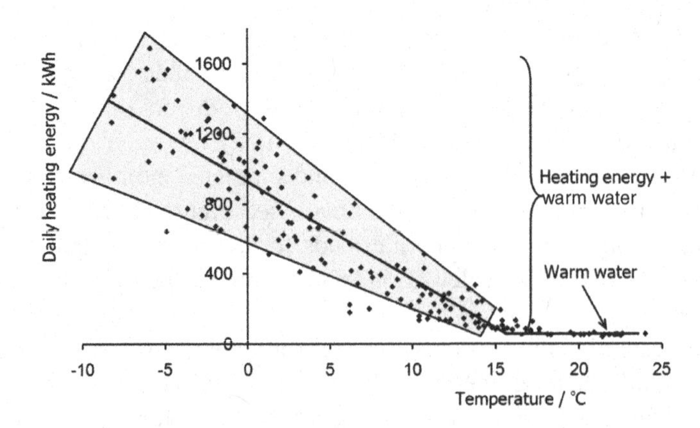

Figure 6.35 Daily heating energy of the Town Hall public building in Ostfildern

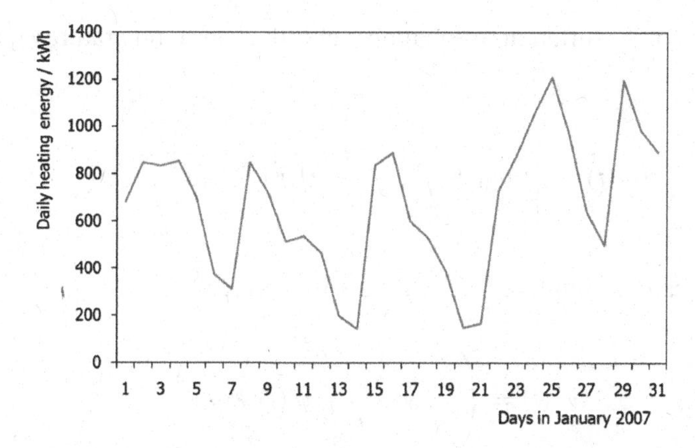

Figure 6.36 Measured daily heating energy for the Town Hall in January 2007. The weekend drop can be clearly seen

the effective storage mass Φ_c arise (see Figure 6.38). Inputs from solar gain, internal gains, heating or cooling sources sum up to a total hourly gain $I\,(t)$:

$$\begin{aligned}
\Phi_c &= C \cdot \dot{\Theta}_1 \\
\Phi_e &= \frac{1}{R_e}\,(\Theta_1\,(t) - \Theta_e\,(t)) \\
\Phi_{12} &= \frac{1}{R_{12}}\,(\Theta_1\,(t) - \Theta_2\,(t)) \\
I\,(t) &= \Phi_c + \Phi_e + \Phi_{12} \\
I(t) &= \dot{q}_{sol} + \dot{q}_{\text{int}} + \dot{q}_{heat} + \dot{q}_{cool}
\end{aligned} \tag{6.18}$$

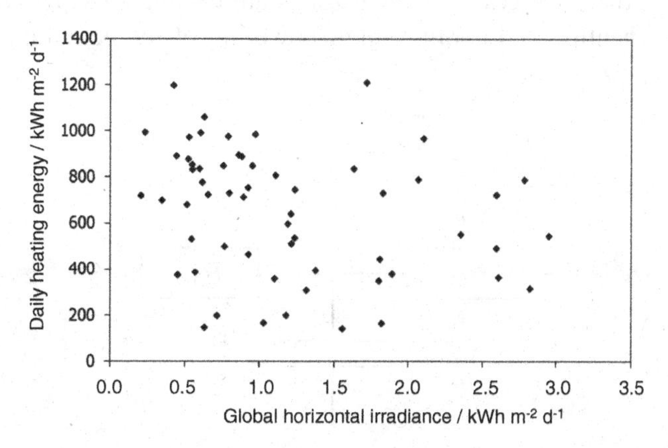

Figure 6.37 Daily heating energy in the Town Hall building as a function of daily global irradiance

The solution of the differential equation yields the zone temperature Θ_1 as a function of time:

$$\Theta_1(t) = \left[\Theta_{1,0} + \int_0^t \left(\frac{\psi(t)}{\tau(t)} e^{\int \frac{1}{\tau(t)} \cdot dt}\right) dt\right] e^{-\int \frac{1}{\tau(t)} \cdot dt} \tag{6.19}$$

The equation can be simplified if the time τ is constant:

$$\Theta_1(t) = \left(\Theta_{1,0} + \frac{1}{\tau}\int_0^t \psi(t)\, e^{\frac{t}{\tau}}\, dt\right) e^{-\frac{t}{\tau}} \tag{6.20}$$

Where the term $\psi(t)$ stands for

$$\psi(t) = \frac{\tau}{C}\left(I(t) + \frac{1}{R_e}\Theta_e(t) + \frac{1}{R_{12}}\Theta_2(t)\right)$$

$$\tau = R_{ges}\, C = \frac{R_e\, R_{12}}{R_e + R_{12}} C$$

If the zone temperature is kept at a user-defined setpoint, the heating or cooling loads required can thus be determined. The model was validated using VDI 6020. Inputs are U-values, g-values and air change rates. No details of the construction are required and there are no numerical stability problems.

The model was implemented in the simulation environment INSEL using a modular structure. Each zone is calculated with a separate block, and earth-connected parts of the building and the interaction between zones are calculated separately. Schedules for air changes, heating and cooling, night lowering of temperature setpoints, etc.,

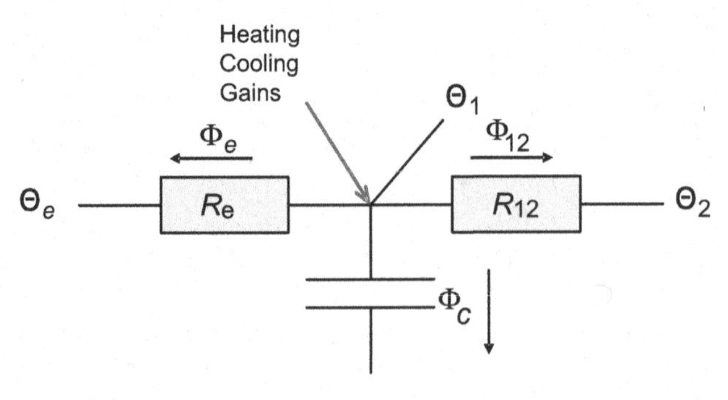

Figure 6.38 Temperature nodes, thermal resistances, heat capacity and heat flows in a simple nodal building model

can be defined for each zone. For the example of the Scharnhauser Park Town Hall, the building was simulated using three different zones, one of which connects to the ground. Temperature setpoints, air change rates and internal gains were defined by schedules. For example, the air change was set at $0.25\,h^{-1}$ during weekends and varied between 0.3 and $0.5\,h^{-1}$ depending on the time during weekdays.

The complete model gives a good agreement between heating energy consumption and simulated demand on a daily basis (see Figure 6.39). Typically the difference between measured consumption and simulation value is between 10 and 20%. On a monthly basis, the simulated values overestimate the measured values by 5%. Typical weekly errors are below 5%, but if weeks with special occupancy occur, the weekly error is near 10%. The implemented online simulation procedure allows the user to follow the building's summer and winter performance on a daily evaluation base and to optimize the performance. Overestimation of the heating energy consumption occurs on weekdays with less occupancy than usual, which is not yet considered in the model. More information about the real occupancy of the building adds to the accuracy of the demand simulation. Currently, work is ongoing to improve the estimation of occupancy rate using the building's electricity consumption. The measured water consumption cannot be used for occupancy estimation, as the building operates a fountain in summer, which distorts the user estimation based on sanitary use (see Figure 6.40). The model is now used for automated energy control and as a fault detection system in the Town Hall. The boundary conditions for meteorological data are obtained from a weather station in the vicinity via the Internet.

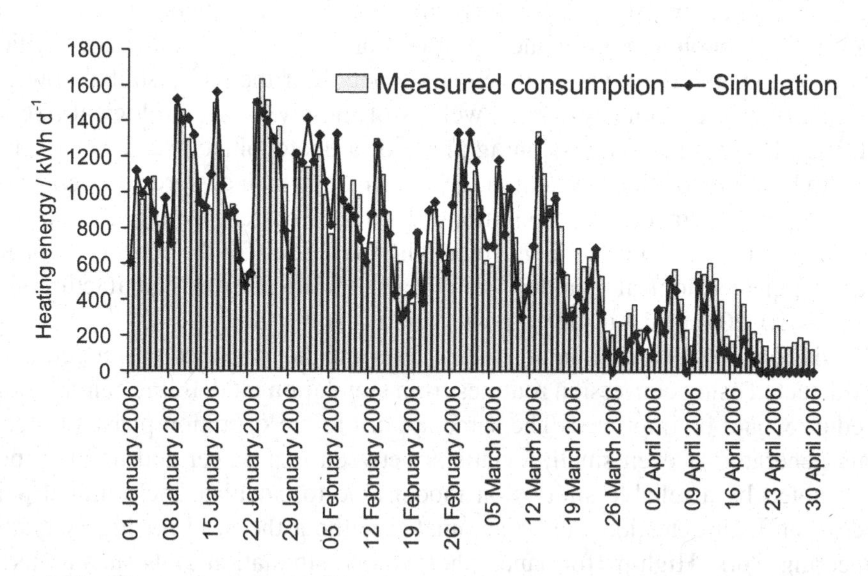

Figure 6.39 Measured and simulated heating energy consumption using a simple dynamic model

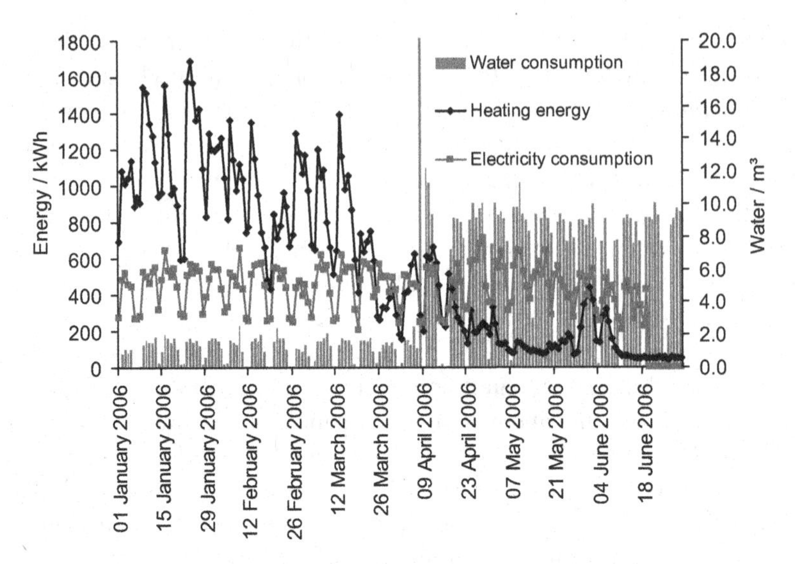

Figure 6.40 Measured daily heating, electricity and water consumption as a basis for internal load calculations

6.3 Online Simulation of Renewable Energy Plants

As the second application of the new area of online simulation, renewable energy plants, here photovoltaic power plants, are simulated. Again, photovoltaic system simulation tools are mostly used for the planning and yield prediction of power plants. However, the economic performance of grid-connected photovoltaic plants critically depends on the electricity yield, which is a function of the real installed power and functionality of the whole system as well as of the given meteorological boundary conditions. During the commissioning phase of a photovoltaic power plant it is important to have photovoltaic system models available which can precisely convert the measured power to standard test conditions. Even if the current–voltage characteristics is measured during commissioning, a model is still needed for the conversion process. At varying meteorological boundary conditions, deviations from manufacturers' data as large as 10–20% of the installed power are difficult to detect.

If deviations between the manufacturers' data and the measured power could be reliably detected using conversion routines from simulation models, the commissioning procedure would be improved. The same applies to the operation phase of a photovoltaic generator: if even small deviations between real power and nominal power can be noted by a reliable simulation model able to calculate the nominal performance, then fault detection can be automated without the need for highly qualified engineering work. High-performance photovoltaic simulation tools such as INSEL, which is used in this work, provide a cheap and reliable possibility to detect faults or power losses in photovoltaic plants (Luther and Schumacher, 1991; Eicker *et al.*, 2005).

Today, simulation tools are mostly used during the design phase to predict the yield of the PV plant. During plant operation, monitoring of the energy production is often considered as sufficient for supervision of the plants. Mostly, only the plant AC power is monitored, as the data is easily available from the inverter. Some attempts have been made to develop low-cost I–V curve sequencers to measure the module performance in more detail (van Dyk *et al.*, 2005), but such systems are not standard practice. Advanced monitoring systems use data loggers, which regularly transmit the measured values via a modem or Internet connection to a server, where the data is visualized and significant faults can be detected by analysing the measured performance (Beyer *et al.*, 2004).

Simulation has been used in the analysis of monitoring results to determine the loss mechanisms, but mostly not as an online tool (Reinders *et al.*, 1999). Only within the PVSat project with its satellite-based meteorological data is a full simulation model used, also based on the INSEL environment to calculate the theoretical yield of the plant once per month (Stettler *et al.*, 2005). As the error of satellite-based hourly irradiance measurements is over 20%, deviations of system yield from the potentially possible yield can only be detected above this error range (Lorenz *et al.*, 2005).

In this work, the role of simulation tools for the commissioning of PV plants and their subsequent operating period is analysed. During the commissioning period, it is important to determine accurately the real installed power and to check the correct functioning of all components. As the measurements are taken under varying irradiance and temperature conditions, models are needed to transfer the measured power to standard test conditions. Different methods of PV generator parameter extraction for the simulation models are analysed.

If the PV power plant has a longer lasting follow-up, it is important to implement automatic routines for fault detection. Here a new methodology is demonstrated for online remote simulation and fault detection. The required communication technology for the data exchange is explained and online simulation of a grid-connected PV generator is carried out.

6.3.1 Photovoltaic System Simulation

An accurate physical model for most PV modules is the two-diode equation for the current density j, which requires six fitted parameters to describe the two exponential functions for the Shockley and recombination diodes (factors c_s and c_r and series resistance r_s), the photocurrent with factors c_0 and c_T and the current loss through the parallel resistance r_p. The diode factors are 1 for the Shockley diode term and 2 for the recombination term:

$$j = j_{ph}(c_0, c_T) - j_s(c_s)\left[\exp\left(\frac{q(V + jr_s)}{1kT}\right) - 1\right]$$
$$-j_r(c_r)\left(\exp\left[\frac{q(V + jr_s)}{2kT}\right) - 1\right] - \frac{V + jr_s}{r_p} \qquad (6.21)$$

To determine the fitted parameters, current–voltage curves (I–V curves) at different irradiance and temperature levels are required, which are not available for the majority of modules on the market. Work has also been done on using nonlinear regressions on dark I–V curve measurements at different temperature levels to determine all parameters except the photocurrent, which has to be obtained from light measurements (King *et al.*, 1996). Again the data is usually not available from the manufacturers.

If measured I–V curves are given at one or more irradiance and temperature levels, INSEL provides a block for the determination of all parameters. The fitted algorithm is based on linearizing the equation system for all measured current density–voltage (j, V) data point sets. This is achieved by setting the series resistance r_s to a fixed value and then solve the overdetermined linear equation system with five unknowns. A genetic algorithm is then used to vary the series resistance and solve the equation system again, until the global minimum quadratic error between the measured and simulated curve is obtained.

As an alternative, a new method was developed in Stuttgart to obtain analytically six parameters for the one diode equation from the manufacturer's data sheet. In the one-diode model, the diode factor α is now a free parameter, and there is only one coefficient c_s needed for the exponential term. To determine all six parameters, the short-circuit current, the maximum power point (MPP) voltage and current, the open-circuit voltage and the temperature coefficients for current and voltage are required. The new feature in the implemented algebraic equations is the assumption that an infinite shunt resistance is no longer required. INSEL now contains the one-diode parameters for more than a thousand PV modules:

$$j = j_{ph}(c_0, c_T) - j(c_r)\left[\exp\left(\frac{q(V + jr_s)}{\alpha kT}\right) - 1\right] - \frac{V + jr_s}{r_p} \qquad (6.22)$$

The PV generator is operated with an MPP tracker, which varies the voltage on the I–V curve until the product of current and voltage is maximized. The inverter is modelled using the Schmidt–Sauer equation requiring only three partial load efficiencies for the model parameter extraction (Schmidt and Sauer, 1996). The energy dissipation is made up of the losses which are independent of input power (internal current supply and magnetization losses p_{own}), the losses in the semiconductor switches v_{switch} which are linearly dependent on the power output, and the ohm cable losses r_{ohm} which rise as the square of the AC performance. The inverter efficiency can be represented to good accuracy by a second-order polynomial and the efficiency equation is solved for the input DC power p_{DC}

$$\eta = \frac{p_{AC}}{p_{DC}} = \frac{p_{DC} - \left(p_{own} + v_{switch}\, p_{AC} + r_{ohm}\, p_{AC}^2\right)}{p_{DC}}$$
$$= 1 - \frac{p_{own}}{p_{DC}} - v_{switch}\eta - r_{ohm}\eta^2 p_{DC} \tag{6.23}$$

$$\eta = \frac{1 + v_{switch}}{2r_{ohm}\, p_{DC}} \pm \sqrt{\frac{(1 + v_{switch})^2}{(2r_{ohm}\, p_{DC})^2} + \frac{p_{DC} - p_{own}}{r_{ohm}\, p_{DC}^2}} \tag{6.24}$$

The three loss coefficients p_{own}, v_{switch} and r_{ohm} can be calculated from three efficiency values η_1, η_2 and η_3 given by most manufacturers in technical data sheets for power ratios of, for example, $p_1 = 10\%$, $p_2 = 50\%$ and $p_3 = 100\%$ of the rated power.

The input to the model is the DC power. Parameters for 500 inverters are provided in the INSEL database.

6.3.2 Communication Strategies for Simulation-Based Remote Monitoring

To allow precise system simulation of a PV power plant, monitoring data for the meteorological boundary conditions such as irradiance and ambient temperature has to be available. Irradiance data should best be available in the module plane, or can be calculated from horizontal data to high accuracy. To detect faults in the power plant, DC voltage and current and AC power should be monitored. If possible the module temperature should be measured, but it can also be calculated from ambient temperature and more precisely from temperature and wind speed. All the data need to be transferred from the data logging equipment to a server, from which clients can access the data, visualize it and simulate performance.

There are different server technologies with varying degrees of response time: ftp servers only transfer on request complete data files, which are produced at a set time interval, for example every hour. Alternatively, a web server provides HTML pages with included XML data, which are actualized in a given time interval. If monitoring data is stored locally in a database, which is often the case in BMS, an elegant method is to access the database server via open database connectivity communication (ODBC).

For real-time access to the measured data, socket servers such as DataSocket from National Instruments write the data to the computer memory, where it is continuously actualized. Other protocols such as OPC provide the same functionality. The simulation environment INSEL with its graphical interface VEE already includes the DataSocket server and client blocks, where monitored data can be easily written to the memory (server) and read from the server for the simulation process (client or simulation agent).

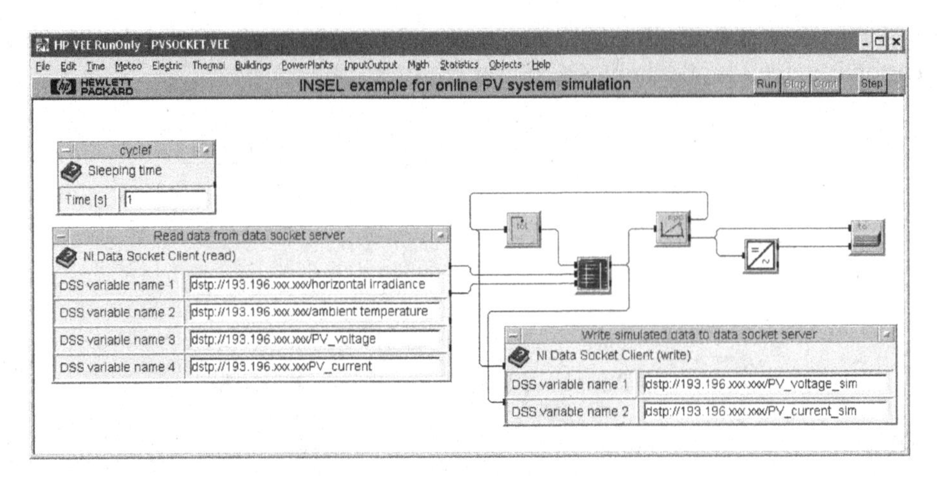

Figure 6.41 INSEL block diagram for a grid-connected PV system in the graphical editor, where measured data is read from a data socket server and simulated data is written back

To avoid data losses in case of communication problems with the socket connection using computers on different sites, it is advisable to run a local client on the server computer, which just writes hourly or daily data files. In the demonstration application all the measured data is written to a data socket server in Offenburg from a LabVIEW visualization. The Stuttgart data socket client accesses the real-time data each second, performs a complete system simulation for the given meteorological boundary conditions and writes back the simulated results to the data socket server – all within a millisecond (see Figure 6.41). If the graphical interface from Agilent VEE is used as the data acquisition software, all tasks such as data logging, visualization, communication and simulation are combined within the same INSEL simulation environment. If other software packages such as LabVIEW are employed for monitoring purposes, all INSEL blocks for simulation and communication can be easily integrated as functions from dynamic link libraries. VEE is mainly used in the INSEL simulation environment to construct simulation models freely using the available blocks within the menu structure. In addition, powerful visualization tools are provided. The visualization scheme can easily integrate the simulated values from the online remote simulation and issues alarms if significant deviations are detected between simulation and measurement.

6.3.3 Online Simulation for the Commissioning and Operation of Photovoltaic Power Plants

Experimental Set-up for Remote PV System Simulation

To test and demonstrate the method, a small grid-connected PV power plant was monitored at the University of Applied Sciences in Offenburg and remotely simulated at the University of Applied Sciences in Stuttgart. The PV generator consists of six 90 W_p modules from Photowatt at a tilt angle of 25°. The module is made of 72 polycrystalline silicon solar cells of 101.25 × 101.25 mm, coated with silicon nitride anti reflection material. The system is coupled to a 500 W Würth Solergy inverter.

PV Module Characteristics

In the first step, six one-diode model parameters and the inverter characteristics had to be determined using the data sheet information from the 90 W Photowatt PW 1000 module and the Würth Solergy WE 500 NWR inverter data.

The INSEL block pvdet1 was used to calculate the one-diode parameters from the manufacturers' data. The values extracted from the Photowatt module data sheet are summarized in Table 6.10 below. Using the extracted parameters for the simulation model, the performance of the PV system was calculated, written to the server and visualized.

The simulated performance was in general 15% higher than the measured performance (see Figure 6.42). This means that the manufacturers' performance values do not correspond to the module power of the system installed.

As a consequence, I–V curve measurements were carried out on the generator. A Stella field tester was used at high irradiance levels of about 900 W m^{-2}. A two-diode curve fitting routine implemented as a block in INSEL (pvfit2) was then applied to all three measured curves to extract the module parameters of the two-diode equation, which are summarized in Table 6.8.

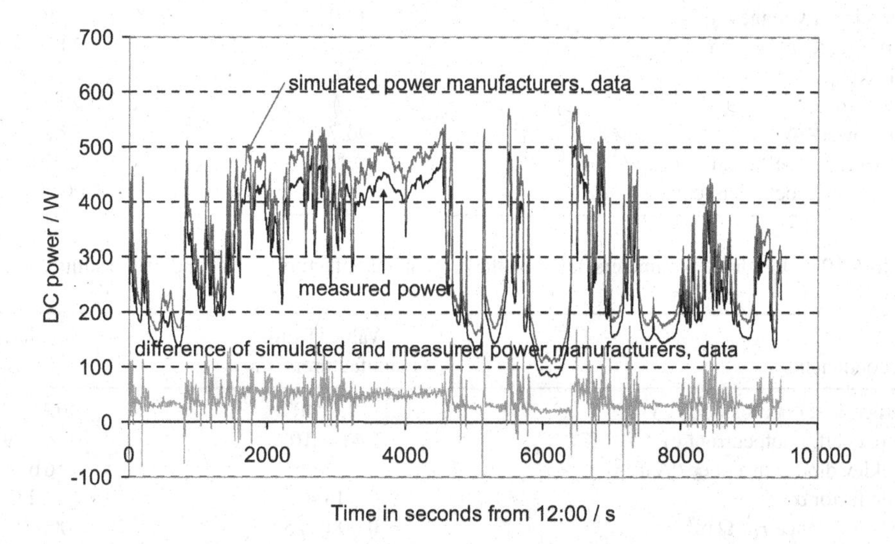

Figure 6.42 Measured and simulated power using the manufacturers' data sheet information

Table 6.8 Two-diode parameters extracted from measured I–V curves

Parameter	Simulated values
Photocurrent coefficient c_0 / V^{-1}	0.1613
Temperature coefficient photocurrent c_T/ V^{-1} K^{-1}	0.000 411 51
Shockley diode current parameter c_s / A m^{-2}	17 139
Recombination diode current parameter c_r / A m^{-2}	1.713
Series resistance r_s / Ω m^2	0.000 371 3
Shunt resistance r_p / Ω m^2	0.048 05

The two-diode equation was then used to calculate standard test conditions (STC) for the module and the results were compared with the data sheet information from the manufacturers. The reduction of MPP power from the 90.7 W manufacturers' information to 77.2 W from the online simulation correspond to the 15% power reduction mentioned above (see Table 6.9).

Using these STC values calculated by the two-diode model as the new data sheet information, the one-diode model parameters were again extracted (see Table 6.10). They are much more consistent than the original data, where the shunt resistance was negative and the photocurrent temperature coefficient was far too small.

The two-diode curve fitted to three measured I–V curves is shown in Figure 6.43. The two-diode curve under STC conditions and the derived one-diode curve at STC show very similar behaviour.

Table 6.9 STC conditions calculated with two-diode model versus manufacturers data

Parameter	Manufacturers' data	Simulated values
Open-circuit voltage / V	43.0	42.36
Short-circuit current / A	2.8	2.89
MPP voltage / V	33.6	30.41
MPP current / A	2.7	2.54
MPP power / W	90.7	77.2
Temperature coefficient voltage / V K^{-1}	−0.156	−0.151
Temperature coefficient current / A K^{-1}	0.000 93	0.004 04

Table 6.10 One-diode parameters based on the original manufacturers' data and on the simulated STC curve

Parameter and coefficients	Values from manufacturers' data	Simulated values
Photocurrent coefficient c_0 / V^{-1}	0.2240	0.1557
Temp. coeff. photocurrent c_T / V^{-1} K^{-1}	7.44×10^{-6}	0.000 394
Shockley diode current c_s / A m^{-2}	14 066	16 689
Diode factor α	1.04	1.0232
Series resistance r_s / Ω m^2	0.000 24 59	0.000 361 9
Shunt resistance r_p / Ω m^2	−0.1134	0.048 87

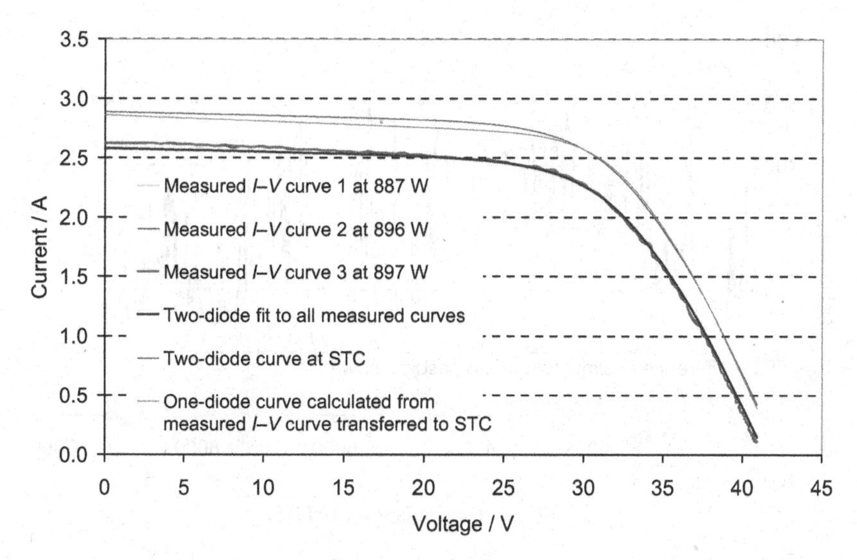

Figure 6.43 Measured I–V curves, two-diode model fitted to the I–V curves, transformation of two-diode curve to STC, and calculation of one-diode parameters from STC conditions

The whole system was then simulated online using the irradiance on the tilted plane and module temperature only as an input. As the process of data transfer and simulation takes only milliseconds, simulations were first carried out on a second basis. However, the deviation between measurement and simulation was rather high, as the measured irradiance information from the pyranometer has a slow time constant of about 20 seconds. For fast-moving clouds, a sharp drop in irradiance leads to only a slow reduction of the measured pyranometer value, whereas the PV plant immediately reduces power. The simulation value is based on the pyranometer measurement and thus overestimates the simulated power. Conversely, in the case of a rapid rise in irradiance, the pyranometer measures too low a value and the PV model underestimates the simulated power, so that the error can be 10% or higher (see Figure 6.44).

If the monitored data is averaged over an interval of 1 minute, the difference between simulation and measurement is nearly zero, because the pyranometer time constant is no longer significant (see Figure 6.45).

The simulation data fits the measured power data excellently even on a day with wildly fluctuating meteorological conditions. The difference in energy output over the measurement period of 3 hours is 0.5%. The mean arithmetic error between measurement and simulation is 2%.

6.3.4 Summary of Renewable Energy Plant Online Simulation

In this work, a methodology was proposed and demonstrated to simulate grid-connected photovoltaic power plants not only in the planning stage, but also online

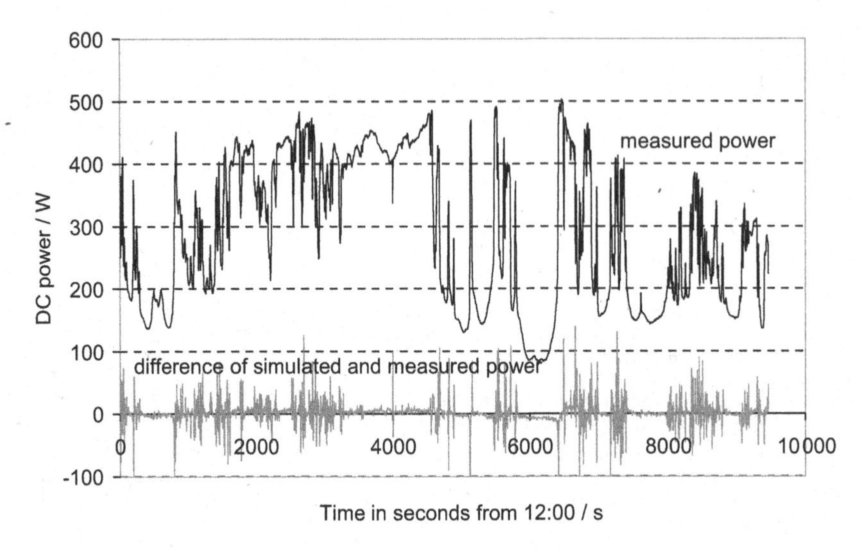

Figure 6.44 Measured and simulated DC power of the MPP tracked PV generator at a measurement time interval of 1 second on the 6.4.2005

using Internet communication. The simulation models used are physically accurate models of the photovoltaic modules, the MPP tracker and the inverter and reproduce real-time measured data within a 2% error. All physical models together with the parameter sets for over 1000 modules and 300 inverters are implemented with the required communication blocks in the simulation environment INSEL.

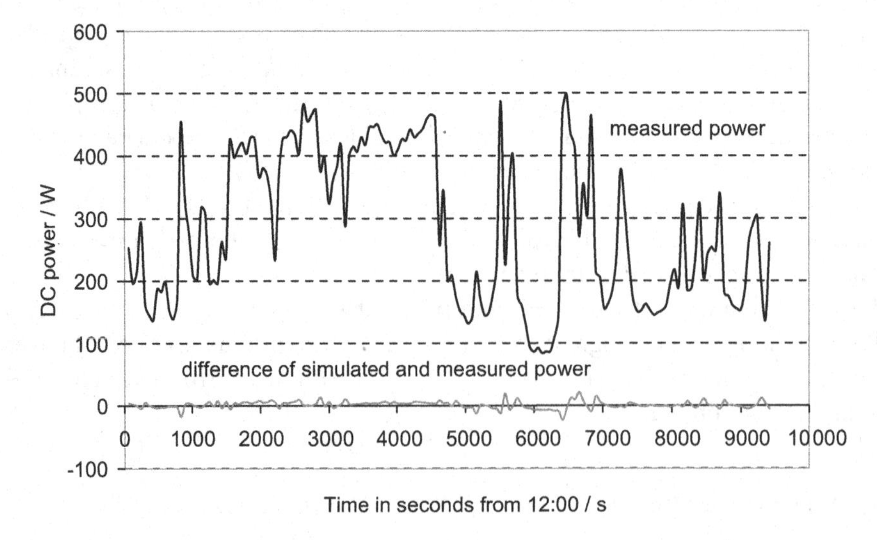

Figure 6.45 Measured and simulated DC power of the MPP tracked PV generator at a measurement averaging time interval of 1 minute on 6 April 2005

Real-time access to monitored data is possible using a range of protocols such as socket communication. The online simulation of a small photovoltaic power plant revealed that the real installed power was 15% lower than given by the manufacturer. Using curve fitting algorithms to measured *I–V* curves, new parameter sets for the generator could be determined, which were then used for online simulation. At a time averaging interval of 1 minute, a mean arithmetic error of 2% could be achieved for online simulation. This allows very precise fault detection plant operation.

The method allows during the nominal power of photovoltaic generators to be derived for any given meteorological conditions and is thus excellently suited for the commissioning procedure. With the parameter sets calculated, fault detection during the operating lifetime is possible within an error range of a few per cent.

7

Conclusions

The energy consumption in buildings has been analysed with a special emphasis on summer performance and low-energy cooling supply. This focus is supported by the ongoing implementation process of the European Energy Performance Directive of Buildings, which contributes to the development of unified and transparent calculation methods not only for heating energy, but also for cooling, lighting and electricity demand. The main factors contributing to the summer performance and thermal comfort in residential and office buildings were studied using laboratory experiments and monitoring results from buildings. In addition, passive and active cooling systems were developed and analysed.

The thermal performance of glazed façades in various construction typologies was characterized in detail, as they determine the external loads of a building. It was shown that the total energy transmittance through single and double façades can be low if sun protection devices are used effectively. Double façades with varying blind locations achieve *g*-values measured under a laboratory solar simulator of about 10%, that is lower than a single façade with closed blinds. Façades implemented in a building have even lower measured *g*-values, as incidence angles especially under summer conditions are high. Secondary gains only play a role in façade systems with low-e coated glazing, if no shading is used. The shading element reduces the room-side surface temperature by about 6 K at irradiance levels around 500 W m^{-2}.

However, there are non-negligible ventilation gains if the ventilation air is taken directly from the façade. Air gap temperature increases were between 2 and 6 K, depending on the free cross-section of the air entry. The measured blind temperature

Low Energy Cooling for Sustainable Buildings Ursula Eicker

was up to 15 K above air inlet temperature. For tilted windows, measured air exchange rates with the room are typically between 0.5 and 2 h^{-1}, if no cross-ventilation takes place. Daily heat gains obtained were then between 36 and 86 Wh $m^{-2} d^{-1}$ for a standard office geometry; that is, of the same order of magnitude as thermal gains from people.

When ventilated photovoltaic façades are employed, it is possible to collect the thermal energy produced in the air gap with thermal efficiencies between 15 and 30%. Despite losses in thermal efficiency, it is recommended to work with gap sizes over 10 cm or to keep flow velocities very low in order to reduce pressure drops and obtain high coefficients of performance between the heat produced and electricity employed. In the best case, COPs of 50 could be achieved. In summer the thermal energy can be used to prewarm air for desiccant cooling systems.

The magnitude of cooling loads in office rooms today is often not well known, although it is a prerequisite for any cooling system design. The range and distribution of internal loads were analysed in different offices in great detail over a 3-year period. Internal loads were as high as 400–500 Wh $m^{-2} d^{-1}$ for a standard two-person office situation, with about 20% due to the people, 5–10% to lighting and the rest to electrical appliances such as computers and printers. Such internal loads are significantly higher than external loads from a shaded façade. Comfort levels for typical German summer conditions are good using purely passive night ventilation strategies. Measured air exchange rates – mainly caused by cross-ventilation – were high in the case study building with average summer values of 9.3 h^{-1}. To remove daily loads of about 8–10 kWh at this air change rate, the temperature difference between room and outside air must be 6 K for 9 hours of night ventilation. As this is not always the case in warm summers, for example in the year 2003, here nearly 10% of all office hours were above 26°C.

Mechanical night ventilation leads to better control of night air flows, but the electricity consumption is significant. In a well-designed passive building in Tübingen, measured COPs of the night ventilation were about 4.0 with maximum values of 6.0. Also, the rather low air exchange rates of the monitored mechanical ventilation systems of about 2 h^{-1} were far from sufficient to decrease significantly the room air temperatures. A much better performance was obtained from earth heat exchangers, both air and water based. Here the average annual COPs were as high as 35 to 50. The only disadvantage is the rather limited cooling contribution, as only fresh air is used for the earth heat exchange and as water-based distribution systems such as activated ceilings cannot deliver high cooling power.

At higher cooling energy demand, active cooling technologies are often required. Different solar thermal cooling technologies were analysed and an overview of existing projects was given. Design and monitoring results from two large desiccant cooling systems were presented. It was shown that under German climatic conditions, the coincidence of full regenerative operation and available solar power was rather low. However, simulation studies demonstrated that auxiliary heating can be nearly completely avoided if the control strategy is adapted to the available solar thermal temperature levels. The COPs were between 0.5 and 1.0, increasing with lower regeneration temperature. This gives an added advantage of reducing auxiliary heating or better use of auxiliary cooling.

New developments in diffusion–absorption cooling and liquid desiccant use were presented for the low cooling power range below 10 kW. The main challenge in the development of diffusion–absorption chillers is the design of an indirectly heated bubble pump, which operates under slug flow conditions. Stable flow conditions can achieved with convection and nucleate boiling with lifting ratios between liquid and vapour of 4–5. Under nucleate boiling conditions, heat transfer is optimized and the highest lifting ratio is obtained. However, as long as the required solution flow is sufficient, lower driving temperature differences between tube wall and fluid boiling temperature are advisable to reduce the external generator temperature. On the low partial pressure side, the main difficulty is to achieve a good performance in the falling film evaporator and absorber. Unequal liquid distribution into the tubes of the evaporator led to low evaporation rates in the first prototypes, which could be subsequently improved by changing the construction between the evaporator top plate and tubing inlet. Much experimental experience was gained with heat exchanger constructions. Both shell-and-tube and plate heat exchangers did not give satisfactory heat recovery results for the solution heat exchange, as solution flow rates are very low. The best results were obtained for coaxial heat exchangers with heat recovery factors of up to 92% (weak solution). The rather slow diffusion rate of the refrigerant into the auxiliary gas atmosphere increases with lower total system pressure, improving the whole performance of the system. With the last prototype, COPs of nearly 0.4 can be achieved.

A liquid desiccant cooling unit was developed and tested, providing sensible cooling of the fresh air in small ventilation units. The process is based on exhaust air drying in a spray-cooled heat exchanger absorber unit using LiCl or CaCl salt solutions. The nearly isothermal drying process of the exhaust air is then followed by a further spray-cooled heat exchanger, where heat is transferred from the warm supply air to the humidified cool exhaust air. For a small system with a volume flow of only 200 $m^3 h^{-1}$ a cooling power of nearly 1 kW was obtained. This system offers interesting applications in the residential building sector.

The role of computer simulation not only for the design, but also for the operation of complex energy plants and buildings was analysed. Especially for solar thermal cooling plant design, it is very important to know the details of the cooling load time series. The correlation between the cooling machine power and required solar thermal collector area is very weak and varies by a factor of 10 depending on full load hours and location. There is a much better correlation between the collector area and cooling energy, varying for example at locations in Spain between 3 and 6 m^2 per megawatt hour of building cooling energy. A good cost-effectiveness can be obtained if the system is operated with an optimized control strategy for partial load conditions, which allows a reduction of the solar collector driving temperature and as a consequence the collector field size. The use of new communication tools offers the possibility to extend computer simulation to the operational control of energy plants and buildings. This will increase the transparency of a building's energy flows and support the optimization process during a building's lifetime. Online comparisons of building energy consumption and simulated demand have been carried out. Good agreement between simulation and measurement could only be obtained if changing internal loads caused by user behaviour could be assessed with reasonable accuracy. Furthermore, online simulations of power plants were carried out, which showed the usefulness of simulation tools during the commissioning and operating time of the plants.

In conclusion, the work aimed to increase knowledge of the summer performance of buildings and of technologies for supplying cooling energy at low primary energy consumption. The potential and limitations of passive and active cooling systems were demonstrated by a range of monitored building projects, laboratory experiments and simulation studies.

References

Adnot, J. (1999) Final report 'Energy efficiency of room air-conditioners'. Study for the Directorate General for Energy DG XVII, contract 4.1031 /D/97.026.

Adnot, J., Waide, P. *et al.* (2003) Final report 'Energy Efficiency and Certification of Central Air Conditioners (EECCAC)', Vol. 2. Study for the D.G. Transportation-Energy (DGTREN) of the Commission of the EU.

Afonso, A., Collares-Pereira, M., De Oliveira, J.C. *et al.* (2003) A solar/gas powered absorption prototype to provide small power heating and cooling. Proceedings of the ISES Solar World Congress, 16–19 June, International Solar Energy Society (ISES), Göteborg, Sweden.

AIT-Spezial, Intelligente Architektur 12 (Feb. 1998) Sonderheft der AIT – Architektur Innenarchitektur Technischer Ausbau (Special Edition on Intelligent Architecture AIT – Architecture, Interior Architecture, Technical Improvement). Verlagsanstalt Alexander Koch GmbH, Leinfelden-Echterdingen, Germany.

Ajib, S., Schultheis, P. (1998) Untersuchungsergebnisse einer solarthermisch betriebenen Absorptionskälteanlage. *TAB Technik am Bau*, Vol. 29(2), pp. 49–54.

Ajib, S., Nilius, A., Karno, A. (2004) Erste Untersuchungsergebnisse eines Versuchsstandes einer niedertemperaturbetriebenen Absorptionskältemaschine (First investigation results at a test stand of a low-temperature driven Absorption Refrigeration Machine). Proceedings of the 14th Symposium Thermische Solarenergie, Ostbayerisches Technologie-Transfer-Institut eV (OTTI), Staffelstein, Germany. pp. 488–493.

Albers, K.-J. (1991) Untersuchungen zur Auslegung von Erdwärme – Austauschern für die Konditionierung der Zuluft für Wohngebäude. *Forschungsberichte des Deutschen Kälte- und Klimatechnischen Vereins*, No. 32.

Albers, J. (2002) Betriebserfahrungen mit solar gestützter Absorptionskälte bei den Bundesbauten in Berlin (Experiences gained in the operation of solar-assisted absorption cooling at the buildings of the German government in Berlin). Proceedings of the 2nd Symposium Solares Kühlen in der Praxis, Fachhochschule Stuttgart-Hochschule für Technik, Germany. Vol. 56, pp. 135–144.

Albers, J., Ziegler, F. (2003) Analysis of the part load behaviour of sorption chillers with thermally driven solution pumps, Proceedings of the 21st IIR International Congress of Refrigeration, 17–22 August, International Institute of Refrigeration (IIR), Washington, DC.

Albring, P. (2001) Anlagen und Systeme der solaren Kälteerzeugung (Plants and systems of the solar cold production). Proceedings of the 1st Symposium Solares Kühlen in der Praxis, Fachhochschule Stuttgart - Hochschule für Technik, Germany. Vol. 53, pp. 105–123.

Altfeld, K. (1985) *Exergetische Optimierung flacher solarer Lufterhitzer*, VDI Fortschrittsberichte Reihe 6, No. 75, VDI Verlag.

Low Energy Cooling for Sustainable Buildings Ursula Eicker

Argiriou, A. (1996) Ground cooling. In Santamouris, M. and Asimakopoulos, D. (Eds) *Passive Cooling of Buildings*, James & James.

Arsenal Research (2007) Solares Kühlen für Büro- und Dienstleistungsgebäude – a study for the Magistratsabteilung 27, Vienna.

Ashford, P. (1998) Assessment of potential for the saving of carbon dioxide emissions in European building stock. Report prepared for EuroACE building energy efficiency alliance.

Atmaca, I., Yigit, A. (2003) Simulation of solar-powered absorption cooling system. *Renewable Energy*, Vol. 28, pp. 1277–1293.

Auer, F. (1998) Bericht über 2. Europäischen Solarkongress in Slowenien (Report to the Second European Solar Congress in Slovenia). Sonnenenergie 5/1998.

Bachofen, F. (1999) European Patent Office, patent number EP 0 959 307 A1, declaration number 9881042.5, declarant POLYBLOCK AG 8404 Winterthur (CH), publication day 24 November 1999, Patent Paper 1999/47.

Balares, C. (2003) Recent activities on absorption cooling. Proceedings of the Workshop on Absorption Cooling – Research & Development, 17–18 January 2003. Arsenal Research, Vienna.

Beccali, M., Butera, F., Guanella, R. *et al.* (2003) Simplified models for the performance evaluation of desiccant wheel dehumidification. International Journal of Energy Research, Vol. 27, pp 17 – 29.

Breembroek, G., Lazáro, F. (1999) International heat pump status and policy review 1993-1996. IEA Heat Pump Centre.

Beyer, H.G., Heilscher, G., Bofinger, S. (2004) A robust model for the MPP performance of different types of PV-modules applied for the performance check of grid connected systems. Proceedings of the Eurosun, Freiburg.

Bi, Y., chen, L., Wu, C. (2001) Measured performance of a solar-ground source heat pump system with vertical double spiral coil ground heat exchanger. *International Journal of Ambient Energy*, Vo. 22(1), pp. 3–11.

Biesinger, A. (2002) Untersuchungen an einem indirekt beheizten Austreiber einer Diffusions-Absorptionskältemaschine (DAKM). Diploma thesis, Fachhochschule Stuttgart-Hochschule für Technik.

BINE Profi-Info (2000) Elektrischer Energiebedarf Beleuchtung, 2/2000.

Binz, A. (2006) Minergie and Minergie P – the Swiss energy labels. POLYCITY Workshop on Sustainable Town Planning and Energy Benchmarking of Buildings, 2–3 February 2006, Basle.

Blum, H.J. (1998) Das innovative Raumkonzept (The innovative indoor climate concept). *Bauphysik* 20/1998, Vol. 3, pp. 81–86.

Bourdoukan, P., Wurtz, E., Sperandio, M. *et al.* (2007) Global efficiency of direct flow vacuum collectors in autonomous solar desiccant cooling: simulation and experimental results. Proceedings of the Building Simulation Conference IBPSA, Bejing.

Bourseau, P., Bugarel, R. (1986) Absorption-diffusion machines: comparison of the performance of NH_3-H_2O and NH_3-HaSCN. *International Journal of Refrigeration*, Vol. 9(4), pp. 206–214.

Bourseau, P., Mora, J.C., Bugarel, R. (1987) Coupling of absorption-diffusion refrigeration machine and a solar flat-plate collector. *International Journal of Refrigeration* Vol. 10(4), pp. 209–216.

Braun, R., Hess, R. (2002) Solar Cooling. Proceedings of the 7th World Renewable Energy Congress, 1–5 July, World Renewable Energy Network (WREN), Cologne.

Breembroek, G., Lazáro, F. (1999) International heat pump status and policy review 1993-1996. IEA Heat Pump Centre.

Brendel, T., Spindler, K., Mueller-Steinhagen, H. (2004) Aufbau einer Versuchs- und Demonstrationsanlage zur solaren Kühlung am Institut für Thermodynamik und Wärmetechnik der Universität Stuttgart (Set-up of a test and demonstration rig for solar cooling at the Institute of Thermodynamic and Heat Technology of the University of Stuttgart). Proceedings of the 31st Deutsche Klima-Kälte-Tagung, Bremen, Germany. Vol. AA.II.1, pp. 145–158.

Brownsword, R.A., Fleming, P.D., Powell, J.C. *et al.* (2005) Sustainable cities–modelling urban energy supply and demand. *Applied Energy* Vol. 82, pp. 167–180.

Buswell, R.A., Haves, P., Wright, J.A. (2003) Model-based conditioning monitoring of a HVAC cooling coil sub-system in a real building. *Building Services Engineering Research and Technology*, Vol. 24(2), pp. 103–116.

Carslaw, H.S., Jaeger, J.C. (1947) *Conduction of heat in solids*. Oxford.

Castro, J. (2003) Development of a prototype of an air-cooled water-LiBr Absorption Cooling Machine for solar cooling using numerical simulation tools. Proceedings of the Workshop on Absorption Cooling – Research & Development, 17–18 January 2003, Arsenal Research, Vienna.

Cattaneo, A.G. (1935) Über die Förderung von Flüssigkeiten mittels der eigenen Dämpfe – Thermosyphon-Prinzip. *Zeitschrift für die gesamte Kälteindustrie*, Vol. 42(1), pp. 2–8, Vol. 42(2), pp. 27–32, Vol. 42(3), pp. 48–53.

Chaudhari, S.K., Patil, K.R. (2002) Thermodynamic properties of aqueous solutions of lithium chloride. *Physics and Chemistry of Liquids*, Vol. 4(3), pp. 317–325.

Chen, J., Kim, K.J., Herold, K.E. (1996) Performance enhancement of a diffusion-absorption refrigerator. *International Journal of Refrigeration*, Vol. 19(3), pp. 208–218.

Chinnappa, J.C.V. (1992) Principles of absorption systems machines. In Sayigh, A.A.M. and McVeigh, J.C. (Eds) *Solar Air Conditioning and Refrigeration*, 1st edn. Pergamon Press.

Churchill, S.W. (1977) A comprehensive correlating equation for laminar, assisting, forced and free convection. *AIChE Journal*, Vol. 10, pp. 10–16.

Claridge, D.E., Haberl, J., Liu, M. *et al.* (1994) Can you achieve 150% of predicted retrofit savings? Proceedings of the ACEEE 1994 Summer Study on Energy Efficiency in Buildings, American Council for an Energy Efficient Economy, Washington, DC.

Clarke, J. A., Cockroft, J., Connera, S. *et al.* (2002) Simulation-assisted control in building energy management systems. *Energy and Buildings*, Vol. 34(9), pp. 933–940.

Conde, M. (2003) Aqueous solutions of lithium and calcium chlorides: property formulations for use in air conditioning equipment design. M. Conde Engineering, Zurich, private communication.

Conde, M. (2004) Processes in air dehydration with porous membranes and salt solutions. M. Conde Engineering, Zurich, private communication.

Dahlem, K.-H. (2000) Der Einfluss des Grundwassers auf den Wärmeverlust erdreichberührter Bauteile (The effect of groundwater on the heat loss of building parts in contact with the ground). *Berichte aus Praxis und Forschung des Fachgebiets Bauphysik*, Bd. 1. Universitätsbibliothek, Kaiserslautern.

Davanagere, B.S., Sherif, S.A., Goswami, D.Y. (1999) A feasibility study of solar desiccant air conditioning system – Part I: psychometrics and analysis of the conditioned zone. *International Journal of Energy Research*, Vol. 23, pp. 7–21.

Dibowski, G., Rittenhofer, K. (2000) Über die Problematik der Bestimmung thermischer Erdreichparameter. *Heizung Lüftung/Klima Haustechnik*, Vol. 51, pp. 32–41.

DKV Statusbericht No. 22 (2002) www.dkv.org.

Duff, W.S., Winston, R., O'Gallagher, J.J. *et al.* (2003) Novel ICPC solar collector/double effect absorption chiller demonstration project. Proceedings of the International Solar Energy Conference, March, ASME, Hawaii.

EAW Company product information (2003) Absorptionskälteanlage Wegracal SE 15 – Technische Beschreibung (Absorption Chiller Wegracal SE 15 – technical description). EAW Energieanlagenbau GmbH, Germany.

Eicker, U. (2003) *Solar Technologies for Buildings*. John Wiley & Sons, Ltd.

Eicker, U., Höfker, G., Seeberger, P. *et al.* (1998) Building integration of PV and solar air heaters for optimised heat and electricity production. 2nd World Conference and Exhibition on Photovoltaic Solar Energy Conversion, Vienna.

Eicker, U., Fux, V., Infield, D. *et al.* (1999) Thermal performance of building integrated ventilated PV façade. Proceedings of ISES 1999, Solar World Congress, Israel.

Eicker, U., Fux, V., Infield, D. *et al.* (2000) Heating and cooling potential combined photovoltaic-solar air collector façade. Proceedings of 16th European Photovoltaic Solar Energy Conference, Glasgow. pp. 1832–1839.

Eicker, U., Pietruschka, D., Schumacher, J. *et al.* (2005) Improving the energy yield of PV power plants through internet based simulation, monitoring and visualisation. 20th European Photovoltaic Solar Energy Conference, Barcelona.

Eklöf, C., Gehlin, S. (1996) TED – a mobile equipment for thermal response test. MSc thesis, Lulea University of Technology, Sweden.

Energy Information Administration (2006) Detailed tables of 2003 Commercial Buildings Energy Consumption Survey, official energy statistics from the US Government.

Engler, M., Grossmann, G., Hellmann, H.-M. (1997) Comparative simulation and investigation of ammonia-water: absorption cycles for heat pump applications. *International Journal of Refrigeration*, Vol. 20(7), pp. 504–516.

Entex (2004) Gaswärmepumpe aus der Schweiz. *Sonne, Wind & Wärme*, Vol. 28(10), p. 34, www.entex-energy.ch.

Eskilson, P., Claesson, J. (1988) Simulation model for thermally interacting heat extraction boreholes. *Numerical Heat Transfer*, Vol. 13, pp. 149–165.

European Commission (2006) Action plan for energy efficiency, KOM (2006) 545.

EWI/Prognos (2005) Studie die Entwicklung der Energiemärkte bis zum Jahr 2030, energiewirtschaftliche.

Faggembau, D., Costa, M., Soria, M. *et al.* (2003) Numerical analysis of the thermal behaviour of glazed ventilated facades in Mediterranean climates. Part II: applications and analysis of results. *Solar Energy*, Vol. 75(2), pp. 29–39.

Fink, C., Blümel, E., Kouba, R. *et al.* (2002) Passive Kühlkonzepte für Büro- und Verwaltungsgebäude mittels luft- bzw. wasserdurchströmten Erdreichwärmetauschern. *Berichte aus Energie- und Umweltforschung*, 35/2002, Bundesministerium für Verkehr, Innovation und Technologie, Vienna.

Florides, G.A., Kalogirou, S.A., Tassou, S.A. *et al.* (2002) Modelling, simulation and warming impact assessment of a domestic-size absorption solar cooling system. *Applied Thermal Engineering*, Vol. 22(12), pp. 1313–1325.

Fux, V. (2006) Thermal performance of ventilated façades. PhD thesis, Loughborough University.

Gao, Z., Mei, V.C., Tomlinson, J.J. (2005) Theoretical analysis of dehumidification process in a desiccant wheel. *Heat and Mass Transfer*, Vol. 41, pp. 1033–1042.

Gassel, A. (2004) Kraft-Wärme-Kälte-Kopplung und solare Klimatisierung. Habilitationsschrift, TU Dresden.

Gee, R., Cohen, G., Greenwood, K. (2003) Operation and preliminary performance of the Duke Solar Power Roof: a roof-integrated solar cooling and heating system. Proceedings of the International Solar Energy Conference, March, ASME, Hawaii.

Gerhardt, H.J., Rudolph, M. (2000) Steuerungsstrategien für Fenster Betätigungen von Hochhäusern (Control strategy for window apertures in high rise buildings). *Heizung Lüftung/Klima Haustechnik*, Vol. 9, pp. 48–52.

Gertis, K. (1999) Sind neue Fassadenentwicklungen bauphysikalisch sinnvoll? Teil 2: Glas-Doppelfassaden (Are new façade developments rational? Part 2: Double glazed façades). *Bauphysik* 21, Vol. 2, pp. 54–66.

Ginestet, S., Stabat, P., Marchio, D. (2003) Control design of open-cycle desiccant cooling systems using a graphical environment tool. *Building Services Engineering Research and Technology*, Vol. 24(4), pp. 257–269.

Glück, B. (1991) Zustands- und Stoffwerte (Wasser, Dampf, Luft), Verbrennungsrechnung. Verlag für Bauwesen Berlin.

Government Information Centre, Hong Kong (2004) http://www.emsd.gov.hk/ emsd/eng/pee/wacs.shtml.

Granados, C. (1997) Solar cooling in Spain – present and future. Workshop Forschungsverbund Sonnenenergie.

Gratia, E., De Herde, A. (2007) The most efficient position of shading devices in a double-skin façade. *Energy and Buildings*, Vol. 39, pp. 364–373.

Green Paper on energy efficiency (2005) Office for Official Publications of the European Communities.

Grossmann, G. (2002) Solar powered systems for cooling, dehumidification and air conditioning. *Solar Energy*, Vol. 72, pp. 53–62.

Guiney, B., Henkel, T. (2003) Solar thermal for cooling, heating and power generation. *Renewable Energy World*, Vol. 6(2), pp. 92–98.

Gutiérrez, F. (1988) Behaviour of a household absorption-diffusion refrigerator adapted to autonomous solar operation. *Solar Energy*, Vol. 40(1), pp. 17–23.

Hamed, A.M., Khalil, A., Kabeel, A.E. *et al.* (2005) Performance analysis of dehumidification rotating wheel using liquid desiccant. *Renewable Energy*, Vol. 30, pp. 1689–1712.

Hansen, C. (1993a) Solare Klimatisierung – Beispiel in Benidorm/Spanien (Solar air-conditioning – example in Benidorm/Spain). *Sonnenenergie*, Vol. 18(2), pp. 8–9.

Hansen, C. (1993b) Solare Klimatisierung (Solar climatisation). *Ki Klima-Kälte-Heizung*, Vol. 21(9), pp. 356–357.

Hartmann, K. (1992) Kälteerzeugung in Absorptionsanlagen (Cold production of absorption plants). *Die Kälte- und Klimatechnik*, Vol. 45(9). pp. 622–634.

Hauser, G., Heibel, B. (1996) Zuluftfassaden – Simulationsmodell und mess-technische Validation (Supply air front – simulation model and validation by measurements). *Gesundheitsingenieur* 117, Vol. 1, pp. 1–8.

Hauser, G., Kaiser, J., Rösler, M. *et al.* (2004) Energetische Optimierung, Vermessung und Dokumentation für das Demonstrationsgebäude des Zentrums für umweltbewusstes Bauen, Universität Kassel, Abschlussbericht zum BMWA Forschungsvorhaben FKZ 0335006Z.

Hausladen, G., Kippenberg, K., Langer, L. *et al.* (1998). Solare Doppelfassaden: Energetische und raumklimatische Auswirkungen (Solar double façades: energetic and space climatic impacts). *Ki Luft- und Kältetechnik*, Vol. 11, pp. 524–529.

Henne, A. (1999) Luftleitungs-Erdwärmeübertrager - Grundlegendes zum Betrieb. Technik am Bau 10, pp. 55–58.

Henning, H.M., Hindenburg, C., Erpenbeck, T. *et al.* (1999) The potential of solar energy use in desiccant cooling cycles. Proceedings of the International Sorption Heat Pump Conference (ISHPC), Munich.

Henning, H.-M. (2001) Energetische und wirtschaftliche Aspekte solar unterstützter Klimatisierung (Energetic and economical aspects of solar assisted air-conditioning). Proceedings of the 1st Symposium Solares Kühlen in der Praxis, Fachhochschule Stuttgart - Hochschule für Technik, Germany. Vol. 53, pp. 22–42.

Henning, H.-M. (2004a) *Solar-assisted air-conditioning in buildings – a handbook for planners*. Springer Verlag.

Henning, H.-M. (2004b) Solare Klimatisierung – Stand der Entwicklung Fraunhofer Institut für Solare Energiesysteme ISE Tagung Solares Kühlen, Wirtschaftskammer Österreich, 7 May, Vienna.

Hensen, J., Bartak, M., Drkal, F. (2002) Modeling and simulation of a double skin facade system. *ASHRAE Transactions*, Vol. 108(2), pp. 1251–1259.

Hering, E., Rolf, M., Stohrer, M. (1997) *Physik für Ingenieure*, 6th edu. Springer- Verlag.

Herold, K.E. (1996) Diffusion-absorption heat pump. Final Report for Gas Research Institute, GRI-96/0271. Gas Research Institute, USA.

Hicks, T.W., Neida, B. (2000) An evaluation of America's first energy star buildings. Proceedings of the ACEEE 2000 Summer Study on Energy Efficiency in Buildings, American Council for an Energy Efficient Economy, Washington, DC.

Hindenburg, C. (2002) Anlagenplanung und Betrieb einer offenen sorptionsgestützten Klimaanlage. Proceedings of the 2nd Symposium Solares Kühlen in der Praxis, HfT Stuttgart.

Höfker, G. (2001) Desiccant cooling with solar energy. PhD thesis, Institute of Energy and Sustainable Development, De Montfort University, Leicester.

Holter, C., Meißner, E. (2003) Solar cooling system in Pristina/Kosovo. Proceedings of the ISES Solar World Congress 2003, 16–19 June, International Solar Energy Society (ISES), Göteborg, Sweden.

Höper, F. (1999) Optimierte Anlagenschaltung zur solaren Kühlung mit Absorptionstechnik (Optimized design for solar cooling with absorption refrigeration units). *Ki Luft- und Kältetechnik*, Vol. 35(8), pp. 397–400.

IEA (2008) *Promoting Energy Efficiency Investments – Case Studies in the Residential Sector*. IEA.

IEA Heat Pump Centre (2002) HPC-IFS2: Closed loop ground-coupled heat pumps. http://www.heatpumpcentre.org.

Infield, D., Mei, L., Eicker, U. (2004) Thermal performance estimation for ventilated PV facades. *Solar Energy*, Vol. 76, pp. 93–98.

Jachan, C. (2003) Hygienischer Tauglichkeitsnachweis und Optimierung der bau-physikalischen Performance von Gebäuden in Passivbauweise (Hygienic, suitability verification and optimization of building physical performance in buildings constructed in passive design), Dissertation, Technische Universität Wien.

Jain, P.C., Gable, G.K. (1971) Equilibrium property data equations for aqua-ammonia mixtures. *ASHRAE Transactions*, No. 2180, pp. 149–151.

Jakob, U. (2006) Investigations into solar powered Diffusion-Absorption Cooling Machines. PhD thesis, De Montfort University, Leicester.

Jakob, U., Eicker, U., Taki, A.H. *et al.* (2003) Development of an optimised solar driven Diffusion-Absorption Cooling Machine. Proceedings of the ISES Solar World Congress 2003, 16–19 June, International Solar Energy Society (ISES), Göteborg, Sweden.

Kanoğlu, M., Bolattürk, A., Altuntop, N. (2007) Effect of ambient conditions on the first and second law performance of an open desiccant cooling process. *Renewable Energy*, Vol. 32(6), pp. 931–946.

Karagiorgas, M., Mendrinos, D., Karytsas, C. (2004) Solar and geothermal heating and cooling of the European Centre for Public Law building in Greece. *Renewable Energy*, Vol. 29(4), pp. 461–470.

Karbach, A. (1998) Solare Kühlung (Solar cooling). In VDI Berichte. *Kälteversorgung in der technischen Gebäudeausrüstung*, 1st edn. VDI Verlag. No. 1432, pp. 85–95.

Karbach, A., Fischer, H., Niebeling, D. (1997) Solare Kühlung - Erneuerbare Energien möglichst effizient nutzen (Solar cooling - use renewable energy possible effectively). *Heizung Lüftung/Klima Haustechnik*, Vol. 48(10), pp. 47–50.

Kast, W. (1988) *Adsorption aus der Gasphase, Ingenieurwissenschaftliche Grundlagen und technische Verfahren*, VCH Verlag.

Kautsch, P., Dreyer, J., Hengsberger, H. *et al.* (2002) Thermisch-hygrisches Verhalten von GlasDoppelFassaden unter solarer Einwirkung. (Thermal-hygric behaviour of glazing-double-façades under solar influence). Berichte aus Energie- und Umweltforschung 36/2002, Bundesministerium für Verkehr, Innovation und Technologie, Vienna.

Keizer, C. (1979) Absorption refrigeration machine driven by solar heat. Proceedings of the 15th IIR International Congress of Refrigeration, Venice, Italy. pp. 861–868.

Keller, B. (1997) *Klimagerechtes Bauen*. B.G. Teubner Verlag.

Kim, D.S., Machielsen, C.H.M. (2002) Evaluation of air-cooled solar absorption cooling systems. ISHPC '02, Proceedings of the International Sorption Heat Pump Conference, Shanghai, 24–27 September.

Kim, D.S., Wang, L., Machielsen, C.H.M. (2003) Dynamic modelling of a small-scale NH_3/H_2O absorption chiller. Proceedings of the 21st IIR International Congress of Refrigeration, 17–22 August, International Institute of Refrigeration (IIR), Washington, DC.

Kimura, K. (1992) Solar absorption cooling. In Sayigh, A.A.M. and McVeigh, J.C. (Eds) *Solar Air Conditioning and Refrigeration*, 1st edn. Pergamon Press, pp. 13–65.

King, D.L., Dudley, J.K., Boyson, W.E. (1996) PVSIM – a simulation program for photovoltaic cells, modules and arrays. 25th PVSEC, Washington, DC. pp. 1295–1297.

Klopfer, H. *et al.* (1997) *Lehrbuch der Bauphysik*, 4th edn. Teubner Verlag.

Kohlenbach, P., Medel y Molero, S., Schweigler, C. *et al.* (2004) Weiterentwicklung und Feldtest einer kompakten 10kW H_2O/LiBr Absorptionskälteanlage (Further development of a compact 10kW H_2O/LiBr absorption chiller). Proceedings of the 3rd Symposium Solares Kühlen in der Praxis, Fachhochschule Stuttgart - Hochschule für Technik, Germany. Vol. 65, pp. 145–158.

Kornadt, O., Lehmann, L., Zapp, F.J. (1999) Doppelfassaden: Nutzen und Kosten (Double layer façade: benefit and costs). *Bauphysik* 21, Vol. 1, pp. 10–19.

Kouremenos, D.A., Stegou-Sagia, A., Antonopoulos, K.A. (1994) Three dimensional evaporation process in aqua-ammonia absorption refrigerators using helium as inert gas. *International Journal of Refrigeration*, Vol. 17(1), pp. 58–67.

Krüger, D., Hennecke, K., Schwarzbözl, P. *et al.* (2002) Parabolic trough collectors for cooling and heat supply of a hotel in Turkey. Proceedings of the 7th World Renewable Energy Congress, 1–5 July, World Renewable Energy Network (WREN), Cologne.

Kumar, R., Kaushik, S.C., Srikonda, A.R. (2003) Cooling and heating potential of earth-air tunnel heat exchanger (EATHE) for non-air-conditioned buildings. *International Journal of Global Energy Issues*, Vol. 19(4), pp. 373–386.

Labs, K. (1989) Earth coupling. In Cook, J. (Ed.) *Passive Cooling*. MIT Press.

Lamarche, L., Beauchamp, B. (2007) A new contribution to the finite line source model for geothermal boreholes. *Energy and Buildings*, Vol. 39, pp. 188–198.

Lamp, P., Ziegler, F. (1997) Solar cooling with closed sorption systems. Proceedings of the Workshop Solar Sorptive Cooling, 16–17 October, Forschungsverbund Sonnenenergie (FVS), Hardthausen, Germany. pp. 79–92.

Lang, W. (2000) Typologische Klassifizierung von Doppelfassaden und experimentelle Untersuchung von dort eingebauten Lamellensystemen aus Holz zur Steuerung des Energiehaushaltes hoher Häuser unter Berücksichtigung der Nutzung von Solarenergie (Classification of double glazed façades and experimental investigations of integrated lamella systems for the energy regulation of high buildings by consideration of solar energy). Dissertation, TU München.

Lavan, Z., Monnier, J.-B., Worek, W.M. (1982) Second law analysis of desiccant cooling systems. *Journal of Solar Energy Engineering*, Vol. 104, pp. 229–236.

Lävemann, E., Keßling, W., Röhle, B. *et al.* (1993) Klimatisierung über Sorption. Endbericht Phase I des Forschungsvorhabens Nr. 032 9151 B des BMFT, Sektion Physik der Ludwig-Maximilians-Universität München.

Lee, K. (2002) Untersuchung zur Einsatzmöglichkeit von Doppelfassaden bei hohen Verwaltungsgebäuden mit Glasfassaden im extrem gemäßigten Klimagebiet (Application of double façades in high administration buildings with glazing facades in extremely moderate climates). Dissertation, TU Berlin.

Li, X., Chen, Z., Zhao, J. (2006) Simulation and experiment on the thermal performance of U-vertical ground coupled heat exchangers. *Applied Thermal Engineering*, Vol. 26, pp. 1564–1571.

Loewer, H. (1978) Solar-Kühlung in der Klimatechnik (Solar refrigeration for air conditioning installations). *Ki Klima-Kälte-Ingenieur*, Vol. 6(4), pp. 155–162.

Lokurlu, A., Richarts, F., Krüger, D. *et al.* (2002) Wärme- und Kälteversorgung eines Hotels mit Parabolrinnenkollektoren (Heat and cooling supply of a hotel with parabolic trough collectors). Proceedings of the 12th Symposium Thermische Solarenergie, Ostbayerisches Technologie-Transfer-Institut eV (OTTI), Staffelstein, Germany. pp. 235–239.

Lorenz, E., Betchke, J., Drews, A. *et al.* (2005) Automatische Ertragsüberwachung von Photovoltaikanlagen auf der Basis von Satellitendaten: Evaluierung des PVSAT-2 Verfahrens, 20. Symposium Photovoltaische Solarenergie, Staffelstein, pp. 66–71.

Luther, J., Schumacher, J. (1991) INSEL – a simulation system for renewable electrical energy supply systems, Proceedings of the 10th European Photovoltaic Solar Energy Conference, Lisbon. pp. 457–460.

Maclaine-Cross, I.L. (1988) Proposal for a desiccant air conditioning system. *ASHRAE Transactions*, Vol. 94(2), pp. 1997–2009.

Mei, L., Infield, D., Eicker, U. *et al.* (2003) Thermal modelling of a building with an integrated ventilated PV-facade. *Energy and Building*, Vol. 35(6) pp. 605–617.

Meissner, E., Podesser, E., Enzinger, P. *et al.* (2004) Projekterfahrungen mit solaren Kältemaschinen: 17kW NH_3-H_2O Absorptionskälteanlage für Weinkühlung beim Weingut Peitler in der Steiermark/Österreich (Project experiences with solar cooling machines: 17kW NH_3-H_2O absorption cooling machine for the Peitler winery in the Steiermark/Austria). Proceedings of the 3rd Symposium Solares Kühlen in der Praxis, Fachhochschule Stuttgart - Hochschule für Technik, Germany. Vol. 65, pp. 59–70.

Mei, L., Infield, D., Eicker, U. *et al.* (2002) Parameter estimation for ventilated photovoltaic façades. *Building Services Engineering Research & Technology*, Vol. 23(2), pp. 81–96.

Mendes, L.F., Collares-Pereira, M., Ziegler, F. (1998) Supply of cooling and heating with solar assisted heat pumps: an energetic approach. *International Journal of Refrigeration*, Vol. 21(2), pp. 116–125.

Mei, L., Infield, D. (2002) Cooling load calculations. Final publishable progress report of EU AIRCOOL project ERK6-CT1999-00010, published by University of Applied Sciences in Stuttgart, ursula.eicker@hft-stuttgart.de.

Mei, V.C., Chen, F.C., Lavan, Z. *et al.* (1992) An assessment of desiccant cooling and dehumidification technology. Report prepared by the Oak Ridge National Laboratory, Contract No. DE-AC05-840R21400.

Meyer, S. (2001) Wirkung eines hybriden Doppelfassadensystems auf die Energiebilanz und das Raumklima der dahinterliegenden Räume (Influence of a hybrid double façade system on the adjacent room climate). Dissertation, TU Cottbus.

Ministry for Transport and Buildings, Germany (2000) Entwurf der Energieeinsparverordnung. Tagungsbeitrag zum 10. Symposium thermische Solarenergie, Staffelstein, May.

Mößle, P. (2000) Erfahrungen mit der solarunterstützten Klimatisierung (Experiences with the solar assisted air-conditioning). Proceedings of the Dresdner Colloquium – Solare Klimatiserung, ILK Dresden, Germany.

Munters, C.G. (1960) Air conditioning system. US Patent 2,926,502.

Narayankhedkar, K.G., Maiya, M.P. (1985) Investigations on triple fluid vapour absorption refrigerator. *International Journal of Refrigeration*, Vol. 8(11), pp. 335–342.

Nick-Leptin, J. (2005) Political framework for research and development in the field of renewable energies. International Conference or Solar Air Conditioning, Staffelstein.

Niebergall, W. (1981) Sorptions-Kältemaschinen (Sorption cooling machines). In Plank, R. *Handbuch der Kältetechnik*, reprinted 1st edn. Springer Verlag. Vol. 7, pp. 105–114, 249–271, 443–445 and 285–313.

Noeres, P., Pollerberg, Cl., Dötsch, Chr. *et al.* (2004) Solare Kühlung mit Parabolrinnen-DSKM – Erfahrungen mit solarthermischer Kälteerzeugung. Drittes Symposium Solares Kühlen in der Praxis, Stuttgart.

Öberg, V., Goswami, D.Y. (1998a) A review of liquid desiccant cooling. *Advances in Solar Energy*, Vol. 12, pp. 413–470.

Öberg, V., Goswami, D.Y. (1998b) Experimental study of the heat and mass transfer in a packed bed liquid desiccant air dehumidifier. *Journal of Solar Engineering*, Vol. 120, November, pp. 289–297.

Oesterle, E., Koenigsdorff, R., Mösle, P. *et al.* (1998) Zukunftsorientiertes Energiekonzept für das Projekt 'Stuttgart 21', Bericht zum Planungsgebiet A1 (Forward-looking energy concept for the project 'Stuttgart 21', Report to the planning area A1). Stuttgart 21, Heft 11, Landeshauptstadt Stuttgart, Amt für Umweltschutz, pp. 47–51.

Ohn, J. (1995) Kölner Pilotprojekt: Verwaltungsgebäude solar gekühlt (Cologne pilot project: administrative building solar cooled). *Heizung Lüftung/Klima Haustechnik*, Vol. 46(12), pp. 58–59.

Olsson, C.O. (2004) Prediction of Nusselt number and flow rate of buoyancy driven flow between vertical parallel plates. *Journal of Heat and Mass Transfer*, Vol. 126, pp. 97–104.

O.Ö. Energiesparverband (1996) Energiekannzahlen und -sparpotentiale im Lebensmittel Einzelhandel.

Pahud, D., Kohl, T., Mégel, T. *et al.* (2002) Langzeiteffekt von Mehrfach-Erdwärmesonden. Schlussbericht März 2002, DIS-Projekt 39690, Bundesamt für Energie, Switzerland.

Panaras, G., Mathioulakisa, E., Belessiotisa, V. (2007) Achievable working range for solid all-desiccant air-conditioning systems under specific space comfort requirements. *Energy and Buildings*, Vol. 39(9), pp. 1055–1060.

Park, M.S., Vliet, G.C., Howell, J.R. (1994a) Coupled heat and mass transfer between a falling desiccant film and air in cross-flow: Part I – Numerical model and experimental results. HTD Vol. 275, Current Developments in Numerical Simulation of Flow and Heat Transfer, ASME, pp. 81–90.

Park, M.S., Vliet, G.C., Howell, J.R. (1994b) Coupled heat and mass transfer between a falling desiccant film and air in cross-flow: Part II – Parametric analysis and results. HTD Vol. 275, Current Developments in Numerical Simulation of Flow and Heat Transfer, ASME, pp. 73–79.

Park, M.S., Howell, J.R., Vliet, G.C. *et al.* (1994c) Numerical and experimental results for coupled heat and mass transfer between a desiccant film and air in cross-flow. *International Journal of Heat and Mass Transfer*, Vol. 37, Suppl. 1, pp. 395–402.

Pennington, N.A. (1955) Humidity changer for air conditioning. US Patent 2,700,537.

Petukhov, B.S. (1970) In *Advances in Heat Transfer*, Vol. 6, ed. T.F. Irvine and J.P. Hartnett, Academic Press.

Pfafferott, J. (2003) Evaluation of earth-to-air heat exchangers with a standardised method to calculate efficiency. *Energy and Buildings*, Vol. 35, pp. 971–983.

Pfafferott, J., Herkel, S., Wambsganß, M. (2004) Design, monitoring and evaluation of a low energy office building with passive cooling by night ventilation. *Energy and Buildings*, Vol. 36, pp. 455–465.

Piechowski, M. (1999) Heat and mass transfer model of a ground heat exchanger: theoretical development. *International Journal of Energy Research*, Vol. 23, pp. 571–588.

Plank, R., Kuprianoff, J. (1960) *Die Kleinkältemaschine* (*The household refrigeration machine*), 2nd edn. Springer Verlag, pp. 360–397.

Podesser, E. (1982) Kühlung mit Sonnenenergie (Cooling with sun energy). *Ki Klima-Kälte-Heizung*, Vol. 10(1), pp. 29–32.

Price, L., de la Rue du Can, S., Sinton, J. *et al.* (2006) Sectoral trends in global energy use and GHG emissions. LBNL Report 56144.

Pulselli, R.M., Simoncini, E., Pulselli, S. *et al.* (2007) Energy analysis of building manufacturing, maintenance and use: EM-building indices to evaluate housing sustainability. *Energy and Buildings*, Vol. 39, pp. 620–628.

Quinette, J.-Y., Albers, J. (2002) Solar air-conditioning by absorption. Proceedings of the Industry Workshop Solare Klimatisierung/Solar Assisted Air Conditioning of Buildings. aircontec – International trade fair for air-conditioning technology at Light+Building, Fachinstitut Gebäude-Klima FGK, Germany. www.sorptionsgestuetzte-klimatisierung.de.

Rafferty, K. (2004) Direct-use temperature requirements - a few rules of thumb. *Geo-Heat Center Quarterly Bulletin*, Vol. 25, No. 2.

Reinders, A., van Dijk, V., Wiemken, E. *et al.* (1999) Technical and economic analysis of grid-connected PV systems by means of simulation. *Progress in Photovoltaics*, Vol. 7, pp. 71–82.

Reol, N. (2005) És sostenible la construcció sostenible?. Proceedings of II Jornada Empresa i Medi Ambient, Barcelona, 15 November.

Sack, N., Kuhn, T., Beck, A. (2001) Entwicklung einer Referenzmethode zur kalorimetrischen Bestimmung des Gesamtenergiedurchlassgrades von transparenten und transluzenten Bauteilen. Abschlussbericht Projekt REGES, BMFT 0329798A.

Saelens, D., Blocken, B., Roels, S. *et al.* (2005) Optimization of the energy performance of multiple-skin façades. Ninth International IBPSA Conference, Montreal, 15–18 August.

Safarik, M., Weidner, G. (2004) Neue 15kW H_2O-LiBr Absorptionskälteanlage im Feldtest für thermische Anwendungen (New 15kW H_2O-LiBr Absorption Cooling Machine in field test for thermal applications). Proceedings of the 3rd Symposium Solares Kühlen in der Praxis, Fachhochschule Stuttgart - Hochschule für Technik, Germany. Vol. 65, pp. 159–171.

Safarik, M., Gramlich, K., Schammler, G. (2002) Solar absorption cooling system with 90°C - latent heat storage. Proceedings of the 7th World Renewable Energy Congress, 1–5 July, World Renewable Energy Network (WREN), Cologne.

Salsbury, T.I., Diamond, R.C. (2001) Fault detection in HVAC systems using model based feedforward control. *Energy and Buildings*, Vol. 33, pp. 403–415.

Saman, W.Y., Alizadeh, S. (2001) Modelling and performance analysis of a cross-flow type plate heat exchanger for dehumidification/cooling. *Solar Energy*, Vol. 70(4), pp. 361–372.

Saman, W.Y., Alizadeh, S. (2002) An experimental study of a cross-flow type plate heat exchanger for dehumidification/cooling. *Solar Energy*, Vol. 73(1), pp. 359–371.

Sanner, B., Rybach, L. (1997) Oberflächennahe Geothermie – Nutzung einer allgegenwärtigen Ressource. *Geowissenschaften*, 15, Heft 7.

Santamouris, M. (2005) A common evaluation protocol of buildings. EU FP6 Ecobuildings Symposium, Deutsches Technikmuseum Berlin, 22–23 November.

Schirp, W. (1990) Gasbeheizte Diffusions-Absorptions-Wärmepumpe (DAWP) für Wohnraumbeheizung, Brauchwassererwärmung und Wohnraumkühlung. *Ki Klima-Kälte-Heizung*, Vol. 18(3), pp. 113–118.

Schmidt, H., Sauer, D.U. (1996) Wechselrichterwirkungsgrade. Praxisgerechte Modellierung und Abschätzung, Sonnenenergie 4, pp. 43–47.

Schölkopf, W., Kuckelkorn, J. (2004) Verwaltungs- und Bürogebäude – Nutzerverhalten und interne Wärmequellen. OTTI Kolleg Klimatisierung von Büro- und Verwaltungsgebäuden, Regensburg.

Schumacher, J. (1991) Digitale simulation regenerativer elektrischer Energieversorgungssysteme. Dissertation, Universität Oldenburg.

Schumacher, J. (2004) The simulation environment INSEL. www.insel.eu.

Schwarz, C., Lotz, D. (2001) Gas-Wärmepumpen – Absorber: Einsatz im Ein- und Zweifamilienwohnhaus. Proceedings of the Fachtagung Heizen - Kühlen - Klimatisieren mit Gas-Wärmepumpen und -Kälteanlagen, 14 November, Fulda, Germany. ASUE, pp. 35–43. www.asue.de.

Shah, R., Rasmussen, B.P., Alleyne, A.G. (2004) Application of a multivariable adaptive control strategy to automotive air conditioning systems. *International Journal of Adaptive Control and Signal Processing*, Vol. 18, pp. 199–221.

Shaviv, E., Yezioro, A., Capeluto, I.G. (2001) Thermal mass and night ventilation as passive cooling design strategy. *Renewable Energy*, Vol. 24, pp. 445–452.

Simader, G., Rakos, C. (2005) *Klimatisierung, Kühlung und Klimaschutz: Technologien, Wirtschaftlichkeit und* CO_2 *Reduktionspotentiale*. Austrian Energy Agency.

Sklar, S. (2004) New dawn for distributed energy. *Cogeneration and On-Site Power Production*, Vol. 5(4), pp. 115–121.

Smirnov, G.F., Bukraba, M.A., Fattuh, T. *et al.* (1996) Domestic refrigerators with absorption-diffusion units and heat-transfer panels. *International Journal of Refrigeration*, Vol. 19(8), pp. 517–521.

Srikhirin, P., Aphornratana, S. (2002) Investigation of a diffusion absorption refrigerator. *Applied Thermal Engineering*, Vol. 22(11), pp. 1181–1193.

Steemers, K. (2003) Energy and the city: density, buildings and transport. *Energy and Buildings*, Vol. 35, pp. 3–14.

Stettler, S., Toggweiler, P., Wiemken, E. *et al.* (2005) Failure detection routine for grid connected PV systems as part of PVSAT2 project. 19th European Photovoltaic Solar Energy Conference & Exhibition, Barcelona.

Stierlin, H. (1964) Neue Möglichkeiten für den Absorptions-Kühlschrank (New chances for the absorption refrigerator). *Kältetechnik*, Vol. 9, pp. 264–270.

Stierlin, H.C., Ferguson, J.R.(1990) Diffusion Absorption Heat Pump (DAHP). *ASHRAE Transactions*, Vol. 96, pp. 1499–1505.

Storkenmaier, F., Harm, M., Schweigler, C. (2003) Small-capacity water/LiBr absorption chiller for solar cooling and waste-heat driven cooling. Proceedings of the 21st IIR International Congress of Refrigeration, 17–22 August, International Institute of Refrigeration (IIR), Washington, DC.

Stürzebecher, W., Braun, R., Garbett, E. *et al.* (2004) Solar driven sorption refrigeration systems for cold storage depots. Proceedings of the 3rd International Conference on Heat Powered Cycles (HPC 2004), 11–13 October, Larnaca, Cyprus. Paper no. 2117.

Truschel, S. (2002) Passivhäuser in Europa. Diploma thesis, University of Applied Sciences, Stuttgart.

Tzaferis, A., Liparakis, D., Santamouris, M. *et al.* (1992) Analysis of the accuracy and sensitivity of eight models to predict the performance of earth-to-air heat exchangers. *Energy and Buildings*, Vol. 18, pp. 35–43.

Uchihara, M., Yamamoto, T., Takahashi, S. (2002) Model based control of air conditioning systems for painting. *Toyota Technical Review*, Vol. 52, No. 1.

Ürge-Vorsatz, D., Novikova, A. (2008) Potentials and costs of carbon dioxide mitigation in the world's buildings. *Energy Policy*, Vol. 36, pp. 642–661.

US Department of Energy (2002) US Lighting Market Characterization, Volume I: National Lighting Inventory and Energy Consumption Estimate. Final Report, Building Technologies Program, www.eren.doe.gov/buildings/documents/.

van Dyk, E.E., Gxasheka, A.R., Meyer, E.L. (2005) Monitoring current voltage characteristics and energy output of photovoltaic modules. *Renewable Energy*, Vol. 30, pp. 399-411.

Velázquez, N., Best, R. (2002) Methodology for the energy analysis of an air cooled gas absorption heat pump operated by natural gas and solar energy. *Applied Thermal Engineering*, Vol. 22, pp. 1089–1103.

Verein Deutscher Ingenieure (2000) VDI Guideline 4640: Thermal use of the underground. http://www.vdi.de.

Vollmer, K. (1999) Thermische Charakteristik und Energieeintrag von hinterlüfteten PV-Fassaden (Thermal characteristics and energy gains of ventilated PV façades). Diploma thesis, University of Applied Sciences, Stuttgart.

Voss, K., Herkle, S., Pfafferott, J. (2007) Energy efficient office buildings with passive cooling - results and experiences from a research and demonstration programme. *Solar Energy*, Vol. 81, pp. 424–434.

Wardono, B., Nelson, R.M. (1996) Simulation of a double effect LiBr/H_2O Absorption Cooling System?. *ASHRAE Journal*, Vol. 38(10), pp. 32–38.

Watts, F.G., Gulland, C.K. (1958) Triple-fluid vapour-absorption refrigerators. *Journal of Refrigeration*, July and August, pp. 107–115.

Willers, E., Neveu, P., Groll, M. *et al.* (1999) Dynamic modelling of a liquid absorption system. ISHPC '99, Proceedings of the International Sorption Heat Pump Conference, Munich, 24–26 March.

Wolkenhauser, H., Albers, J. (2001) Systemlösungen und Regelungskonzepte von solarunterstützten Klimatisierungssystemen Teil 1 – Kaltwassersysteme (System solutions and control strategies of solar assisted air-conditioning systems part 1 – cold water systems). *Heizung Lüftung/Klima Haustechnik*, Vol. 52(12). pp. 41–49.

Yamaguchi, Y., Shimoda, Y., Mizuno, M. (2007) Transition to a sustainable urban energy system from a long term perspective: case study in a Japanese business district. *Energy and Buildings*, Vol. 39, pp. 1–12.

Zeng, H., Diao, N., Fang, Z. (2003) Heat transfer analysis of boreholes in vertical ground heat exchangers. *International Journal of Heat and Mass Transfer*, Vol. 46, pp. 4467–4481.

Zhang, Q., Murphy, W.E. (2003) Heat transfer of a multiple borehole field for ground heat pump systems. International Congress of Refrigeration, Washington, DC.

Zia, H., Devadas, V. (2007) Energy management in Lucknow City. *Energy Policy*, Vol. 35, pp. 4847–4868.

Ziegler, F. (1998) *Sorptionswärmepumpen*. Forschungsberichte des Deutschen Kälte- und Klimatechnischen Vereins Nr. 57, Stuttgart.

Ziegler, F. (1999) Recent developments and future prospects of sorption heat pump systems. *International Journal of Thermal Sciences*, Vol. 38, pp. 191–208.

Zimmermann, M. (2003) *Handbuch der passiven Kühlung*. Fraunhofer IRB Verlag.

Zimmermann, M., Andersson, J. (1998) Low energy cooling - case studies buildings. EMPA ZEN, Dübendorf, Switzerland.

Zöllner, A. (2001) Experimentelle und theoretische Untersuchungen des kombinierten Wärmetransports in Doppelfassaden (Experimental and theoretical examinations of the combined heat transfer in double façades). Dissertation, TU München.

Index

Absorption cooling
ammonia, 116
COP, 116
cycles, 115
Diffusion absorption, 156
double effect, 121
double effect COP, 199
power range, 113
simulation models, 201
water/lithium bromide, 115
Air collector
fin efficiency, 146
performance, 143
Air exchange
measurement, 67
wind influence, 68

Building management system
communication infrastructure, 228
function, 226
socket servers, 241
Building simulation
dynamic model, 234
VDI 6020, 236
Buildings
U-values, 3

Collector efficiency, 112
Compression chiller
COP, 111
Control
energy savings, 197
model based control, 198
Cooling load, 208
internal load, 66
ventilation gains, 42
Cost
investment solar cooling, 123
passive buildings, 4

Desiccant cooling, 129
Contact matrix absorber, 178
control strategy, 131, 132, 139
COP, 140
dehumidification, 135
Desiccant rotor, 178
Heat exchanger absorber, 179
humidifier performance, 138
Liquid desiccant, 175
regeneration temperatures, 135
Diffusion absorption cooling, 156
History, 157
Double façade
building measurements, 40
overview, 23

Earth heat exchanger
annual COP, 89
exit temperatures, 90
ground to air, 88
horizontal brine system, 92

Earth heat exchanger (*Continued*)
pressure drop, 88
vertical, 97
Electricity consumption
air conditioning, 13
lighting, 13
office buildings, 10
residential buildings, 7
Energy consumption
buildings, 2
electricity, 3
fans, 62, 64
Germany, 2
hot water, 6
office heating, 9
Energy reduction coefficient
measured values, 36

g – value
building measurements, 39
definition, 30
Geothermal cooling
direct coupling, 83

Heat exchanger
cross flow, 179
rotating heat exchanger, 148

Night ventilation
eboek building, 75
mechanical, 75
power consumption, 75
SIC building, 74

Passive buildings
Lamparter office, 62
Passive cooling
daily loads, 61
simulation, 71
Photovoltaics
façade, 47
heat gain, 56
online simulation, 245
temperature levels, 49
two diode model, 240
POLYCITY project, 229
town hall, 230
Pressure drop
photovoltaic façade, 55

Soil temperatures, 90
Solar cooling
collector area, 122, 199
collector cost, 220
collector yield, 215
control strategy, 213
cooling tower electricity consumption, 198
history, 118
investment cost, 220
storage volume, 215
Solar simulator
irradiance distribution, 32
Standards
DIN V 18599, 233
EN 832, 230
Sun shading
energy reduction coefficient, 22

Temperature stratification, 69